MARINE POLLUTION

ADVISORY COMMITTEE ON POLLUTION OF THE SEA

YEAR BOOK 1990

edited by

Patricia Dent

Published on behalf of ACOPS by

PERGAMON PRESS

Member of Maxwell Macmillan Pergamon Publishing Corporation

OXFORD · NEW YORK · BEIJING · FRANKFURT
SÃO PAULO · SYDNEY · TOKYO · TORONTO

U.K.	Pergamon Press plc, Headington Hill Hall, Oxford OX3 0BW, England
U.S.A.	Pergamon Press, Inc., Maxwell House, Fairview Park, Elmsford, New York 10523, U.S.A.
PEOPLE'S REPUBLIC OF CHINA	Pergamon Press, Room 4037, Qianmen Hotel, Beijing, People's Republic of China
FEDERAL REPUBLIC OF GERMANY	Pergamon Press GmbH, Hammerweg 6, D–6242 Kronberg, Federal Republic of Germany
BRAZIL	Pergamon Editora Ltda, Rua Eça de Queiros, 346. CEP 04011, Paraiso, São Paulo, Brazil
AUSTRALIA	Pergamon Press Australia Pty Ltd., P.O. Box 544, Potts Point, N.S.W. 2011, Australia
JAPAN	Pergamon Press, 5th Floor, Matsuoka Central Building, 1–7–1 Nishishinjuku, Shinjuku-ku, Tokyo 160, Japan
CANADA	Pergamon Press Canada Ltd., Suite No. 271, 253 College Street, Toronto, Ontario, Canada M5T 1R5

First edition 1990

Library of Congress Cataloging-in-Publication Data
International Council for the Exploration of the Sea.
Advisory Committee on Pollution of the Sea.
Marine pollution: yearbook 1990/Advisory Committee on Pollution of the Sea. — 1st ed.
p. cm.
"Published on behalf of ACOPS" — Colophon.
1. Marine pollution.   I. Title.   GC1085.I52   1990
363.73′94′09162—dc20   90–43161

British Library Cataloguing in Publication Data
ACOPS yearbook 1990
1. Marine pollution
I. Advisory Committee on Pollution of the Sea
363.7′394   GC1085

ISBN 0 08 040809 5

Typesetting by Calne Valley Services and printed in Great Britain by B.P.C.C. Wheatons Ltd, Exeter

Foreword

A lthough my present duties date from the beginning of 1990, I became aware of ACOPS and its contributions in the maritime field during my term as Chairman of the IMO Council.

The standing in which the ACOPS Yearbook is held has increased steadily over the years, principally deriving, I believe, from its value as an up-to-date and readily accessible summary of policy and legal developments relating to the marine environment.

The Yearbook reinforces ACOPS' policy-forming role in the field of marine environment protection and I have no hesitation in commending it to the many entities, whether in the public or private sector, engaged in this very important area of endeavour.

W.A. O'NEIL
Secretary-General
International Maritime Organisation

Message from The Rt Hon Lord Callaghan KG
President of ACOPS

As the 1990s begin to unfold we are already able to see that mankind will be facing considerable challenges in all aspects of the environment; yet with this the decade also brings hope for a brighter future for our planet. To those of us who initiated the NGO environmental movement shortly after the last war, it is a matter of considerable satisfaction to see environmental issues become the top priority of action programmes in intergovernmental agencies and governments alike.

Let there be no doubt that this new approach amongst members of the international community was very much overdue. In the past, response to environmental problems has almost always been remedial. This must cease to be the case, as the only satisfactory long-term policy is to incorporate 'environment friendly' policies into economic planning strategies. However, along with my colleagues at ACOPS I disagree with those who would halt economic development, as such a dogma would in particular seriously affect countries in the Southern Hemisphere.

It is important to note that a healthy environment is an economic, as well as social, necessity. I therefore welcome the activities of bodies such as the Organisation for Economic Cooperation and Development that attempt to calculate not only the cost of actions to control and reduce pollution but also the hidden cost to a society that does not take such action. The financial institutions also have a new role to play.

Once again I have to record that in a number of areas, internationally agreed legislation has not been followed up by adequate domestic enforcement through legal measures, fiscal policies and provision of the necessary technical infrastructure. In other cases bureaucratic indifference and complete lack of political leadership have resulted in inaction. It is essential that governments and intergovernmental agencies should append their signatures to international agreements and carry out their obligations. If this is not done we shall be inundated with unenforceable legislation which will give the false impression of having solved the problems it addresses but will lead to public cynicism when it fails to do so.

Whilst it is for governments, working in particular through agencies, to take action, non-governmental organisations such as ACOPS are increasingly being recognised as playing an important role. We all have to accept that higher environmental standards will involve higher costs of even basic necessities than we are presently accustomed to. NGOs should therefore encourage, through public awareness campaigns, the adoption of necessary environmental measures.

As a parliamentarian, I am naturally happy to see an increased commitment to environmental protection expressed by the European Parliament and the US Senate and House of Representatives. This culminated in the setting up of Global Legislators for a Balanced Environment (GLOBE see 3.24) which hopes soon to include members of the Japanese Diet and the Soviet Parliament. I should like to record the pleasure with which we received a message and comment from Mr Carlos Pimenta, Portugal's former Environment Minister, a distinguished Member of European Parliament, of GLOBE and, also, of ACOPS.

I cannot conclude without placing on record my grateful thanks to Dr Viktor Sebek for his creative leadership and also to his small but devoted staff for their efforts to which this Yearbook bears testimony, and without which ACOPS could not survive. We have also been extremely fortunate in persuading Lord Clinton-Davis, a former European Commissioner, to resume his connection with ACOPS and to become our Chairman.

There is much work for ACOPS to do in the years ahead, and on behalf of everyone I wish it every success.

James Callaghan

Callaghan of Cardiff
President of ACOPS

Message from Carlos Pimenta
Member of the European Parliament's Environment Committee

The world as we know it is facing global rupture. The environment around us is reaching the limits of its capacity to adapt to the pollution created by the human race. Our climate is changing because of our over-use of fossil fuel and of ozone-damaging substances; we are destroying the diversity that nature created at a rate of several thousand species per year through the destruction of natural habitats such as the tropical forests and coastal zones; we dirty the environment with household and toxic waste and endanger health by transporting waste, especially toxic waste, to the cheapest country for disposal.

Without the conviction and support of the human inhabitants of this planet, there could be no new approach to these environmental problems, and it is a new approach that we need.

First, economic thinking must be changed. We must no longer use our resources without taking into account environmental damage, and all our industries should, as a priority, aim to reduce the amount of waste produced to the minimum. An extension of the 'polluter pays' principle is needed as well as new government fiscal policies which will make companies focus and re-evaluate their thinking on environmental damage. Fiscal policy should no longer be based only on revenue or sales volume.

Second, thinking on energy use must change so that it is based on a rationalisation of need. The northern more-developed countries should bias their energy policies towards promoting energy conservation, rather than increasing power supply, as at present. At a global level, research into new and renewable sources of energy must continue and be expanded. A global pooling of knowledge on technology already developed should feature in each country's energy policy. A change in lifestyle for most people is inevitable but this does not necessarily mean that life will be uncomfortable.

Third, the link between North/South developments and environmental problems needs to be acknowledged. Debt relief and trade must be linked with environmental policies.

Fourth, the lack of an internationally-binding legal framework and of an executive body able to regulate the problems and take decisive action is causing the exponential growth of environmental damage. The forthcoming 1992 United Nations Environmental Conference in Brazil is the place to set up new global laws and organisations. Failure to do this will be unforgivable.

Success depends on the seizing of as many opportunities as possible to talk at international level. The ACOPS conference on toxic waste, which took place in London in October 1989, is a good example of seizing an opportunity since the delegates not only addressed the problem of waste; through discussion of international experts, the whole environmental question and the lack of global answers became apparent.

The establishment of a strong and permanent communication at a global level, in order to find global solutions to environmental problems, would greatly help the work of the political institutions involved in deciding environmental policy, provided that these institutions can overcome national prejudices and work together at global level.

Political cooperation at global level was also the main objective of the setting up GLOBE, Global Legislators Organisation for a Balanced Environment. This informally links some members of the European Parliament with those members of the US Congress and Japanese Senate who are interested in global environmental change.

The nineties will be decisive years as far as environmental policy is concerned. We need global solutions for the global problems facing us and we must be bold enough to change the way we relate to our planet.

Carlos Pimenta
Member of the European Parliament's
Environment Committee

Contents

Preface

Green must have already been termed the catchword of the nineties. Everyone is going 'green', and many successfully, due largely to the way in which the public has responded to the message coming through the popular media. Organisations, both national and international, governmental and non-governmental, are growing, and nowhere more so than in the field of marine pollution. Keeping abreast of events, organisations and meetings is a full-time occupation. Indeed, can the decision-makers and politicians of today keep up with the urgent need, and the demand, that we clean up our act?

The object of this Yearbook is not necessarily to give a 'total' picture of events and organisations involved in maritime environmental affairs; rather, the aim is to give a balanced review of what is happening. And from the way the Yearbook has grown in size, a lot certainly is happening.

Many people have expressed concerns about the legacy of pollution that we are leaving to our children: littered and oiled beaches, seas devoid of fish and other marine life, blooms of poisonous algae. So we asked the children from two primary schools — Shotley CP school, Suffolk, which views across the Orwell Estuary the busy ports of Felixstowe and Harwich, and Shipbourne CP School in the heart of Kent — to paint their picture of marine life and pollution. The results give an insight into how the next generation is growing up in an environment in which pollution is an everyday (but still undesirable) problem. You will already have glanced at some of these paintings on the cover. Let us hope that tomorrow's decision-makers will not still be clearing up after us.

This edition of the Yearbook spans an 18-month period, from June 1988 to the beginning of 1990; future editions of the Yearbook will appear annually. It has been a mammoth task to put together, one that would have not been possible without the cooperation and assistance of many individuals and organisations all over the world. To name them all here would be impossible, but I would like to give special thanks to Dr David George at the Natural History Museum, London, for letting us reproduce some of his outstanding underwater photos, and also Professor Eddie Brown and colleagues at Cardiff Law School, UWIST, for updating the legal sections.

Patricia Dent
Editor

Introduction

I regard it as a great privilege to have been elected as Chairman of ACOPS in January 1990 for the second time. My first appointment was in 1983 but I was compelled to relinquish the Chair in December 1984 on my nomination as a member of the Commission of the European Communities.

Since 1983, ACOPS has become a truly international non-governmental organisation. This is exemplified by the fact that we now have a team of international Vice Presidents of the highest reputation, many of whom have held or still hold high public office, throughout most of the maritime regions of the world. Since the last Yearbook was published in 1988, our newly elected Vice Presidents include Professor Elisabeth Mann Borgese, who chairs Canada's International Council for Ocean Development and is the President of the International Ocean Institute in Malta; Mr. Chandrika Prasad Srivastava, who recently relinquished the position of Secretary General of the International Maritime Organisation after years of distinguished and devoted service; and Professor Gennady Polikarpov, a member of the Ukrainian Academy of Sciences. We are indeed delighted to have them on board.

ACOPS is also proud to have added to its membership a host of prominent individual members, in particular, a number of Members of the European Parliament who have displayed a keen and practical interest in environmental affairs over a long period of time.

There can be no doubt that over the last few years there has been a remarkable change in public perception of the significance of environmental issues. This is true not only in Europe, but throughout the world, including the developing countries. Today, the environment is right at the top of the political agenda. Perhaps the most significant examples of this are the chapter devoted to environment in the European Communities' Single European Act — the first time that this has occurred since the enactment of the Treaty of Rome; the continuing debates concerning environmental problems at the summits of world leaders, such as those in Paris in July 1989 and at the forthcoming Hous-

ton summit in July of this year; the convening of the 1992 United Nations Conference on the Environment, which just about coincides with the 20th anniversary of the historic Stockholm meeting which resulted, among other things, in the establishment of the United Nations Environment Programme.

As a member of the Commission of the European Communities, from 1985 to 1989, and charged with the specific responsibility for transport, environment, and nuclear safety, I was able to witness first hand the transformation of environmental policies, not only in the European Community, but in global terms, and in this context there can be no doubt whatsoever that the European Community is playing an increasingly important role. The great benefit, of course, of European Community legislation, over and above most international conventions, is that they are enforceable through the European Court of Justice. Although too many Member States commit too many breaches of environmental law, the ultimate sanction of being brought before the European Court of Justice is remarkably effective.

One of the most critical elements in the development of environmental policy is a more knowledgeable public opinion, aware of actions that may be taken to enforce better environmental behaviour at political and legal levels. To this end, greater openness on the part of governments and industry is essential. I am pleased therefore that ministers have reached a common position in connection with the draft directive on freedom of access to environmental information, although I fear that there are still too many exceptions based on confidentiality, a much over-used concept to perpetuate unnecessary restrictions on transparency of information.

Other very important initiatives have been taken by the present Delors Commission. For example, the proposal to establish a European Environmental Agency; the draft directive on strict liability for environmental damage caused by waste, and indeed, a whole host of legislative proposals

2

on which DGXI are currently working. This is greatly encouraging, but the fact remains that even now DGXI receives a relatively small proportion of the Commission's entire budget — and this is simply not good enough.

Transport and disposal of toxic waste has become one of the key environmental issues, and has been singled out for the agenda of the 1992 UK Conference on the Environment. It is, of course, a multi-faceted problem, and, while it is of considerable importance to the protection of the marine environment, it is by no means confined to this. In this regard, ACOPS has played a highly significant role, having been involved in meetings leading to the adoption of the Basle Convention and then having arranged, in conjunction with a series of intergovernmental agencies (such as the IMO and UNEP) and many governments, its own conference (6.2.1) in order to pave the way towards finding solutions to issues that remained unresolved at Basle and indeed subsequently, and which can only be tackled by the secretariat and the conference of parties, which cannot itself be set up until the convention enters into force.

To stimulate governments into signing and ratifying the Convention has therefore become a major priority as far as ACOPS is concerned, and to achieve this objective, I and my Vice Presidential colleagues have availed ourselves of every possibility to engage in discussions with heads of intergovernmental agencies, governmental officials, in our respective countries and regions. We believe that it is incumbent upon environmental NGOs to do whatever is possible to promote progressive endeavours to protect the environment which are undertaken by the international community.

Of course, the work of ACOPS over the last two years has been extremely extensive, and by no means confined to the issue of toxic waste. We have seized many opportunities to participate in the work of the IMO and the UNEP and we shall continue so to do in the year ahead.

Certainly our activities would not have been possible without financial assistance from intergovernmental agencies, governments, local authorities, voluntary bodies — especially the Wildlife Fund for Nature UK — and industry. As far as the production of the Yearbook is concerned, we would wish to express our special thanks to Mr Robert Maxwell who has generously underwritten the cost of this Yearbook.

I would also wish to congratulate the editor, Patricia Dent, who has devoted such enormous efforts to continue the Yearbook as a work of considerable scholarship, and to all the contributors who have helped to make this possible.

ACOPS is proud of its Yearbook and proud of its work. As Chairman a special vote of thanks is due from me to my predecessor, Lord Campbell of Croy, who has worked indefatigably for ACOPS for so long; also to Dr Viktor Sebek, our Executive Secretary, Mrs Jennie Holloway, our Legal Advisor, my Vice-Chairman, Wardley Smith, and, of course, to our President, Lord Callaghan, who has been such a constant source of guidance and support.

Lord Clinton-Davis
Chairman

Control of the Transboundary Movements of Hazardous Waste

by Javier Perez de Cuellar
Secretary General
United Nations

The Advisory Committee on Pollution of the Sea is to be commended for convening a Conference in 1989* to assist in the rapid implementation of the Basle Convention on the Control of Transboundary Movements of Hazardous Wastes and their disposal. The subject of that meeting was especially important as transborder disposal of hazardous wastes is one of the major problems contributing to the environmental crisis that threatens the future of our planet.

It is encouraging that the signatories to the Basle Convention have recognised the immediate need for the control of the transboundary movement and disposal of hazardous wastes and have called for the development of technologies that will help to prevent the generation of such wastes. Clearly, the Basle Convention represents an important step towards the type of international cooperation required to address the problem of transboundary movements of hazardous waste. Other steps, I am sure will follow.

I am pleased that the signatories to the Basle Convention have brought to the attention of the International Maritime Organisation the need for a review of the existing rules under the 1972 London Dumping Convention, with a view to recommending additional measures that may be needed to control and prevent dumping of hazardous and other wastes at sea. They have also recognised the urgent need to develop rules on liability and compensation for damage resulting from the transboundary movements and disposal of hazardous and other wastes. Furthermore, the relevant international organisations have also been called upon to review the existing rules, regulations and practices with respect to the transport of hazardous wastes by sea, with a view to recommending any additional measures needed to assist states in protecting and preserving the marine environment.

As you are aware, the protection and preservation of the marine environment is one of the underlying principals of the United Nations Convention on the Law of the Sea, which establishes the global framework for the legal control of marine pollution and the conduct of maritime activities. The Convention has laid down the basic obligation of states to protect and preserve the marine environment and to "act so as not to transfer, directly or indirectly, damage or hazards from one area to another."

The problem of trade in hazardous wastes is not limited to any particular area of the world or any particular level of national development. All countries, whether exporters or recipients of hazardous wastes, developed or developing, North or South, have a vested interest in controlling it. By seeking to put an end to the unlawful dumping and movement of hazardous wastes and strictly control its movement, the parties to the Basle Convention are helping to protect the environment. By fostering international cooperation and promoting concerted action through competent international organisations, they also contribute to improving relations among states and eliminating mistrust and suspicion.

Javier Perez de Cuellar
Secretary General, United Nations

---

*This text was a Message to ACOPS' International Conference on Protecting the Environment from Hazardous Substances, London 3–5 October 1989.

1. Outline of Events

1.1 PAST EVENTS

1988

JUNE   **Wellington, New Zealand** Convention on the Regulation of Antarctic Mineral Activities was adopted (see 3.22).

OCTOBER
3–7   **IMO, London** 11th Consultative Meeting of LDC (see 3.11).
24–28   **IMO, London** 56th Session of the Maritime Safety Committee (see 3.2.4).

NOVEMBER
21–15   **IMO, London** 61st Session of the IMO Council (see 3.2.2).

DECEMBER
31   Annex V of MARPOL 73/78 entered into force (see 7.3.1).

1989

MARCH
13–17   **IMO, London** 27th Session of Marine Environment Protection Committee of IMO (see 3.2.3).
22–24   **Basle**   UNEP Diplomatic Conference on Transport and Disposal of Toxic Waste (see 3.3.12).
22   Convention on the Control of Transboundary Movements of Hazardous Wastes and their Disposal (Basle Convention) came into being (see 3.3.12).

25   *Exxon Valdez* ran aground in Prince William Sound, Alaska, spilling 250,000 barrels of crude oil (see 2.1.4).

APRIL
3–12   **IMO, London**   57th Session of Maritime Safety Committee of IMO (see 3.2.4).
17–28   International Conference on Salvage (see 3.2.6).
28   International Convention on Salvage adopted.

MAY   **Nairobi** 15th Session of UNEP Governing Council (see 3.3.11).
12–18   **Athens** 19th Session of GESAMP (see 3.8).
31–2 June   **Port of Spain** First CARICOM Ministerial Conference on the Environment (see 3.17).

JUNE
5–9   **IMO**, London   62nd Session of IMO Council (see 3.2.2).
12   **San Diego** 41st Annual Meeting of International Whaling Commission (see 3.15).

JULY
14–16   **Paris** 15th Summit of the Seven (see 3.23).

SEPTEMBER
28–29   **IMO, London** 61st Session of Legal Committee of IMO (see 3.2.5).

OCTOBER
3–6   **Athens** Sixth Ordinary Meeting of Contracting Parties to Barcelona Convention (see 3.3.1).
17   **IMO, London** 28th Session of MEPC (see 3.2.3).
9–20   **Paris**, 15th Antarctic Treaty Consultative Meeting (see 3.22).

6

9–20	**IMO, London** 16th Session of IMO Assembly (see 3.2.1).
20	**IMO, London** 63rd Session of IMO Council.
23–27	**IMO, London** 12th Assembly of IOPC Fund. (see 3.10).
30–3 Nov	**IMO, London** 12th Consultative Meeting of Contracting Parties to LDC (see 3.11).

DECEMBER

22	UN General Assembly resolves to convene a Conference on Environment and Development in 1992 (see 3.25).

1990

MARCH

7,8	**The Hague** Third International Conference on the Protection of the North Sea (see 4.2).
12–16	**IMO, London** 29th Session of MEPC (see 3.2.3).

APRIL

29	Amendments to SOLAS entered into force (see 3.2.4).

MAY

21–25	**IMO, London** 58th Session of Maritime Safety Committee of IMO.

JUNE

11–15	**IMO**, London 64th Session of IMO Council.

August

27–31	**Rotterdam** Pacem in Maribus XVIII, on Ports and Harbours (IOI)

September

24–28	**IMO, London** 13th Assembly of IOPC Fund.

FORTHCOMING EVENTS

OCTOBER

4–12	**Copenhagen** 78th Statutory Meetng of ICES (see 3.20).
22–24	**Isle of Man**, UK Irish Sea Conference.
29–2 Nov	**London**, IMO 13th Consultative Meeting of the LDC (see 3.11).

NOVEMBER

19–23	**London, IMO** International Conference on Oil Pollution Response (see 3.2.7).

1991

JANUARY

1	Amendment 25-89 to IMDG Code enters into force (see 3.2.4).

FEBRUARY

18	Amendments to Annex V of MARPOL 73/78 are due to enter into force.

MAY

13–17	**IMO, London** 59th Session of Maritime Safety Committee of IMO.

OCTOBER

8–11	**Cairo** 7th Ordinary Meeting of Contracting Parties to Barcelona Convention.

1992

MAY–JUNE	**Brazil**, United Nations Conference on Environment and Development (see 3.25).

2. Overview of Major Pollution Incidents

2.1 OIL POLLUTION

2.1.1 Oil Pollution Trends

In the *1987/88 ACOPS Yearbook* we reported that there had been a reversal of the downward trend in maritime oil spills seen in previous years. Figures for the years 1988 and 1989 indicate that this upward trend is continuing.

Between 1984 and 1987 the Tanker Advisory Service (TAC) in the USA reported low figures on the total accidental oil spills. The number of incidents of oil spill reported by TAC in 1988 increased by 1, from 12 to 13; however in 1989 the total number of incidents reported jumped to 29, an increase of over 100% (see 6.6.4). Increases in tons of oil spilled in 1988 and 1989 were also reported by TAC: 178,265 tons in 1988 and 162,969 tons in 1989. This compares to only 8,700 tons in 1987.

The International Tanker Owners Pollution Federation (ITOPF, see 6.4.6) attended a total of 14 oil spills around the world in 1988; this increased to 24 in 1989, and included the *Exxon Valdez* incident in Alaska, which occupied three ITOPF staff for the spring and summer of 1989. This increase relates in part to attendance at smaller spills than previously.

In UK waters the ACOPS surveys of oil pollution (see 8.1) indicate that beach pollution in 1988 had decreased, but oil spillage in ports was up by 17% on 1987. Reported slicks in the open sea also increased, perhaps due to increase in surveillance flights.

In 1989, preliminary interpretation of the ACOPS oil pollution survey data show a 37% increase in oil spill incidents and reports, and the number of reports increased from 559 to 764. This may in part be an indication of increased public awareness and the many relatively small incidents at sea that have been reported. Reports of oil in the open sea have also increased: 1989 showed an increase of 44%; however, the increase in reports of oil on the coast is only 20%. Incidents of oil pollution in UK ports increased by 22% in 1989.

The Gulf War

On 8 August 1988, the UN Secretary General, Javier Perez de Cuellar, announced that Iran and Iraq had accepted a cease-fire in the eight-year Arabian Gulf War. The cease-fire began on 20 August. According to INTERTANKO, a total of 250 seamen have been killed during attacks on shipping, and some 343 tankers were hit.

2.1.2 Follow-up on Previous Oil Incidents

Amoco Cadiz

On 21 February 1989 awards against Amoco for pollution claims following the *Amoco Cadiz* disaster of 1978 (see previous *Yearbooks*) were adjusted. The additional sum awarded against Amoco came to Frs 116 million; the original judgment was Frs 261 million.

The adjusted ruling was made by Frank McGarr, the retired judge who had made the original ruling. The sum was adjusted for two main reasons: firstly because clean-up equipment used by the French Government was rented, not bought, and hence a resale deduction was not valid; secondly, oyster growers affected by the disaster successfully argued that their compensation did not take into account the long-term effects from oil persisting in sandbanks.[1]

Texaco Refinery, Panama

The Smithsonian Tropical Research Institute of Balboa Panama, has examined the effects of a spill from a Texaco refinery in Panama in 1986 on a complex of mangroves and coral reefs (*ACOPS Yearbook 1986/87*, p6). Both the coral and mangrove were consequently damaged. These ecosystems had been surveyed before the spill occurred. Before the spillage plants and animals covered the roots of mangroves in the study area, including algae and invertebrates, such as sponges, hydroids and other sessile organisms. In channels on the sea-bed the roots supported oysters and barnacles as well as some mussels. After the accident oysters and barnacles in the channels and rivers disappeared from the reefs for 15 months. The effect on some organisms may be serious.

2.1.3 Major Spills involving Ships[2]

Barge

On 24 August 1988 an Eastern Carriers barge suffered a tank fracture and spilled an estimated 212,000 gallons of cargo oil into Chesapeake Bay and Indian Creek on the eastern seaboard of the USA. The incident caused minimal environmental damage, as the oil spilled was light and conditions were hot and sunny, helping evaporation. Response to the spill was also rapid.

Odyssey

On 10 November 1988, the 65,000-ton tanker *Odyssey*, carrying about 1 million barrels of North Sea crude oil split in two during a storm, caught fire and sank in the North Atlantc Ocean. All 27 crew members were lost.

Aoyagi Maru

On 12 December 1988 the Japanese fishing vessel *Aoyagi Maru* struck a reef after losing power while transferring fish in the Bering Sea. More than 52,000 gallons of fuel oil leaked from the vessel.

Nestucca

On 23 December 1988 the tug *Ocean Service* collided with the barge it was towing, the *Nestucca*, on Puget Sound, Portland, USA. Waves breaking over the barge contributed to spillage of oil from the barge, and it was decided to tow the barge out to sea, away from the beaches and sensitive areas. About 70,000 gallons of oil were lost. Oil from the barge washed ashore along a three-mile stretch of beach in Grays Harbor.

For the next three weeks oil continued to float inshore from where the vessels had been towed out to sea. Both beaches and marshes in Washington and in Canada were affected; approximately 20 tons of oil reached the coast of Vancouver Island. Major clean-up initiatives were necessary in both countries. Approximately 900 oiled birds were collected in the area, and another 600 died as a result of oiling.

Bahia Paraiso

On 28 January 1989, the 400-ft Argentine vessel *Bahia Paraiso* ran aground about 1.1 miles from Palmer Station in the Antarctic. The vessel was carrying 250,000 gallons of diesel fuel, and also jet fuel, compressed gas and other drums of unspecified petroleum. Amongst the 316 people on board were 81 tourists. All were safely transferred to Palmer Station. Oil spilled from the listing vessel, at an estimated rate of 2 gallons per minute from open hatchways. After two days the list of the grounded vessel increased. It then lifted and floated free for several hours until finally turning over against the shore.

Much Antarctic wildlife was at risk from the spill, including a colony of 24,000 penguins. Krill were washed up on a short stretch of beach; scientists expect the effect of the oil on the krill will have serious repercussions up the food chain. On 9 February hatches on the *Bahia Paraiso* were sealed to prevent further losses of oil. Attempts to salvage the ship were abandoned in March with the onset of winter. Fears were that the fuel remaining on board would eventually leak out as the ship breaks up.

Humboldt

On 27 February 1989 the Peruvian vessel *Humboldt* ran aground in the Antarctic off Fildes Bay, King George Island. Initial reports were of a 140-yard oil slick floating towards the coast.

Exxon Houston

The 73,212 tonne *Exxon Houston* broke her moorings whilst discharging crude oil at the Hawaiian Independent Refinery on 2 March 1989. The tanker grounded near Oahu and spilled 33,600 gallons of the oil. A two-mile stretch of beach was contaminated.

Exxon Valdez

On 24 March 1989 the tanker *Exxon Valdez* ran onto a reef in Prince William Sound, Alaska. See 2.1.4 for special report.

Hawaii

At the end of March 1989 fuel oil hit the islands of Molokai and Lanai in Hawaii. At the time, the origin of the oil was unknown. However, analysis of the oil identified the US passenger vessel *SS Independence* as the source.

Kanchenjunga

On 26 April 1989 the Indian tanker *Kanchenjunga* grounded six miles from port in Jeddah, Saudi Arabia. The vessel struck a reef in the Red Sea, spilling an estimated 3.9 million gallons of oil. About 20,000 barrels of the oil were blown onto the shoreline north of Jeddah.

Shamrock Ocho/Orange Coral

On 2 May 1989 the *Shamrock Ocho* and the *Orange Coral* collided in the Jurushima Strait, Japan. Approximately 6,000 gallons of oil were reported spilled, causing a 10-mile slick. According to the Maritime Disaster Prevention Center in Tokyo, the slick mostly affected fishermen.

World Prodigy

On 23 June 1989, the Greek tanker *World Prodigy* grounded four miles south of Newport, Rhode Island, USA. About 420,000 gallons of oil were spilled. Response by the US Coast Guard was quick, although they decided against using dispersant, because of the threat this could pose to shellfish. One of the most sensitive spots affected was Narragansett Bay, a rich breeding ground for numerous species of shellfish and fish. The spawning season had just begun when the spill occurred. According to local officials, the ship had entered they bay without a local pilot on board, an action which violates state law.

Presidente Rivera
On 24 June 1989, the Uruguayan tanker *Presidente Rivera* ran aground near Marcus Hook, Pennsylvania, USA, spilling nearly 300,000 gallons of fuel oil into the Delaware River. The oil affected three states and impacted 21 miles of mostly industrial shoreline.

Puppy P
On 29 June 1989 the Maltese tanker *Puppy P* collided with a Panamanian bulk carrier, *World Quince*, in the middle of the Arabian Sea. The *Puppy P* reportedly spilled over 1 million gallons of furnace oil which later polluted the west coast of India between Colaba Point and Worli.

Marao
On 14 July 1989 the Portuguese tanker *Marao* struck a breakwater as it entered the port of Sines, Portugal, spilling about 8,000 gallons of crude oil. Winds carried the 20-kilometre oil slick onto tourist beaches at the height of the season.

Fiona
On 17 September 1989 the Maltese tanker *Fiona* collided with the Liberian registered *Phillips Oklahoma* outside the Humber River, near Hull, England. *Phillips Oklahoma* spilled about 500 tons of its crude oil cargo. The Marine Pollution Control Unit began spraying dispersant on the oil within four hours of the accident. Winds blew the slick away from the shore.

Pacificos
On 4 October 1989 the Cyprus steam tanker *Pacificos* spilled close to 10,000 tons of crude oil off the coast of East London, South Africa.

Mercantil Marica
On 21 October 1989 the Brazilian bulk carrier *Mercantil Marica* ran aground in the Sogne Fjord, north-west of Bergen, Norway. Diesel fuel leaked immediately, but the cargo of bunker fuel leaked minimally. Oil affected about 20km of beach and some oiled birds were recovered. Environmental impact was expected to be minimal.

Milos Reefer
On 14 November 1989 the Greek refrigerated cargo ship *Milos Reefer* spilled at least 237,000 gallons of fuel into the Bering Sea. The spill was a result of the vessel running aground in the Bering Sea National Wildlife Refuge off St Matthew Island. The spill threatened wildlife, including walruses and birds. Authorities were worried about the

possibilities of rats making their way from the boat to the island, where they would pose a threat to birds' nests.

Kharg 5
On the morning of 19 December 1989, the Iranian tanker *Kharg 5* caught fire approximately 100 nautical miles from the Moroccan coast. See 2.1.5 for full report.

Aragon
On 29 December 1989 the Spanish supertanker *Aragon* leaked 350,000 gallons of crude oil into the North Atlantic Ocean north-east of the island of Madeira. Officials said that no clean-up was necessary as the slick as moving north-westerly away from the Madeira Islands. However,

An oil cleaning expert uses a suction pump to attempt to collect some of the oil that reached the shore of Funchal Island (Madeira on 9 February 1989. The oil was believed to have come from the Aragon *(see 2.1.3). Photo: Popperfoto/Reuter*

by February 1990, oil, reported to have originated from the *Aragon* polluted the shores of the island of Funchal (Madiera).

Porto Santo
A crude oil slick 20km long and 3km wide drifted onto beaches on the island of Porto Santo on 15 January 1990. The slick was reported to be moving on towards other Portuguese islands south of Porto Santo, and clean-up crews concentrated on breaking it up. The origin of the oil remained unkown, although officials thought it unlikely to

have come from the Spanish tanker *Aragon*, as that slick had completely disappeared by 2 January. Experts thought the oil more likely to be fresh, possibly from a tanker cleaning out its tanks in the area.

American Trader
On 31 January 1990 a six-ton anchor swung loose from the 80,000 ton *American Trader*, crashed into the ship's side and caused a three-foot gash. The tanker spilled 300,000 gallons of oil which spread across nearly 20 square miles off the Californian coast. Beaches around Los Angeles were affected by the oil, as were many sea birds and migrant birds.[3]

2.1.4 Exxon Valdez

Early on Good Friday 24 March 1989 the tanker *Exxon Valdez* ran aground in Prince William Sound having just loaded some 170,000 tons of oil in the port of Valdez, Alaska. The oil had originated in the oilfield of Prudhoe Bay, on the north coast of Alaska and then traversed the 1300 kilometre Trans-Alaska pipeline to Valdez. According to evidence, given to the National Transportation Safety Board enquiry into the accident, a pilot had navigated the ship through the Valdez narrows and had then left the ship. The Captain, Joseph Hazelwood, ordered a course change from the regular track to avoid icebergs from the Columbia Glacier to the north. He gave instructions for the ship to return to the regular traffic lane when abreast of Busby Island; he then left the bridge. However, the vessel, on auto pilot, continued in a straight line for 11 minutes until it hit Bligh Reef, at a speed of 12 knots. The rocky pinnacles of the reef cut into the hull holing eight cargo tanks and three ballast tanks. Eleven million gallons of oil (36,000 tons) were lost. The US Coast Guard Strike Force

Map of Prince William Sound, Alaska, showing the area most affected by the Exxon Valdez spill. From The Pilot

A worker tests a heated water spray hose on oil covered rocks, Smith Island, Prince William Sound. This was one of the many methods used to attempt to clean oil from the beaches in the wake of the Exxon Valdez disaster (see 2.1.4). Photo: Popperfoto/ Reuter

who quickly air lifted pumps to the vessel transferred the remainder of the oil to other tankers.

When the oil companies set up their terminal at Valdez, the State Government appreciated that a spill might occur so the Alyeska Service Company was set up with enough trained men and equipment to enable them to respond within 5 hours of an incident being reported. In this case it took 12 hours to mobilise and travel the 28 miles to the *Exxon Valdez*; half of the equipment was out of order, while some lay under snow.

Prince William Sound is one of the world's richest fishing grounds. It is also a treasured wilderness area, encircled by wildlife refuges and national parks and forests, so the spill soon became a matter of national concern.

Within a week the oil had moved out of Prince William Sound, and into the Gulf of Alaska; after a month the oil had swept round Kodak Island and into the Lower Cook Inlet, while strong winds pushed the oil into the fjords of the Kenai Peninsula.

Chaos reigned — "the oil must be removed" was the battle cry. The problem of cleaning oil from a rocky coastline is immense; if environmental damage is to be avoided, doing nothing is probably the best, 'cleaning' the worst. Cleaning may be useful on some sheltered rocky shores where oil would otherwise remain for years. Removal of the 'bulk' or 'loose oil' is potentially useful as it reduces re-oiling. This was one aim of at least some of the shore cleaning activities. Pressure hoses fed with hot water were used to remove the oil — and everything else — from the rocks. Organisms were not necessarily removed by the hot water, but some were killed. Cold water hoses were then used to wash the oil down into the sea where booms and skimmers were supposed to collect the oil for disposal. This they failed to do, so the next tide and onshore wind re-deposited the oil on to the rocks. At peak 10,000 people were employed on this thankless task, using water jets, rakes and shovels and paper towels! They were paid some $16.69/hour (about £10). At the end of July Exxon Corporation estimated the total cost of cleanup as US $1,280 million; it was costing some $40 million per week. The Exxon Chairman said this included the cost of paying 10,000 clean-up workers, renting and operating 1,000 vessels and 70 aircraft. The figure also included money to reimburse the Federal Government and States which have contributed to the clean-up.

In the conditions prevailing efforts directed towards dispersing the oil before it reached the shore would have been the best response. This breaks up the oil into tiny droplets which mix with the water and eventually are bio-degraded. In many areas of the world and in some States in America the use of dispersant is forbidden. However, just two weeks before the spill, Alaska environmental officials had authorised the use of dispersants. The onscene Commander did not think there was enough wave motion to disperse the oil so before using it he had demanded tests. The authorities clearly did not have the modern dispersant — as used in the UK — which does not require mixing energy; nor had they 'breaker boards' which the UK has used so successfully with the earlier dispersants. Much time was lost and finally dispersants were not used. It has been interesting to read of a biological treatment, seeding oiled beaches with a special fertiliser to stimulate bacterial degradation. This was suggested by the American Zobell in 1969 and tried with little success many times in the last 20 years.

The most important and as yet unanswered question is what will be the long-term effect on the environment. The whole area was described as one of the world's last untouched wildernesses though this is totally hyperbolic; fishing, routine oil industry operations and past mining must have had a considerable impact. However, much had never been surveyed and little was known about the wildlife. Rather more was known about fish as the fisheries are very rich and fishing is a major industry worth many millions of dollars each year. It was said that all the fish would be killed, that the spawning grounds ruined and the salmon would go away and not return. Great efforts, apparently successful, were made to protect the salmon hatcheries. The herring fishery makes most of its money from roe — several spawning areas were said to be badly oiled. Salmon enters the area to spawn in the rivers. There are five salmon hatcheries which provide between 50 and 60% of the Prince William Sound peak salmon harvest worth as much as 35 million dollars a year. The young salmon feed on the immense plankton blooms which occur in April and May. It has been reported that the annual bloom shortly after the spill was very high.

The bird population suffered greatly and immense numbers of dead birds, mostly auks, were counted — some 27,000 up to mid July. The total bird population is not known. There is also a large mammal population of seaotters, seals, sea lions and even whales, which frequent the bays. Large numbers of seaotters have been found dead or dying. The rare American bald eagle is also at risk: 120 have been reported dead out of a decreasing population of some 5,000.

This spill can be compared with that of the *Torrey Canyon*, when some 70,000 tons of crude oil were lost on the coast of Cornwall, a prosperous inshore fishing ground. It is interesting to note that no reduction in the weight of fish landed at Cornish ports occurred as a result of that spill, in fact, the total landing 1967, was greater than on either the two previous or the two following years.

In mid August the State of Alaska filed a civil lawsuit against Exxon and the Alyeska Pipeline Service Company. The suit filed in Anchorage Supreme Court seeks payment for the *Exxon Valdez* oil spill. The State also wants to ensure that Exxon and Alyeska provide clean-up of more than 600 square kilometres of water and shoreline which were polluted. The State claims Exxon is responsible for the grounding because it failed to select and supervise the ship's officers adequately.

In addition, by congressional decree, federal and state agencies are authorised to make a damage assessment which could result in a financial claim against the spiller sufficient to cover the costs of complete restoration. Exxon, it has been said, will countersue charging that the State itself was responsible for some of the damage because it had argued against the use of dispersants — a charge denied by the State.

The effect of this legal activity is scientists are unable freely to publish or discuss their results, as their employers are interested parties.[41]

2.1.5   Kharg 5

In the early hours of 19 December 1989 the Iranian MT *Kharg 5* (284,000 DWT) suffered a series of explosions following heavy weather damage in the Atlantic Ocean, off the north-west coast of Africa. In the following days, about 70,000 tonnes of the Iranian crude oil cargo was spilled into the sea, most of it more than 100 nautical miles from the Moroccan coastline.

The vessel's owners, the National Iranian Tanker Company, promptly contracted with the Dutch salvage company, Smit Tak, who responded by sending tugs as well as a fire-fighting team and equipment to the scene. However, despite getting the tanker under tow within about two days, it was driven by storms to within 15 nautical miles of the Moroccan coast, although only small amounts of cargo were being lost at the time. Eventually the salvors succeeded in towing the vessel away from the coast and resumed their search for a suitable sheltered location for a ship-to-ship transfer of the remaining cargo. Since none of the countries in the region was willing to offer a safe haven to the *Kharg 5*, the damaged tanker had to be towed some 1,500 miles south, towards the calmer waters of the equatorial regions of the Atlantic Ocean, where a successful ship-to-ship transfer operation was eventually completed in early February, some 50 days after the original accident. The salvors' actions not only saved the tanker and some 200,000 tonnes of cargo, but they also averted a major oil pollution incident.

Much of the crude oil spilled from the vessel in mid-December evaporated and dispersed naturally in the open waters of the Atlantic Ocean, leaving small amounts of emulsified residue and sheen which approached the Moroccan coast. Attempts to disperse these residues with chemicals applied from ships and aircraft were ineffective and so booms and skimmers, obtained from the UK, were deployed to protect sensitive coastal resources, including oyster beds in a lagoon at Oualidia. In the event, a combination of the prevailing northeasterly wind and the southerly-flowing Canary current prevented the highly-scattered patches of emulsified residue from reaching the coast, and carried them further out to sea where they dissipated naturally.

Because of pronouncements of environmental disaster, made by a number of prominent persons, this incident attracted widespread media attention that was out of all proportion to the real threat posed to the coastal resources of Morocco. The actual environmental impact is believed to have been small.[5]

2.1.6   Other Offshore Oil Spills

Piper Alpha
On 6 July 1988 a series of explosions and fires on Occidental Petroleum Corporation's *Piper Alpha* platform in the North Sea caused massive oil spills and loss of life. A total of 167 of the 230 people on board the platform or in a nearby rescue boat were killed. It took more than three weeks for experts to halt the fires at the wells, while work on installing first-stage cement plugs in the 36 wells associated with the platform took until 15 August to complete. The platform was situated about 120 miles north-east of Aberdeen, Scotland and the oil released into the water from the accident dispersed and sank. [6]

2.1.7   Land-based Oil Spills[2]

Mersey River
On 19 August 1989 a Shell pipeline on the Mersey River ruptured, spilling 150 tons of thick crude oil. Initially the oil was taken out with the tide, but later an abnormally high tide and strong winds carried the oil up the river as far as Warrington. Thousands of birds and their habitats were affected by the oil which resisted response to dispersant.

In February 1990 the Liverpool High Court fined Shell Oil (UK) £1 million after pleading guilty to spilling 156 tons of crude oil. Shell also spent £1.4 million on clean-up and damage claims in response to the pollution.

Shell admitted that corrosion, brought on by seawater, had caused a six-inch split in the 16-year-old pipe. The company was criticised for not informing the rivers authority of the incident for 2½ hours.[7]

US Virgin Islands
On 20 September 1989 — the day hurricane Hugo hit — fuel oil spilled from a storage tank on the island of St Croix, US Virgin Islands. About 17,000 gallons of oil that escaped containment contaminated three miles of beach.

Birds feed on a sandbank in the Mersey Estuary as the oil from the ruptured Shell pipeline flows past them (see 2.1.7). Photo: The Independent

Only a small number of birds were affected. The leak was one of the devastating effects of the hurricane.

Amundsen-Scott South Pole Station
An intermittant leak in the fuel system at the US Amundsen-Scott South Pole Station in Antarctica was discovered in late September 1989. The pipeline reportedly leaked about 44,000 gallons of diesel fuel before it was stopped. The fuel has percolated down to a permeable layer of snow, and is expected to move down a further 50 metres where it will be halted by more compacted firn.

McMurdo Station
On 11 October 1989, as much as 50,000 gallons of jet fuel, diesel fuel and petrol leaked from flexible tanks at Williams Field on the Ross Ice Shelf, Antarctica. The fuels became a congealed mass which was scraped up using bulldozers and returned to the US National Science Foundation McMurdo Station nearby. Environmental impact was expected to be minimal.

Arthur Kill, New York
On 2 January 1990 an Exxon pipeline ruptured and spilled more than 500,000 gallons of heating oil into Arthur Kill, the waterway separating Staten Island, New York and New Jersey, USA. It was thought that a boat collided with the pipeline in about 7 foot of water, outside the regular shipping channel. Within 36 hours about 80% of the oil had evaporated, gone into solution or sunk, leaving about 100,000 gallons visible in the water. It was expected to degrade quickly. A major concern for environmentalists was the treat to Pralls Island, a bird sanctuary.

On 9 January 1990, Exxon officials disclosed that workers had failed to detect the leak until six hours after it began because of a mechanical malfunction. Moreover, the leak detection system had been malfunctioning for the previous 12 years.

By mid January environmental damage was being assessed, but initial reports were that the spill had had serious impact. About 400 dead birds had been recovered and 86 live oiled birds had been treated.

2.2 CHEMICAL AND OTHER HAZARDOUS SUBSTANCES

2.2.1 Follow-up on Previous Chemicals Incidents

Brigitta Montanari
We reported in the *1987/88 Yearbook* (p 9) on attempts to salvage the Italian gas tanker *Brigitta Montanari* which went down in 270 ft of water off the Dalmation coast in the Adriatic in 1984. Salvage of the vessel was deemed important because it was carrying 1,300 tons of vinyl chloride monomer (VCM). The long-term effects of VCM are more serious than the short-term, namely it is carcinogenic; it is also very unstable, forming explosive mixtures when it comes into contact with air. In August 1987 VCM was de-

tected in the air and water around the wreck, indicating that it posed a direct contamination threat.

On 8 May 1988 the vessel was lifted to 55 metres and pulled to a sheltered bay and raised to 30m. Between 11 and 14 June 1988 the VCM was transferred to another Italian tanker. All 700 tons remaining in the tanks of the vessel were recovered safely. The *Brigitta Montanari* was then refloated and towed to the scrapyard.[8]

2.2.2 Chemicals in the Sea

Tributyl Tin (TBT)
According to the UK junior Agriculture Minister Donald Thompson, curbs on the use of TBT in anti-fouling paints for small boats and the fish-farming industry in May 1987 had already produced significant positive effects by in 1988. In the Crouch Estuary TBT levels in shellfish had fallen by two-thirds, while in other estuaries monitored in England concentrations had fallen by 'over half'.[9]

Seals

In the summer of 1988 seals hit the headlines in North Sea states, as a virus devastated the population (see 2.2.2). Photo: Don McPhee

Between April and September 1988 the plight of North Sea seals hit newspaper headlines throughout the North Sea states. Seals were dying by the hundred, from a mysterious disease which was found to be closely related to the virus that causes cannine distemper in dogs. By September Dutch scientists had developed a vaccine but administering it proved difficult.

In October a *Siren* report estimated that 14,000 common seals had been washed up in Denmark, Sweden, Norway, FRG, Holland and the UK. This is 23% of the estimated seal population of the north-western hemisphere. It was predicted that up to 80% of the seal population could be killed by the epidemic. Grey seals, whose main habitat is in the UK, were also affected.

Scientists and activists linked the disease with pollution in the North Sea, in particular PCBs which are believed to suppress the immune system.

More than 11,000 tonnes of heavy metals and 5 million tonnes of treated sewage are dumped in the North Sea every year. The Baltic Sea takes 10,000 tonnes of phosphorus and 1.2 million tonnes of nitrous wastes yearly.[10]

It was stated in the House of Commons in July 1989 that the Eastern North Sea received 768,000 tonnes of nitrogen from the Continent as opposed to 79,000 tonnes from Britain. The Continent discharges 14.6 tonnes of mercury compared with 1.9 tonnes from Britain. For cadium the figures are 31.6 and 5.8 tonnes. While industrial waste is only 1/6th of the 1.2 million tonnes from the Continent and of the waste incinerated in the North Sea is only 2% of the total.

Billingham

ICI is to spend up to £35 million to forecast river and sea pollution from their Billingham factory (N.E. England) which discharges 300,000 tons of acid ammonium sulphate each year. 165,000 under licence in the North Sea and 60,000 in the estuary of the river Tees. The plant is due for completion in mid 1990s. The waste will be converted into sulphuric acid used elsewhere in the plant.

Karin B

The saga of toxic waste and the *Karin B* was closely followed by the popular media in the summer of 1988. The saga began in Koko, Nigeria, in June 1988, when the Nigerian Government invited the US Environmental Protection Agency, Harwell Laboratory and Friends of the Earth to a site at Koko, Bendal State. FoE's account was of some 3,500 tonnes of hazardous waste contained in drums on the outskirts of Koko. The waste was in 10,000 drums and 30 shipping containers which had been brought from Italy over a nine-month period. Some 20 types of waste were listed on a manifest, including PCBs, biocides and drugs, solvents, acids, residues and industrial solids. Almost half the containers were found to be leaking, others were piled unprotected against heat or storm.

Although most of the containers seemed to be Italian in origin, others were from Norway and West Germany. The Italian Government hired Ambiente, a subsidiary of the Italian state-owned ENI, to find a disposal outlet for the waste. The cargo of wastes was placed on the *Karin B* and a second vessel, the *Deep Sea Carrier*. Whilst the ships were at sea, however, public pressure caused the Italian Government to back-track and press for the cargo to be off-loaded in another country. A British company, Leigh Interests, offered to analyse and re-package the cargo, and possibly dispose of it in the UK. However, the company had not counted on the measures the British Government could call upon to prevent the *Karin B* from offloading. These included the Dangerous Substances in Harbour Areas Regulations 1987, and the Control of Pollution Act 1974.

The British authorities succeeded in turning away the *Karin B*, amidst wide media coverage; French and Dutch authorities did likewise. At this point the Italian authorities relented and ordered the *Karin B* to sail for Italy.

The incident prompted Member States of the European Communities to move towards implementing EC transfrontier shipment Directives which were supposed to have been tranposed into state law by October 1985 and to re-define the meaning of 'hazardous waste' in existing legislation.

This and the Venezuelan incident (reported in the *ACOPS Yearbook 1987/88*, p.8) serve to illustrate how vital and timely the Basle Convention on the Control of Transboundary Movements of Hazardous and Toxic Waste and their Disposal (see 3.3.12) is.[11]

Piper Alpha

Four PCB transformers were housed on the *Piper Alpha* platform which exploded on 6 July 1988 (see 2.1.6). Occidental, owners of the oil rig, were unable to confirm the fate of the transformers and their fluid contents. A preliminary assessment of other chemicals on the platform was made by Occidental, and the company concluded that "chemicals that were likely to have been released into the marine environment will have been rapidly dispersed and diluted". Traces of PCBs were later detected in the mud of the sea surrounding the platform.[2]

Perintis

In March 1989 the cargo ship *Perintis* sank about 35 miles south-east of Brixham, Devon, in the English Channel. It was carrying in its cargo a number of highly toxic pesticides: 1 tonne Permethrin; 0.6 tonne Cypermethrin; on the deck were large containers of Lindane. The deck cargo of Lindane broke loose but all containers were recovered, except for one, carrying 0.5 tonne Lindane. This container sank.

The drums containing the Permethrin and Cypermethrin were located inside the sunken ship and were retrieved. In April 1989, the missing container of Lindane had still not been located. [12]

Maasgusar

On 14 March 1989 the chemical tanker *Maasgusar* exploded and caught fire south of Tokyo. The vessel was carrying a cargo of 25,700 tons of methanol, acrylonitrile and other chemicals. A total of 23 crew were reported missing.[2]

Tampa, Florida

On 16 May 1989 a 420,000-gallon storage tank gave way at a storage lot in Tampa, Florida, USA, discharging a total of nearly 132,000 gallons of phosphoric acid. Local US Coast Guard reported that about 21,000 gallons of the acid flowed down to the Sparkman Channel before it was diverted into a retention pond. Prop wash was used to restore the pH balance of the channel; no large fish kills were observed.[2]

Canadian scientists report surprisingly high levels of toxic chemicals in snow taken from remote areas in the Arctic. The pollutants include polychlorinated biphenols, organochlorine insecticides and polycyclic? aromatic hydrocarbons.

2.3 RADIOACTIVE POLLUTION

2.3.1 Incidents

Muroroa

According to a report made public on 10 November 1988 by explorer Jacques-Yves Cousteau, nuclear test explosions have caused cracks and rock slides in Muroroa atoll. Scientists have also found traces of iodine 131, a faintly radioactive substance, in Muroroa lagoon. The French have already moved their nuclear test site to Fangataufa, a neighbouring island which has a stronger base.

In early September 1989 a medical researcher in Australia claimed that medical statistics in the territory of French Polynesia were being manipulated to make the rate of cancer in the region appear lower that it really is.[13]

2.3.2 Disposal of Nuclear Waste

UK plans to dump nuclear submarines at sea came to a halt on 2 November 1989 when the London Dumping Convention ruled that military vessels still fell within the brief of the Convention. Options of drilling holes in the sea-bed and dropping submarine parts into a sea-bed 'grave' were also ruled out by the LDC meeting (see 3.11).[14]

This gannet, found on a UK beach has fallen foul of plastic rope. Plastics will generally float until they are washed ashore or entangle in something that will pull them underwater. Photo: TBG

2.4 PLASTIC WASTE

QE2

On 5 July 1989 the *QE2* was labelled by the *Today* newspaper as the 'Dustbin of the Sea'. This was after two sailors, disgusted by the practice of dumping bags full of refuse overboard at night, captured the action on video. The

entry into force of Annex V of MARPOL 73/78 on 31 December 1988 clearly prohibits the dumping of garbage and other wastes from ships into the sea (see 7.3.2).[15]

Fish Nets

Over the past two years, the often devastating effects that fish nets of certain design can have on marine wildlife has been picked up by the popular media.

In the past diving birds and sea mammals have been able to avoid fishing nets because these were made of hemp, cotton or flax and their fibre size made them acoustically or visually detectable. Discarded nets would also sink and eventually rot. Today nets are made of very thin and extremely strong plastic filaments. They are often not detected by birds and sea mammals until they become entangled, and they don't rot when discarded, and only sink when they have gathered enough debris to pull them to the bottom.

Chris Mead of the British Trust for Ornithology has analysed the recovery of ringed auks in the Channel and southern North Sea area. He compared historic causes of death with recent deaths (June 1987–June 1989) and found the percentage of deaths due to entanglement in nets had increased from 4.8% to 37% for guillemots; the figures for razorbills had risen from 6.5% to 26%.[16]

It has been estimated that, in the Pacific, the Japanese squid fishing fleet sets a total of 30,000 miles of monofilament nets every night during the seven-month season. Up to 100,000 marine mammals and half a million birds die in it annually. Up to 10% of North Pacific fur seals are killed each year in discarded fishing nets. Annex V of MARPOL (see 7.3.1), prohibiting the dumping of plastics, came into force at the beginning of this year. This Annex includes discarding of fishing nets. However, any fishing boat from a country not party to the Convention can still dump plastic waste and nets outside territorial waters. Many countries, Pacific states in particular, have banned the use of monofilament nets in their waters and zones (see 3.19).[17]

2.5 ENDANGERED SPECIES

Olive Ridley Turtle

The Olive Ridley turtle has only two main breeding grounds in the world: one in India, the other on the Pacific Beaches at San Augustinillo in Mexico. It is at this second site that the *Independent on Sunday* reported massive slaughter of the turtles in March 1990. Turtle killing is not in itself illegal in Mexico; there is a quota for their capture. Turtle skins are sent to Japan where they are made into shoes and handbags; turtle eggs are a delicacy and a purported aphrodisiac.

According to fishery ministry officials in Mexico the quota of 20,000 Olive Ridley turtles for each breeding season is not being exceeded. The *Independent on Sunday* report puts the actual figure slaughtered this season at 70,000. The Olive Ridley turtle is an endangered species. The Mexican Government is presently considering signing an international agreement banning trade in endangered species.[18]

Birds

According to the British Trust for Ornithology, Arctic terns have not bred successfully in the British Isles since 1982. Some 25 years ago there were 20,000 pairs; in 1989 there were an estimated 1,000 pairs. Eric Mortensen of the BTO points out that Shetland seabirds had a disastrous breeding season in 1988, and that the same was true in the Faeroes. He advocates a complete ban on hunting; but also that the most urgent need is for an investigation into the state of the birds' food supplies. [19]

2.6 SEWAGE

2.6.1 Algal Blooms

A variety of toxic substances are produced by many phytoplankton species, mainly pelagic dinoflagellates. Factors influencing the growth, or 'bloom', of these species include temperature, sunlight and particularly the availability of nutrients such as phosphorus and nitrogen compounds. In fresh water lakes, phosphates are the major limiting factor. Common land-based sources of nutrients that may cause algal blooms are domestic wastes (principally sewage), agricultural run-off of excess fertilisers, animal wastes from intensive livestock units, aquaculture and some industrial effluents and atmospheric decomposition. Some of the toxic substances produced by these algae are highly poisonous to man and other animals; drinking from contaminated freshwater lakes where concentrations are high have resulted in deaths of domestic animals. In sea water, toxic substances are diluted in the water, however, they are accumulated by filter feeders, mainly bivalve molluscs.

The effects of these toxins on man have been recognised for some time; between 1969 and 1983 some 200 cases of paralytic illness resulted from the eating of mussels contaminated by algal blooms.

In the summers of 1988 and 1989 toxic algal blooms developed off southern Scandinavia. In May/June 1988 blooms were particularly virulent in the Skagerrak and the Kattegat; about 600 tonnes of caged fish were killed, along with bottom-dwelling organisms. The bloom coincided with a virus epidemic which killed thousands of seals on the coasts of Germany, the Netherlands and Britain. The algae behind these mortalities was *Chrysochromulina polylepis*; surprisingly, it was the first time it had been observed in algal blooms and was not previously known to exert toxic effects.

The blooms appear to have been caused by a combination of meteorological, hydrographic and human influences, according to a European Commission report. [20] High rainfall in winter months had caused high run-offs of nitrogen; high winds mixed the nitrogen into deeper layers of water where nitrogen levels were already elevated due to rising inputs from farmland, sewage works and the atmosphere. Light winds in May/June kept nitrogen levels at lower sea levels high and this, combined with high temperatures gave the right conditions for *C. polylepis* to bloom.

Public pressure helped move a number of North Sea governments to act: the West German Government announced a £5 billion programme to install nitrogen and phosphorus stripping equipment in sewage works; Danish and Swedish authorities introduced similar programmes, as well as curbs on fertiliser use. [21]

In August 1989 another alga, *Primnesium parvum,* bloomed in fjords near Stavanger, Norway, causing the death of at least 500 tonnes of salmon. [22] Also in August 1989, an orange tide of algae appeared off the west coast of Cornwall. The algae caused irritation to swimmers and may have caused the death of lugworms and crabs through depletion of oxygen. Meanwhile on 7 August, Belgium's public health department warned people not to swim in the North Sea because of high concentrations of *Salmonella* in the water, caused again by the hot weather. [23]

Brown-yellow algae bloomed along the Italian Adriatic in the region of Rimini, according to a report in the *Guardian* on 8 August 1989. Dead fish, crabs and mussels were washed up along the coast, although the slime was declared safe to swim in.

Warm weather in the spring of 1990 gave rise to a need to ban consumption of mussels (*Mytilus edulis*) on the north-east coast of England for a short period.

These incidents indicate the need for regular monitoring programmes to ensure public safety, and to minimise disruption to shell fish harvesting.

On 5 March 1990 UK Environment Secretary, Christopher Patten, announced the dumping of raw sewage sludge by the UK into the North Sea would be phased out by the end of 1998. This is three years later that other North Sea states are demanding. The announcement was made on the eve of the Third North Sea Conference (see 4.2) and represented a complete turnround in policy from four months previously. [24]

2.6.2 Blue Flag Awards

The European Blue Flag Campaign is sponsored by the Commission for the European Communities and a number of other sponsors throughout the Community. It was launched during 1987, the European Year of the Environment, at the initiative of the Foundation for Environment Education in Europe.

The main objectives of the Blue Flag Campaign is protection of the marine and coastal environment, including beaches. The attribution of a Blue Flag to beaches and ports denotes a high standard of environmental quality, and the provision of basic facilities and information. The European Blue Flag for Beaches is awarded annually. Applicant bathing beaches have to fulfil a number of criteria, one of the most important being water quality. The European Blue Flag for Ports is awarded to marinas or ports with mixed use, but not to commercial ports. Initial awards have been made for three years.

In 1989 European Blue Flags were awarded by the European Jury to 567 beaches and 126 leisure ports/marinas. This was a big increase over 1988. Table 1 lists the number of beaches/ports in each country that applied for an award and the number of Blue Flags awarded. [25]

Table 1   Blue Flag Awards 1989

Country	No. of Beach applicants	Beaches awarded Blue Flag	Ports awarded Blue Flag
Belgium	39	24	-
Netherlands	ng	7	5
Denmark	100	89	37
UK	41	22	3
FRG	28	14	11
Ireland	36	22	-
France	204	125	18
Italy	78	17	10
Portugal	170	107	3
Spain	285	120	32
Greece	*	6	8

ng: information not given.

* Greek authorities only monitored beaches where there is a risk of pollution; hence many Greek beaches that could expect to qualify for a Blue Flag award cannot compete due to lack of data on water quality.

Notes and References
1. *Lloyds List* 22 February 1989.
2. Information from *Oil Spill Intelligence Reports* (*OSIR*) or *Lloyds List*, unless otherwise stated.
3. *OSIR*; *Guardian* 10 February 1990
4. Compiled from *National Geographic* (January 1990) and *New Scientist* reports.
5. Report by Ian White, ITOPF.
6. *OSIR*.
7. *OSIR*; *Guardian* 24 February 1990.
8. *Medwaves*, Issue 16/1/1989.
9. HC *Written Answers* 1 December 1988.
10. *Siren* no. 38; October 1988.
11. *ENDS Report* no. 161, p9; no. 163, p8.
12. *Metocean Update*, Issue 9, May 1989.
13. *Guardian*, 11 November 1988; *New Scientist*, 16 September 1989.
14. *Guardian* 3 November 1989.
15. *Today* 5 July 1989.
16. *BTO News* no. 163, 1989, p1.
17. *Guardian*, 27 October 1989.
18. *Independent on Sunday*, 4 March 1990.

As the Yearbook goes to press, news is coming in of another tanker casualty. On 9 June 1990 the Norwegian oil tanker Mega Borg caught fire in the Gulf of Mexico. Much of the oil that escaped was consumed by the intense heat. Oyster beds and shrimp nurseries were put at risk by the event. Photo: Walter Frerck/Popperfoto

19. *BTO News* no. 163, p8.
20. Water Pollution Research Report 10, HMSO.
21. *ENDS Report* no. 163, p4.
22. *ENDS Report* no. 175, pp5–6.
23. *Guardian* 8 August 1989.
24. *Guardian* 6 March 1990.
25. *The European Blue Flag* 1989, CEC.

3. Activities of the United Nations and other Global Intergovernmental Agencies and Meetings

Foreword

by AJ Gabaldón
Former Minister for Environment
Venezuela

During the year under review, the environment has become a top global issue. When the leaders of seven major industrial nations and the EEC met in Paris in July 1989 for the 15th Annual Economic Summit (see 3.23), the environment was brought to the centre stage in the world economy. This 'green summit' highlighted a clear commitment to policies based on sustainable development.

Many other political leaders and heads of major international organisations have now expressed such a commitment. Consequently, many international agencies have significantly strengthened their environmental programmes: locally, nationally and at the regional and international levels.

The global environment, in particular the atmosphere, ozone layer, the oceans and the greenhouse effect, have frequently made headlines during the past year. With this renewed interest, integrated ocean management also received new impetus. The UN Office for Ocean Affairs and Law of the Sea emphasised protection and preservation of the marine environment (see 3.1) and the IMO launched the 'Strategy for the Protection of the Marine Environment', which focuses on pollution from ships and disposal of wastes at sea. Preparations are underway for the UN Conference on the Environment and Development (UNCED-92; see 3.25) where oceans is being considered as one of the major issues to be addressed.

Various events are being staged during 1990 and 1991 to further discussions at the regional level. Additionally, a Second International Symposium on Integrated Global Ocean Monitoring (IGOM-II) is now being planned by several UN agencies for April 1991, to be held in Leningrad. The Symposium will review the latest experiences with respect to ocean research, interpretation, management, observation and formulation of integrated ocean studies.

All releveant international programmes are expected to be represented.

Several important initiatives for joint environmental management are also developing. Take, for example, Latin America and the Caribbean. The Sixth Ministerial Meeting on the Environment in Latin America and the Caribbean, held in Brazil in March 1989, stated the need for an unprecedented level of international cooperation at the global, regional and sub-regional levels. This cooperation should be based on the principle that each state has the sovereign right to manage freely its own natural resources and that high priority should be accorded to free access to scientific information and to the transfer of non-polluting technologies. Also, the members of the Permanent Mechanism for Political Consultation and Coordination (Argentina, Brazil, Colombia, Mexico, Peru, Uruguay and Venezuela) and the Member Countries of the Treaty for Amazonian Cooperation (Bolivia, Brazil, Ecuador, Guyana, Peru, Uruguay and Venezuela), recognised that efforts to achieve environmentally sound development must be redoubled. Moreover, the Caribbean Community (CARICOM) established a Standing Committee of Ministers on the Environment which held its first meeting in Port of Spain, Trinidad and Tobago, in June 1989.

Additionally, the United Nations Development Programme and the Inter-Americam Development Bank launched the drafting of a regional environmental agenda for Latin America and the Caribbean in order to provide regional response to the Brundtland Report. Through its regional office in Mexico, UNEP is working on a Regional Action Plan for the Environment. The Economic Commission for Latin America and the Caribbean (ECLAC) is coordinating the regional contribution to the process leading up to the UNCED-92, mentioned above, which will be convened in Brazil in June 1992. For the Wider Caribbean, the Caribbean Conservation Association (CCA) is developing several strategies for joint action among NGOs in preparation for UNCED-92, and UNEP's Caribbean Environment Programme (CEP) has been steadily developing, providing the framework for cooperation on the joint management of marine and coastal resources.

CEP is focusing on the assessment and control of marine pollution and the management of specially protected areas and wildlife (see 3.3.4). Its long-term goal is to achieve sustainable development of marine and coastal resources in the Wider Caribbean Region through effective integrated management that allows for increased economic growth.

The importance of regional arrangements is increasingly being recognised by international organisations as a more effective way of addressing environmental issues based on a world economy which is becoming increasingly multipolar. Regional arrangements are proving very successful, particularly with respect to the implementation of marine programmes. They allow for a high level of participation and wide responsibility of states in the management of marine resources that are of common interest. They are destined to become an essential component in the development of the whole 'new ocean regime'.

When analysing the UNEP Regional Seas Programme, the Brundtland Report recognised that the political strategy behind the programme and the requirement that management and financing be undertaken by particpiating countries had clearly been crucial to its success. The Carib-

bean Environment Programme is illustrative in that all the states contribute and actively participate in one way or another, including smaller and poorer ones. Certainly, some of the Regional Seas Programmes have clearly demonstrated against the perception of the passive attitude of poor countries expecting rich ones to provide resources. They support the principle that the greater the participation states have in the conduction and financing of programmes, the greater their responsibility will be.

The need to strengthen international joint action for sustainable development of ocean resources is becoming greater as the gap between rich and poor nations continues to widen. There is no other option than to build carefully around and expand upon the most efficient programmes. In my view, the management of the oceans in a multilateral context could provide the opportunity for cooperation among nations that will set the basis for addressing more difficult tasks which lie ahead. I hope that the coming years will see a net development in this sense.

Dr AJ Gabaldón
Former Minister for Environment
Venezuela

3.1 UN OFFICE FOR OCEAN AFFAIRS AND THE LAW OF THE SEA (OALOS)

OALOS is the focal point of the United Nations' activities in the field of marine affairs and the law of the sea. The primary responsibility of the Office is to provide advice, assistance and information to states on all matters related to ocean affairs and the law of the sea, including the protection and preservation of the marine environment; and to promote the widespread acceptance of the United Nations Convention on the Law of the Sea (UNCLOS).

Given the real and potential dangers to the oceans posed by land-generated and other forms of pollution, the Convention devotes considerable attention to this issue. It gives Coastal States the power to enforce national standards and anti-pollution measures within their territorial waters. Outside those waters, however, States must enforce 'generally accepted international rules and standards'. States are also called upon to cooperate at the regional level in order to harmonise their anti-pollution measures and standards.

Recent Activities

In recent years, the international community has become increasingly aware of the danger posed to the marine environment by a variety of activities. On 1 November 1988, the General Assembly of the United Nations expressed its deep concern at the current state of the marine environment and requested the Secretariat to prepare a special report on recent developments related to the protection and preservation of the marine environment in light of the relevant provisions of the Convention on the Law of the Sea.

OALOS prepared the special report (United Nations Document A/44/461 and Corr.1) in response to the request by the General Assembly. The report provides an overview of the Convention as embodying a global framework of new environmental law, a survey of the current state of the marine environment, and, on the basis of this survey and assessment, it identified a number of major areas on which future action should be focused.

Among other things, the report recommended increased international cooperation to control pollution from or through the atmosphere, cooperation to deal with oil spill disasters, and regulation of pollution by dumping. The report also emphasised the importance of wider adherence to existing instruments such as the Convention on the Law of the Sea, which the World Commission on Environment and Development called "the most significant initial action that nations can take in the interests of the oceans' threatened life-support system".

Upon a further request made by the General Assembly at its 1989 Session, OALOS is currently preparing an updated and expanded report on the protection and preservation of the marine environment as a contribution to the proposed 1992 United Nations Conference on Environment and Development (see 3.25).

Among the other activities of the Office in 1989 was its coordination of an Ad Hoc Inter-agency Consultation on Ocean Affairs, convened in Geneva (12–14 July) and chaired by the United Nations Under-Secretary-General for Ocean Affairs and the Law of the Sea, Mr Satya N Nandan.

The marine environment was among the major topics addressed at the consultations, where it was agreed that there was a general need to promote understanding of the important relationship between the ocean environment and the global environment and to encourage scientific research in that respect.

The consultations also concluded that preparations for the 1992 United Nations Conference on Environment and Development should take special account of the role of the oceans for the wider environment as well as address marine environment questions more directly.

Further, OALOS is currently preparing for publication a compilation of all multilateral and bilateral treaties, conventions and agreements having to do with marine pollution. The Office also maintains a compilation of all relevant national legislation in this area.

As part of its activities to promote the protection of the marine environment, OALOS actively participates in and supports the work of the Joint Group of Experts on the Scientific Aspects of Marine Pollution (GESAMP), which is made up of a number of international organisations concerned with the development of marine sciences and/or the protection of the environment.

Sea-bed Mining

The exploitation of mineral resources of the deep sea-bed beyond the limits of national jurisdiction, an important component of the Convention on the Law of the Sea, has been viewed as one of the potential sources of marine pollution. Therefore, the Convention gives the International Sea-Bed Authority, the institution established to administer sea-bed mining, broad discretionary powers to assess the potential environmental impact of a given sea-bed mining venture, recommend changes, formulate rules and regulations, establish a monitoring programme and issue emergency orders to prevent serious environmental damage.

To assist in this process, OALOS has prepared for consideration by the Preparatory Commission for the International Sea-Bed Authority and for the International Tribunal for the Law of the Sea a working paper containing draft regulations on the protection and preservation of the marine environment from activities in the international sea-bed area.

The draft regulations were submitted to the Preparatory Commission at its session held at Kingston, Jamaica 5–30 March 1990. The regulations deal with matters such as procedure for submission and content of an environmental impact statement of a proposed sea-bed mining venture, compliance by a contractor with the obligation to preserve and protect the marine environment and contingency plans for dealing with situations that result in serious harm to the marine environment.[1]

3.2 INTERNATIONAL MARITIME ORGANISATION (IMO)

3.2.1 The Assembly

The 16th Session of the IMO Assembly, which is held biennially, took place in London on 9–20 October 1989.

A resolution was approved which directs the Marine Environment Protection Committee (MEPC) to develop a global convention on oil spill preparedness. In the resolution, the Assembly gave guidance to MEPC that the convention should provide a "framework for international cooperation for combating major oil pollution incidents taking into account the experience gained within existing regional arrangements". The resolution also calls upon the Committee to address:

* the establishment of an international information centre and national response centres;

* the requirement of shipboard contingency plans;

* the pre-positioning of oil spill response equipment; and

* the maintenance of a worldwide inventory of response equipment.

The Secretary-General of IMO was asked to identify the requirements for establishing an international information centre under the auspices of the IMO. Under the timetable approved, the global convention would be adopted by November 1990, and a working group would meet during the MEPC meeting in March with a week of preparatory session held in June to finalise the treaty.

Another resolution approved by the Assembly calls upon nations to implement more effectively the existing IMO conventions.

Under Agenda item 2, the Consideration of the Reports and Recommendations of the Maritime Safety Committee, ACOPS submitted a paper giving a brief account of its Conference on Protecting the Environment from Hazardous Substances and the Conclusions of a meeting of Vice-Presidents of ACOPS. A copy of this paper is contained in Annex I.

3.2.2 Council

The IMO Council is composed of 32 Member States elected by the Assembly for two-year terms. The Council is the executive organ of IMO and is responsible, under the Assembly, for supervising the work of IMO. The IMO Council normally meets twice a year. Between the biennial sessions of the Assembly, the Council performs all functions of the Assembly, except that of making recommendations to governments on maritime safety and pollution prevention; these are reserved for the Assembly. The Council reports to the Assembly on its activities at the annual Session of the Assembly.

As at 31 July 1989, IMO membership stood at 133, two more states having become members during the preceding 12 months.

The 61st Session of the Council was held 21–25 November 1988. At the Meeting the Council took the decision to halve the 1989 programme of meetings because of anticipated budgetary deficit. The savings affected mainly the work of the Maritime Safety Committee.

The 62nd Session of the Council as held 5–9 June 1989. On 7 June 1989, the Council elected Mr William O'Neil of Canada to be the Secretary-General, a four-year term which began on 1 January 1990. He succeeded Mr CP Srivastava of India who had been IMO's Secretary-General since 1974. Mr O'Neil has been associated with IMO since 1972 and was elected as chairman of the Council in 1979.

3.2.3 Marine Environment Protection Committee (MEPC)

The 28th Session of MEPC was held in London on 17 October 1989.

The IMO Secretary-General, Mr C P Srivastava, addressed the Committee pointing out that the protection of the environment is now at the head of the political agenda. He also referred to the draft resolution tabled at the 16th Session of the Assembly which requests the Committee to develop, as a matter of urgency, a draft international convention providing the framework for international cooperation for combating major oil pollution incidents.

William O'Neil from Canada takes the helm at IMO.

Consideration and Adoption of the Proposed Amendments to Annex V of MARPOL 73/78

The Committee considered and adopted the proposed amendments to regulation 5(1) of Annex V of MARPOL 73/78 to designate the North Sea as a special area. It was agreed that the special area requirements should be applied to the North Sea as soon as the amendment entered into force which was deemed to be on 18 February 1991.

29th Session

The 29th Session of MEPC was held at IMO headquarters in London from 12–16 March 1990. Subjects on the agenda included:

- report of the Sub-Committee on Bulk Chemicals;

- uniform interpretation and amendments of MARPOL 73/78 and of the Oil Record Book;

- implementation of Annexes III, IV and V of MARPOL 73/78 and amendments to the IMDG Code to cover pollution aspects;

- prevention of oil pollution from machinery spaces — oily-water separators and monitors and fuel oil sludge;

- prevention of pollution by noxious solid substances in bulk and consideration of possible development of the new Annex VI of MARPOL 73/78;

- technical assistance programme;

- enforcement of pollution conventions: 1, violations of conventions and penalties imposed; 2, casualty investigations in relation to marine pollution;

- Anti-Pollution Manuals: Manual on Chemical Pollution, Section 2 — Search & recovery of packaged goods lost at sea; Manual on Oil Pollution, Section V — Legal aspects;

- identification of particularly sensitive areas, including development of Guidelines for designating special areas under Annexes I, II and V;

- use of Tributyl tin compounds in anti-fouling paints for ships;

- implications of the 1969 Tonnage Convention of MARPOL 73/78;

- prevention of air pollution from ships, including fuel oil quality;

- updating of action plan;

- future work programme.

3.2.4 Maritime Safety Committee (MSC)

The 56th Session of the MSC was held 24–28 October 1988 at IMO headquarters. At the Session, a series of amendments to the International Convention for the Safety of Life at Sea (SOLAS), 1974 were adopted. The amendments are aimed at improving the safety of passenger ships, and the majority were based on proposals put forward by the UK in the aftermath of the *Herald of Free Enterprise* disaster. Three amendments all involve regulations dealing with construction — subdivision and stability, machinery

and electrical installations.

Briefly, the amendments are :

- amendment to Regulation 8 of Chapter 11-1, to improve stability of passenger ships in damaged condition. This amendment will only apply to ships constructed after 29 April 1990. A further amendment to Regulation 8 of Chapter 11-1 is concerned with intact rather than damaged stability;

- amendment adding a new Regulation 20-1 which requires that cargo loading doors shall be closed and locked before a ship proceeds, and remain so until berthing. There is special provision for ships whose doors cannot be locked or opened at the berth;

- amendment to Regulation 22 states that a lightweight survey must be carried out on passenger ships at periods not exceeding five years.

These three amendments became law on 29 April 1990.

A number of other amendments were proposed but the Committee decided that further consideration was necessary before any amendments could be adopted.

The Committee expressed concern that only a limited number of countries have reported to IMO on the implementation of measures to prevent unlawful acts against passengers and crews on board ships.

Draft guidelines on management for safe ship operation and pollution prevention were also considered at the Session. They were due to be discussed at the next Session of MSC.

57th Session

The 57th Session of MSC was held 3–12 April 1989. The Committee unanimously adopted Amendment 25-89 to the International Maritime Dangerous Goods (IMDG) Code and agreed that it should enter into force on 1 January 1991. The Amendment will involved publication of a new edition of the IMDG Code. This amendment is important because it extends the application of the Code to marine pollutants. Marine pollutants have been included in the Code to assist implementation of Annex III of the International Convention for the Prevention of Pollution from Ships 1973 as modified by the Protocol of 1978.

The IMDG Code now includes more than 2,500 pages in five volumes. It was agreed that substantial amendments should normally be adopted at intervals of not less than four years.

A draft Assembly resolution on the use of pilotage services in the Euro-Channel and Ij-Channel was approved by the Committee. The channels lead to Rotterdam Europort and the Ijmuiden and are difficult to navigate. The draft resolution notes that, due to risk of grounding and collision, there is a danger of pollution to the North Sea.

The Committee adopted proposed amendments on navigation through the English Channel and Dover Strait. A new traffic scheme off Finisterre was also adopted.

The Committee approved a draft resolution on Guidelines and Standards for the removal of offshore installations and structures on the Continental Shelf and in the EEZ. After January 1998, no structures should be installed unless their design is such that complete removal is feasible when the time comes to abandon them.

The 58th Session of MSC was held at IMO, London 21–25 May 1990. The 59th Session is due to be held on 13–17 May 1991.

3.2.5 The Legal Committee

The 61st Session of the Legal Committee took place in London on 28–29 September 1989. The following issues were discussed:

The Salvage Convention 1989
The Committee took note of the report on the International Conference on Salvage (see 3.2.6) and agreed to urge governments to give consideration to early signature of the Convention, which would in turn accelerate ratification of the Convention in due course.

Matters arising from the 61st and 62nd Session of the Council
The Committee noted that the Council had endorsed the recommendation of the Legal Committee for a conference of one week's duration to adopt a protocol to the 1974 Athens Convention on the Carriage of Passengers and their Luggage by Sea. It was agreed to recommend that the conference be held at IMO Headquarters from 26–30 March 1990, and to request and authorise the Secretary-General to circulate the draft protocol immediately to governments and organisations concerned. It was also decided to recommend to the Conference to adopt the same procedure on voting rights as was adopted by the 1984 Conference under which substantive decisions on amendments or revisions to the conventions were taken by the customary two-thirds vote of all States participating in the conference, but with an additional requirement regarding the concurrence of a specified proportion of the States party to the conventions being amended. One of the intentions of the revisions at the proposed conference was to encourage greater acceptance of the 1974 Convention by revising the limitation amounts to make the Convention acceptable to more States.

Work Programme 1990
The Committee reaffirmed its decision that top priority would be given to work on a possible hazardous and noxious substances (HNS) Convention during 1990. If time allowed some work on the draft convention of offshore mobile craft might be undertaken but it would not be included on the work programme at this stage. Governments and interested organisations would be circulated with the draft convention on offshore mobile craft well in advance of the 63rd Session to be held in September 1990.

Further work on maritime liens and mortgages and related subjects would depend on the recommendations of the Joint Intergovernmental Group of Experts and the decisions of the relevant bodies of IMO and UNCTAD. The subject would not therefore be included in the work programme for 1990. The same applied to the subject of Arrest of Ships.

It was agreed to consider the subject of wreck-removal and related issues during the 63rd Session to be held 17–21 September 1990.

The Legal Committee noted the information from the Council of the International Civil Aviation Organisation (ICAO) on the development of an international regime for the marking of plastic or sheet explosives for the purpose of detection with interest. The Committee would be willing to consider any proposals on legal aspects which may emerge.

62nd Session
The 62nd Session of the Legal Committee was held at IMO Headquarters in London 2–6 April 1990. Items on the agenda included:

* consideration of a possible convention on liability and compensation in connection with the carriage of hazardous and noxious substances by sea (HNS);

* matters arising from the 15th Extraordinary Session of the Council, the 16th Session of the Assembly and the 63rd Session of the Council.

3.2.6 International Diplomatic Conference on Salvage

The principle of maritime salvage is founded on the equitable premise that one who saves or helps to save maritime property is entitled to remuneration for his services in the form of a reward from the beneficiary of the salved property. The word 'salvage' has a dual meaning denoting both the act of saving as well as the reward itself. The reward must be commensurate with the value of the property saved and the amount of effort expended by the salvor.

According to traditional salvage law, in order for a salvage claim to succeed, three elements must be present, namely, danger, voluntariness and success. The act must be voluntary to the extent that it is not carried out merely for the purpose of self-preservation, or pursuant to a pre-existing contractural obligation.

These principles were codified as international law through the Salvage Convention of 1910.

Salvage at sea is a highly specialised task carried out by professional salvors. In practice, professional salvage services are rendered pursuant to a standard form agreement widely known as Lloyd's Open Form (LOF) which is based on the principle of 'no cure no pay'. In recent years, in cases involving tankers and environmental damage, salvors have frequently found themselves at the 'little or no pay' end of the LOF deal, for a variety of reasons. These include instances where coastal states have refused entry of damaged tankers into their waters with the result that vessels have been taken out to sea and sunk or scuttled. Salvors have thus been reluctant to assist on the basis of 'no cure no pay' in cases of tankers posing serious environmental threats whose cargo has little salved value. In other instances, salvors have expended considerable efforts by way of expensive equipment and manpower but have failed to save the property. Under the traditional principles of salvage law, they have not been entitled to reward, even though as a result of their efforts, considerable environmental damage was averted and monies saved to the benefit of several interests.

In 1980, the LOF was revised to include a provision which, while retaining the 'no cure no pay' principle, seemed to provide some limited remuneration to salvors for preventing pollution in cases where expenses reasonably incurred by the salvor exceeded the reward amount otherwise recoverable under the Agreement.

The world maritime community, realising that the 1910 Salvage Convention was now inadequate, embarked on a revision of the international law. The Comité Maritime International (CMI) produced a draft Salvage Convention which the IMO used as the basic text for a Diplomatic Conference.

The International Diplomatic Conference on Salvage was held at IMO headquarters in London from 17–28 April 1989. Attended by delegates from 66 countries, the Conference adopted the International Convention of Salvage. This new Convention, which replaces a convention on the law of salvage adopted in Brussels in 1910, was opened for signature of 1 July 1989 for one year. It requires adoption by 15 states to enter into force.

The main environmental features of the Convention are the following:

- timely and efficient salvage services are recognised as a contributory factor to protection of the marine environment;

- 'damage to the environment' is defined in Article 1 as "substantial physical damage to human health or to marine life or resources in coastal or inland waters or areas adjacent thereto, caused by pollution, contamination, fire, explosion or similar major incidents";

- Article 8 imposes on the owner and master of an endangered vessel the obligation to take timely and reasonable action to arrange for salvage. They are also required to cooperate fully with the salvors and use their best endeavours to prevent or minimise environmental damage;

- Article 8 also affirms that the salvor also is under a duty of care to prevent or minimise damage to the environment;

- Article 9 affirms the coastal state's rights under international law to protect its coasts from pollution damage resulting from a casualty at sea. The coastal state also has a right under this Article to give directions to the salvor in relation to salvage operations;

- Article 14 deals with the so-called 'special compensation' payable in cases where the salvor failed to save the maritime property but was successful in minimising environmental damage. In such cases, the salvor is entitled to payment of all his salvage expenses plus an additional sum commonly referred to as the 'enhancement' amount. This enhanced sum may normally be equal to a maximum of 30% of the salvage expenses but may be increased to 100% at the discretion of the arbitrator. If, on the other hand, the salvor is negligent and has consequently failed to prevent or minimise environmental damage, special compensation may be denied or reduced.

A further important point is that, in consequence of the adoption of a Resolution by the Diplomatic Conference, the special compensation payable under Article 14 is not subject to the rules of General Average. To this end, the Resolution requests the IMO to take the necessary steps to ensure that the York-Antwerp Rules of 1974 are appropriately amended.[2]

3.2.7 International Conference on Oil Pollution Preparedness and Response

An International Conference on Oil Pollution Preparedness and Response is due to be held at IMO headquarters on 19–23 November 1990. The purpose of the conference is to adopt a convention which will provide the framework for international cooperation for combating major oil pollution incidents.

The Assembly took this decision when it adopted resolution A.674 (16). The draft resolution discussed by the Assembly was submitted by a number of IMO Member States, including those which attended a meeting of leading industrial nations in Paris in July 1989. The communique issued after that meeting called upon IMO to develop further measures to prevent oil spills.

The resolution adopted by the Assembly says that the convention should include such matters as:

- establishment of an information centre within, or under the auspices of IMO;

- encouragement for establishing a national response centre in each Member State;

- development of shipboard contingency plans;

- encouragement for establishing pre-positioned oil spill response equipment;

- fostering International cooperation and coordination of research and development efforts in marine pollution response;

- development and maintenance of an inventory of pollution response equipment available on a worldwide basis.

The resolution states that some of the tasks of the proposed information centre should include:

- maintaining a listing of national response centres and their capabilities;

- maintaining an inventory of pollution response equipment available for use in responding to oil spills;

- maintaining a listing of experts in pollution response and salvage;

- receiving, analysing and disseminating oil spill incident reports.

Resolution A.675 (16) also deals with the prevention of oil pollution. It urges governments to implement and enforce the existing international conventions relating to the safety of life at sea, protection of the marine environment and the training of seafarers. It then invites them to submit promptly the outcome of casualty investigations and other related studies so that IMO can consider further action to prevent tanker casualties.

The Maritime Safety and Marine Environment Protection Committees are requested to examine the role of the human element in such incidents and to review the results of studies in different countries on tanker design and other related topics.[3]

3.3 UNITED NATIONS ENVIRONMENT PROGRAMME (UNEP)

Through the Regional Seas Programme, set up in 1974, UNEP has acted as the principal UN agency in the development of a special regime for the seas and oceans bordering some 120 states. The programme was initially based in Geneva, but moved in August 1985 to Nairobi when it underwent reorganisation to become the Programme Activity Centre for Oceans and Coastal Areas (OCA/PAC). The Centre is responsible for coordinating UNEP programmes dealing with living marine resources, such as fisheries, marine mammals and aquaculture, as well as for its previous activities relating to the protection of the marine environment.

The Centre's sub-programme on the global marine environment continues to contribute towards international efforts in assessing marine pollution and promotes the development of standardised methodology and quality control of marine pollution monitoring and research. The Regional Seas Programme continued its past activities with emphasis on harmonising activities among the various regions dealing with similar subjects.

3.3.1 The Mediterranean

The Sixth Ordinary Meeting of the Contracting Parties to the Barcelona Convention for the Protection of the Mediterranean Sea was held in Athens from 3–6 October 1989. The following issues were discussed:

An assessment of the state of pollution of the Mediterranean Sea by organotin compounds, together with measures for the control of pollution by organotin compounds was adopted. The first measures for controlling the use of antifouling paints containing organotin compounds were brought by France in 1982. Since then, other countries have followed suit, the commonest measures being the ban on the use of TBT paints on vessels smaller than 25m and on mariculture structures. Recently, measures have also been introduced for sea-going vessels. Measures at regional and international levels are promoted through the competent organisations. In a recommendation the Contracting Parties agreed:

- from 1 July 1991 not to allow the use in the marine environment of preparations containing organotin compounds intended for the prevention of fouling by micro-organisms, plants or animals; 1, on hulls of boats having an overall length of less than 25m; 2, on all structures, equipment or apparatus used in mariculture;

- to report to the Secretariat on measures taken in accordance with this decision;

- that a code of practice be developed to minimise the contamination of the marine environment in the vicinity

A fish trap at Regga Point, Gozo, Malta. This trap is one of the traditional methods of fishing from the steep rocky coast. It represents a sustainable local fishery.
Photo: D George

of boat-yards, dry docks, etc., where ships are cleaned of old anti-fouling paint and subsequently repainted.

Among other recommendations adopted during the Conference, of particular interest were the implementation of the Protocol on Specially Protected Areas and Historic Sites and the special actions for the Adriatic Sea.

The main themes of the Meeting were the examination of the activities carried out and future works to be accomplished in the framework of the Mediterranean Action Plan and the preparation of the United Nations Conference on Environment and Development to be held in Brazil in 1992.[4]

3.3.2 The Gulf

The first draft of evaluation of the programmes and activities of Regional Organisation for the Protection of the Marine Environment (of the Gulf) (ROPME) since the adoption of the Kuwait Action Plan in 1978 were completed in autumn 1988. The Third Extraordinary Meeting of ROPME was held in Kuwait 31 October–1 November 1988 where a special plan of action was agreed to clean up the seas after the eight years of wars that have been continuing in the region. KAP States are to be assisted by

UNEP and other international organisations in the plan.

The Seventh Task Team Meeting on marine monitoring and research programme of ROPME was held in Kuwait 22–26 January 1989. The meeting discussed the development of guidelines for the preparation of national reports on the state of the marine environment, the status of the Marine Monitoring and Research Programmes and reviewed synthesis reports. The Meeting revised the scope and approach for the second phase. The Task Team also discussed other projects.

The Protocol on pollution resulting from the exploration and exploitation of the Continental Shelf in the ROPME Sea Area was signed at a Special Meeting of Plenipotentiaries in Kuwait, 28–29 March 1989.

At a meeting of experts from countries involved in the Kuwait Action Plan held from 2–3 July 1989 in Kuwait, draft plans were completed for the clean-up of the regional seas in the aftermath of the Iran-Iraq war.[4]

3.3.3 West and Central African Region (WACAF)

The Sixth Meeting of the Steering Committee for the Marine Environment of WACAF and the Second Meeting of Contracting Parties to the Abidjan Convention and Conference on Hazardous Wastes were held in Dakar, Senegal, 23–27 January 1989. Contracting Parties to the Abidjan Convention adopted a short-term workplan following recommendations made by the WACAF Steering Committee. The short-term plan is a springboard to a five-year medium-term plan due to begin in 1991. One of the priorities is to consolidate the regional pollution monitoring, research and control programme, and prepare a regional oil pollution control manual in cooperation with IMO. The Steering Committee requested UNEP's Executive Director to initiate the development of three additional protocols to the Abidjan Convention, dealing with:

- control of transboundary movement of hazardous waste;
- control of land-based pollutants;
- control of pollution by dumping of wastes.

It was recommended that a Regional Coordinating Unit be established as soon as possible.

The Contracting Parties adopted a five-year medium-term programme extending from 1991 to 1995.

The Contracting Parties called on all states to ratify the Convention and its Protocol; at the time of the meeting only 8 of the 21 WACAF Member States had ratified the Convention.

The Action Plan's project for monitoring marine pollution entered its second phase in April 1989. The first phase had shown that pollution was more likely to occur in coastal regions rather than the open sea.

An Intergovernmental Meeting at Ministerial Level was held in Nairobi, Kenya, on 16 May 1989.[4]

3.3.4 The Wider Caribbean

The Fifth Intergovernmental Meeting and Second Meeting of Contracting Parties to the Cartagena Convention of the Caribbean Environment Programme (CEP) was successfully convened in Kingston from 17–18 January 1990. As its most important achievement, the Meeting adopted "the strategy for the Development of CEP", a breakthrough for joint management of marine and coastal resources by all States and Territories in the region. The Strategy seeks to assist with the stimulation, coordination and consolidation of existing national and regional initiatives and provides the framework for building regional consensus.

The Strategy provides the Caribbean Environmental Programme with goals, principles and objectives for its long-term development. Also, for the implementation of CEP over the next five years, six regionally coordinated comprehensive programmes have been outlined and available mechanisms for Strategy implementation have been recognised. The six mutually reinforcing programmes are:

- Integrated Planning and Institutional Development for the Management of Marine and Coastal Resources;
- Specially Protected Areas and Wildlife (SPAW);
- Information Systems for the Management of Marine and Coastal Resources (CEPNET);
- Assessment and Control of Marine Pollution (CEPPOL);
- Education, Training and Public Awareness for the Appropriate Management of Marine and Coastal Resources; and
- Overall Coordination and Secretariat functions.

Work on one of the regional programmes, CEPPOL, was greatly advanced during the IOC/UNEP Regional Workshop to Review Priorities for Marine Pollution Monitoring, Research, Control and Abatement in the Wider Caribbean Region, held in San José from 24–30 August 1989.

Seven years after the adoption of the Cartagena Convention and its Protocol on combating oil spills, the text of a second Protocol Concerning Specially Protected Areas and Wildlife (SPAW) was officially adopted on 18 January 1990. The SPAW Protocol includes three Annexes, listing protected marine and coastal flora (Annex I), fauna (Annex II) and species to be maintained at a sustainable level (Annex III) which will be prepared during the coming months.[5]

3.3.5 Red Sea and Gulf of Aden

A draft project aimed at strengthening PDR Yemen's capability in the research, monitoring and assessment of pollution was initiated in the autumn of 1988.

A meeting between OCA/PAC and the Programme for the Environment of the Red Sea and the Gulf of Aden (PERSGA) was held at OCA/PAC's headquarters in Nairobi, Kenya on 25–26 April 1989. A draft memorandum was drawn up between UNEP and PERSGA detailing future cooperation.

In August 1989 the Egyptian Government agreed to accede to the Regional Convention for the Conservation of the Red Sea and Gulf of Aden Environment upon its ratification by the People's Assembly in November 1989.

The Experts Meeting on Combating Marine Pollution in the Red Sea and Gulf of Aden was held in Alexandria in November 1989. The Meeting proposed the setting up of liaison points to speed up information about the movements of ship suspected of carrying hazardous cargoes. The Meeting also urged States to ratify the Basle Convention and to adopt national contingency plans for combating marine pollution. A recommendation was made for a regionally coordinated plan for surveying and monitoring marine and coastal pollution. UNEP was requested to prepare a detailed programme for monitoring and assessing the marine envi-

This beautiful sponge, Siphonochalina, *on a reef off Sabah, Kalimantan, is threatened by fishing with the use of illegal explosives. Explosives stun the fish which then float to the surface. However, reefs are also destroyed, along with sessile organisms. Habitat for the fish is lost; a non-sustainable method of fishing. Photo: D George*

ronment in the region.[4]

3.3.6 East Asian Seas

The Seventh Meeting of the Coordinating Body on the Seas of East Asia (COBSEA) was held in Yogjakarta, Indonesia, 17–19 July 1988. The Action Plan was reviewed and a new workplan was adopted.

COBSEA also requested that UNEP continue to:

- manage the trust fund until 1991;

- act as Secretariat of the Action Plan.

The Third Meeting of Experts on the East Asian Seas Action Plan was co-hosted by the Association of the South-East Asian Marine Scientists (ASEAMS) in Quezon City, Philippines from 7–10 February 1989. The Meeting reviewed the environmental problems of the region and drew up a priority list. Problems in order of priority were:

- destruction of support systems, particularly coral reefs and mangroves;

- sewage pollution;

- industrial pollution;

- fisheries and over-exploitation;

- siltation and sedimentation.

The Meeting also reviewed and recommended the implementation of four completed EAS projects, prepared by UNEP with the assistance of IOC and the Monitoring Assessment Research Centre (MARC), of the University of London. The projects include the initiation of phase two of the development of management plans, coastal and marine living resources in East Asia. The experts also recommended the inclusion of five new topics for consideration by the team of scientists studying the implications of the expected climate change for the region.

The Eighth Meeting of COBSEA was held in Brunei Dar-es-Salam from 14–16 June 1989. At the Meeting a request was made for re-appraisal of the East Asian Seas Action Plan with view to it being strengthened. A workplan and budget for the 1989–90 period were also decided upon. Evaluations made by IOC and MARC of the four projects were adopted with only slight revisions.[4]

3.3.7 South-East Pacific

The Coordinated Programme for Research and Monitoring of Pollution in the South-East Pacific (CONPACSE) was discussed at a meeting on 29 September 1988, in Bogota, Colombia. Offshoots of CONPACSE include the setting up of environmental marine departments and an Environment Unit at Colombia's National Institute of Environment and Renewable Resources (INDERENA). In Chile, CONPACSE has been the basis of pollution control projects and five oil combating centres have been established. In Ecuador, the Government has been led by the Action Plan to protect areas such as marine reserves, and pass a new law on water pollution. In Panama a national plan has been established to protect and improve the marine environment, while in Peru there is a programme to clean Lima's beaches.

A Draft Protocol to the Lima Convention was approved by legal and technical experts at a meeting in Cartagena, Colombia, 11–14 April 1989. The Draft Protocol details the preservation and management of protected areas of the South-East Pacific.

At a workshop and seminar on Erosion of Coastal Areas of the South-East Pacific held in Guayaquil, Ecuador in November 1989, coastal erosion was pinpointed as being one of the most serious problems affecting the South American East Pacific coastal region. A first draft of an Erosion/Action Map was prepared during the workshop. A detailed report on erosion trends was due to be submitted before the end of 1990.[4]

3.3.8 South Pacific

A Meeting of Regional Experts, convened by the South Pacific Commission (SPC) and UNEP in Sydney, Australia, in January 1989, discussed the elements of SPREP-POL. SPREP-POL is a consolidation of activities related to marine pollution assessment and control, that resulted from an Intergovernmental Meeting on the South Pacific Regional Environment Programme (SPREP) in July 1988.

At a meeting co-sponsored by OCA/PAC and SPREP in Majuro, Marshall Islands, 16–20 July 1989, 15 regional states adopted a programme of research and other activities related to climate change. The activities will be funded by OCA/PAC, coordinated by SPREP and the work done by the Association of South Pacific Environmental Institutions (ASPEI). Plans include:

- sending experts to six countries that asked for help in conducting studies of impact of expected climate change;

- helping national committees in developing an approach to problems;

- assisting Tonga in studying potential impact at specific sites;

- helping Palau adapt coastal zone management projects.

The Fourth South Pacific Conference on Nature Conservation and Protected Areas was held in Port Vila, Vanuatu in September 1989. The Conference was attended by over 100 delegates, and 13 resolutions relating to natural resource conservation were made during the meeting. Concern over the contribution drift-net fishing was making to destruction of the environment was also expressed and the meeting urged a worldwide cessation and urged governments to outlaw this practice. The Conference noted that only four governments had ratified the SPREP Convention and that the Apia Convention has still not entered into force some 12 years since its adoption.

The meeting adopted the Regional Marine Turtle Conservation and Management Programme for implementation within the framework of SPREP.[4]

3.3.9 Eastern Africa

At its Conference of Ministers held in Victoria, Seychelles, 13–14 April 1989, the Indian Ocean Commission set up a Committee on Environment to advise on how to contain the environmental problems of the countries in the region. The Committee will liaise closely with the Committee on Seas of AMCEN (see 3.18).

A Meeting of Experts was held at OCA/PAC headquarters in Nairobi 21–23 June 1989, where a programme of studies to examine the effects of global warming in the region was drawn up.

On 18 August 1989 France ratified the Convention for the Protection, Management and Development of the Marine and Coastal Environment of the Eastern African Region and its two Protocols.

The First Intergovernmental Meeting on the Action Plan for the region was held at UNEP headquarters in Nairobi on 8-9 November 1989. A number of countries confirmed at the meeting that their governments were in the process of ratifying the Convention for the Protection, Management and Development of the Marine and Coastal Environment of the Eastern African Region. The Meeting resolved to press governments to ratify both the Action Plan and Convention.[4]

3.3.10 South Asian Seas

A meeting was held in Geneva in April 1989 to examine the needs of the Maldives and action to cope with climate change and sea-level rise. The meeting was organised by UNDP and was attended by 23 states and 17 financial institutions. During the meeting, the Maldivian Minister for Foreign Affairs emphasised three priorities for future development plans, including protection of the environment.[4]

3.3.11 Other UNEP Activities

The 15th Session of UNEP Governing Council was held in Nairobi in May 1989. At the Session it was decided to give UNEP's oceans and coastal areas programme higher priority and more money. The Council also voted for pre-emptive action to keep the 'oil pipeline' Red Sea clean. The governments resolved that UNEP, with IMO and regional organisations, should assist in developing an inventory of equipment and expertise available in the region to deal with

Lethal litter. A large sack containing sodium hydroxide washed up on a Yorkshire beach. Photo: Tidy Britain Group

oil spills should they occur. The Council also decided that UNEP and IMO should assist ROPME in implementing its action plan for clean-up in the Red Sea after the Iran-Iraq war.[4]

3.3.12 Basle Diplomatic Conference

The UNEP Diplomatic Conference on Transport and Disposal of Toxic Waste was held in Basle on 22–24 March 1989. A Convention on the Control of Transboundary Movements of Hazardous Wastes and their Disposal was approved and signed on the spot by 34 states and the EEC. Half were developing countries.

The Treaty requires only 20 ratifications to enter into force, a deliberately low number so that the Treaty can become international law as quickly as possible. It was thought likely that the Treaty would enter into force by the middle of 1990; however, at May 1990 only three countries had ratified the Convention.

In addition to the 34 signatories, many countries indicated their intention to sign within three to six months, and still others within one year. More than 100 countries and the EEC attended the final negotiations and approved the Basle Convention.

Treaty Terms
The 53-page Basle Convention has 29 articles and six annexes. These are the principle points.

* hazardous wastes are defined in Article 1 and listed in Annex 1;

* a signatory state cannot send hazardous waste to another signatory state that bans import of it;

* a signatory state cannot ship hazardous waste to any country that has not signed the Treaty;

* every country has the sovereign right to refuse to accept a shipment of hazardous waste;

* before an exporting country can start a shipment on its way, it must have the importing country's consent, in writing. The exporting country must first provide detailed information on the intended export to the importing country to allow it to assess the risks;

* no signatory country may ship hazardous waste to another signatory state if the importing country does not have the facilities to dispose of the waste in an environmentally sound manner;

* when an importing country proves unable to dispose of legally imported waste in an environmentally acceptable way, then the exporting state has a duty either to take it back or to find some other way of disposing of it in an environmentally sound manner;

* the Treaty states that "illegal traffic in hazardous waste is criminal";

* the Treaty does not prohibit the export of toxic waste because it is recognised that the greatest flow of traffic is from North to North and, additionally, because an increasing number of Third World countries need to export their own waste initially to the North;

* shipments of hazardous waste must be packaged, labelled and transported in conformity with generally accepted and recognised international rules and standards;

* bilateral agreements may be made by signatory states with each other and with a non-signatory country, but these agreements must conform to the terms of the Basle Convention and be no less environmentally sound;

* the Treaty calls for a total ban on export of waste south of 60°S (ie. into the Antarctic area);

* the Treaty calls for international cooperation involving, among other things, the training of technicians, exchange of information and transfer of technology;

* the Treaty sets up a secretariat (based in Geneva) to supervise and facilitate its implementation;

* the Treaty asks that less hazardous waste be generated and what is generated be disposed of as close to its source as possible.[6]

3.4 UNITED NATIONS DEVELOPMENT PROGRAMME (UNDP)

The UNDP was established in 1965 through a merger of the Expanded Programme of Technical Assistance (1950) and the UN Special Fund (1959). All Member States of the United Nations are Members of UNDP.

The Organisation is responsible to the UN General Assembly, to which it reports through the UN Economic and Social Council. The Council, which meets annually, is the policy-making body of UNDP, and comprises representatives of 48 countries; 27 seats are filled by developing countries and 21 by economically more advanced countries; one-third of the membership changes every year.

UNDP is the world's largest multilateral grant development assistance organisation maintaining a network of 117 offices; 113 of these are located in developing countries. UNDP draws on the expertise of some 35 specialised and technical UN agencies to work in virtually every sector of development. It also works with non-governmental organisations. In 1989 UNDP served 152 developing countries and territories through some 6,900 projects.

The goal in all UNDP's activities is the permanent enhancement of self-reliant development in all nations. UNDP projects are therefore designed to:

* identify and quantify productive resources;

* provide technical training at all levels and in all requisite skills;

* supply equipment and technology in conjunction with training; identify investment opportunities;

* assist in planning and coordinating development efforts.

Most UNDP projects incorporate training for local workers. Developing country governments themselves provide 50% or more of total project costs in terms of personnel, facilities, equipment and supplies.[7]

3.5 WORLD HEALTH ORGANISATION (WHO)

WHO is a UN Agency concerned with environmental health on a global level. WHO's programme in environmental health covers a large range of pollution problems that can have a direct or indirect impact on human health and well-being. Marine pollution problems and health aspects of coastal water quality are among areas of major concern to the Organisation.

WHO collaborates with its Member States in the execution of various programmes and projects, including the establishment of monitoring networks and other information services. Such projects are frequently implemented in cooperation with other specialised agencies of the United Nations such as UNEP.

Expert committees, periodically convened by WHO alone or jointly with other agencies, make recommendations to national and other authorities on matters relating to fish and shellfish hygiene, and on the selection of sites and methods for waste discharge into the coastal environment.

WHO actively collaborates with UNEP's Regional Seas Programme in regional projects, including environmental sanitation, inventories and control of land-based pollution discharges, coastal water quality problems and seafood contamination.

As one of the co-sponsors of the Joint Group of Experts on the Scientific Aspect of Marine Pollution (GESAMP) (see 3.8), WHO established in 1982, in collaboration with the Food and Agriculture Organisation (FAO), UNEP and IMO, a Working Group on the Review of Potentially Harmful Substances. Its main function is to evaluate pollutants on a substance-by-substance basis for their effects on marine biota and human health. During 1988/1989 the Group made the following progress:

- sub-group for the review of nutrients and algal blooms established in cooperation with UNESCO completed its report;

- review of the health risks associated with carcinogenic substances underway.

WHO participates with other United Nations agencies in the Mediterranean Action Plan. Within this project WHO provides assistance to national research centres participating in the monitoring activities. To this end sampling and analytical methods were further developed and tested to serve as regionally mandatory reference methods. In addition, the scientific rationale was developed for environmental quality criteria to be used in the development of emission standards, standards of use or guidelines for substances listed in the protocol on land-based pollution sources and for coastal bathing waters and shellfish.

Within the framework of the Action Plan for the Protection and Development of the Marine Environment and Coastal Areas of the West and Central African Region, a project was developed with other United Nations agencies. WHO has assumed responsibility for activities related to the monitoring of the microbiological quality of recreational waters and seafood.

Laboratories were designated in participating countries, followed by training of personnel and provision of microbi-ological equipment.

Studies have been continued on the health impact of elevated mercury levels in fish in the Mediterranean and in the Indian Ocean. Monitoring of seafood and of fish consumers as well as dietary intake studies were continued and biological monitoring of seafood consumers initiated.[8]

3.6 FOOD AND AGRICULTURE ORGANISATION (FAO)

One of the concerns of the Food and Agriculture Organisation of the United Nations (FAO) is the effects of pollution (marine and freshwater) on living aquatic resources and their habitat and the contamination of food from the seas, rivers and lakes. The activities of FAO relating to marine pollution during 1989/90 were as follows:

Under the guidance of FAO, the IMD/FAO/Unesco/WMO/WHO/IAEA/UN/UNEP Joint Group of Experts on the Scientific Aspects of Marine Pollution (GESAMP) finalised a report on Long-term Consequences of Low-level Marine Contamination. The report concludes that a systematic framework for exposure and effects assessment can provide an analytical tool to show the qualitative relationships between contaminant exposure and biological effects, even at low level and in the longer term. There are undoubtedly difficulties in interpreting complex and ecological responses, and for some exposure/effects cases the relationship may be weak or even doubtful. However, on the basis of the four cases studied, the analysis leads to the conclusion that long-term biological changes can be attributed to contaminant exposure at low level, or as a result of the slow build up of contamination either in the environment or in the target species.

Also under the umbrella of GESAMP, FAO initiated Working Groups on Scientifically Based Strategies for Marine Environmental Protection and Management and on Potentially Harmful Substances — Chlorinated Hydrocarbons. During 1990/91 another task of GESAMP will be to study environmental impacts of coastal aquaculture.

FAO continued to cooperate with the Regional Seas Programme of UNEP in marine pollution monitoring and research projects in the Mediterranean and the West and Central Africa Region; activities related to analyses of metals and organochlorines in marine biota are of direct concern to FAO. A similar project will be initiated during 1990 in Eastern Africa.

During 1989/90 FAO, jointly with IMO and UNEP, assisted the governments of Somalia and Morocco to estimate hazards from oil pollution resulting from tanker accidents.[9]

3.7 INTERGOVERNMENTAL OCEANOGRAPHIC COMMISSION (IOC) OF UNESCO

The Intergovernmental Oceanographic Commission (IOC) established the Global Investigation of Pollution in the Marine Environment (GIPME) Programme in response to Recommendation 90 of the United Nations' Conference on the Human Environment (Stockholm, 5–16 June 1972).

The overall objective of GIMPE is to provide a scientifically sound basis for the respective assessment and regulation of marine contamination and pollution, including sensibly planned and implemented international, national and regional monitoring programmes.

Summary of Activities
Global
The Ninth Session of the Group of Experts on Methods Standards and Intercalibration (GEMSI) was convened in Villefranche, 5–9 December 1988, and reviewed the development of reference methods, associated manuals, data quality assurance programme, experiences from training workshops and intercalibrations, the preparations for the workshop on the use of sediments in marine pollution research and monitoring. The planning of the Open Ocean Baseline Study, a major activity under GEMSI, is in an advanced stage with a cruise by German RV *Meteor* scheduled for the South Atlantic in March–April 1990. Levels of heavy metals and nutrients will be studied in four deep ocean stations. Laboratories from several countries will participate in this first leg of the study.

GEMSI, by the end of 1989, had finalised manuals on the determination of petroleum hydrocarbons in sediments and on the sampling and analysis of chlorinated biphenyls in open ocean waters.

A manual on quality assurance and good laboratory practice in marine pollution monitoring was also finalised in 1989.

The main emphasis of the Group of Experts on Effects of Pollutants (GEEP) has been the preparation and implementation of the Second Workshop on Biological Effects of Pollutants (Bermuda Biological Station, 10 September–2 October 1988), where techniques proven successful at the first Biological Effects Workshop were tested in a subtropic climatic zone together with a few new techniques. Participants from several regions were invited to this combined training and development workshop, and draft protocols and manuals were tested. The Group has also continued its work in preparing a scientific basis for identifying vulnerable areas. An *ad hoc* meeting was convened, in Woods Hole, June 1988, where the draft criteria prepared by the Marine Environment Protection Committee for IMO were considered, and work on specific case studies initiated.

The Fifth Session of GEEP was held at IMO Headquarters in London, 17–20 April 1989.

Discussions focused on the outcome of the 2nd major GEEP International Workshop on the Biological Effects of Pollutants, Bermuda, 10 September–2 October 1988, and the planning of follow-up actions leading to the next major research and training workshop at Xiamen, PRC, in 1991.

The planning of the Sea-going Workshop on Biological Effects of Contaminants (Bremerhaven, FRG, 12–30 March 1990), co-sponsored by ICES and IOC has reached its final stages.

Due to the success of the first GEEP Workshop on Statistical Treatment and Interpretation of Marine Community Data (Piran, Yugoslavia, 15–24 June 1988), FAO, IOC and UNEP organised a second workshop in Athens, Greece, 18–29 September 1989 which also proved very successful.

Preparation of three manuals describing procedures for measuring the biological effects of pollutants has been finalised during 1989.

The Second Session of GESREM was held in Halifax, Canada, 22–25 January 1990. The meeting reviewed activities by various national and international agencies in the field of preparation and supply of standards and reference materials for the use in regional and international marine pollution monitoring programmes.

The Group of Experts drew a workplan consisting of the following main actions:

- guidelines for preparation and application of working reference materials;

- overview of progress toward development of reference materials and documentation of methods for their preparation;

- provision of reference materials to laboratories in developing countries;

- cooperative efforts toward development of new reference materials;

- publicity for and distribution of the revised catalogue of reference materials.

The IOC Marine Pollution Monitoring System (MARPOLMON) is operational as regional components in several regions, mostly in collaboration with other bodies and networks. From the South-East Pacific, Caribbean, West and Central Africa/IOCEA and the Mediterranean, data are regularly delivered to regional data banks and to IOC. These data are used for regional assessments and for the regional reviews on the state of the marine environment. The data cover petroleum hydrocarbon contamination and in some cases trace metals, and physical oceanography parameters. In the Mediterranean a pilot project on monitoring of marine debris (litter) on beaches and in the coastal zone, has been carried out as a joint activity between IOC, FAO and MAP/UNEP.

In the Caribbean IOC jointly with UNEP initiated in early 1990 a new major Marine Pollution Assessment and Control Programme for the Wider Caribbean Region (CEP-POL).

Training is a very important aspect of GIMPE regional and global pollution monitoring activities. It involves courses and workshop developments, laboratory visits, individual training and inter-companion and intercalibration exercises, and data quality control guidance.

Planned IOC Meetings

- Joint Meeting of the Group of Experts on Effects of Pollutants and of the Group of Experts on Methods Standards and Intercalibration, Moscow, USSR, 14–21 October 1990.

- Xth ICSEM/UNEP/IOC Workshop on Marine Pollution of the Mediterranean Sea, Perpignan, France, 18–19 October 1990.

- 7th Session of the Scientific Committee for GIPME, Paris, 14–24 January 1991.

- 16th Session of the IOC Assembly, Paris, 5–22 March 1991.[10]

3.8 JOINT GROUP OF EXPERTS ON SCIENTIFIC ASPECTS OF MARINE POLLUTION (GESAMP)

GESAMP is the Joint Group of Experts on the Scientific Aspects of Marine Pollution, sponsored by IMO, FAO, UNESCO, WMO, WHO, IAEA, UN and UNEP. It is a multidisciplinary body of independent scientists who provide advice on issues raised by any of the sponsoring agencies. Each agency sponsors two to four GESAMP experts. GESAMP operates through a number of working groups which also include experts from outside GESAMP. The results of GESAMP are published as GESAMP Reports and Studies, and these are available free of charge through any of the sponsoring agencies or the Administrative Secretary of GESAMP (IMO).

GESAMP held its 19th Session in Athens, 8–12 May 1989. The report of the Session has been published as GESAMP Reports and Studies No. 37. A summary of the discussion at the 19th Session is given below.

Review of Potentially Harmful Substances
Substances which are being reviewed are carcinogenic, mutagenic and teratogenic substances; chlorinated hydrocarbons; and oil, including used lubricating oils, oil spill dispersants and chemicals used in offshore exploration and exploitation.

With regard to the evaluation of carcinogenic, mutagenic and teratogenic substances, a number of steps have been accomplished for the preparation of a GESAMP study.

With regard to chlorinated hydrocarbons, GESAMP agreed that the concern about such substances in the marine environment was in many cases misplaced, since this group of compounds comprises a very large range of substances, with very different physical, chemical and biological properties. To identify compounds that really merit concern, a hazard assessment approach was agreed, based on physico-chemical properties of the substances, toxicity and on structure-activity relationships.

With regard to oil, including used lubricating oils, oil spill dispersants and chemicals used in offshore exploration and exploitation and spill control agents, GESAMP approved the outline of a work schedule prepared by experts under the leadership of IMO. It agreed that the review should proceed over a two-year period.

Evaluation of the Hazards of Harmful Substances Carried by Ships
The results of this work are used by IMO in establishing discharge and carriage requirements for chemical tankers, and in identifying packaged goods transported at sea as 'marine pollutants'.

So far GESAMP has developed hazard profiles for 2,500 chemicals transported by ships. This work is ongoing in the light of the many new substances proposed for transport as sea either in bulk or as packaged goods.

Interchange of Pollutants between the Atmosphere and the Oceans
GESAMP approved the report of a workshop on the atmospheric input of trace species to the world ocean. The report indicates that atmospheric input dominates riverine input for most trace species. For most synthetic organic species atmospheric input accounted for 90% or more of the combined atmospheric plus riverine input to the global ocean. This is also the case for many dissolved trace metals, for example lead, cadmium and zinc, while atmospheric and riverine input are similar for copper, nickel and iron. Atmospheric input of nitrogen species dominates that from rivers as well. The major fraction of the input of these elements occurs in the Northern Hemisphere.

Coastal Modelling
This work had originally been initiated with a view to establishing a de minimus level of radioactivity of wastes and other matter, below which these could be considered under the terms of the London Dumping Convention as 'non-radioactive', that is, they could be dumped under a general permit. So far the state of the art of coastal (including continental shelf) modelling relevant to waste inputs by sea dumping or land-based discharges has been evaluated. GESAMP agreed that the Working Group should also recommend on the types of models to be used for specific coastal situations. GESAMP also requested that additional case studies be taken into account and that each should specify in unambiguous terms the receiving coastal environment and the pathway of concern. This work will be completed during the intersessional period.

State of the Marine Environment
The report on the state of the marine environment is now available. The main current areas of concern with regard to the oceans are along the coasts, where waters and habitats suffer both from activities taking place in the coastal zone itself and from a number of others carried out inland. Contamination by nutrients with the attendant risk of eutrophication, and microbial contamination of beaches and seafood from sewage disposal are both highly significant problems of immediate relevance to human health and to the well-being of the marine ecosystem.

Scientifically Based Strategies for Marine Environment Protection and Management
GESAMP agreed that for the formulation of strategies a small steering group should prepare a position paper listing the possible elements of strategies with an indication of the approach which may be adopted by GESAMP towards each of these elements.

Comprehensive Framework for the Assessment and Regulation of Waste Disposal in the Marine Environment
At the request of the Consultative Meeting of Contracting Parties to the LDC, GESAMP agreed to consider the development of a common, comprehensive and holistic framework for the regulation and assessment of dumping at sea of all types of wastes, that is, radioactive as well as non-radioactive wastes. GESAMP noted that this was of particular importance to the future work of the Inter-Governmental Panel of Experts on Radioactive Waste Disposal at Sea (IGPRAD) established by the Consultative Meeting.

The terms of reference for a working group established by GESAMP on the above issue are as follows:

- analyse existing regulatory mechanisms, and their underlying concepts and principles, that are currently used at both national and international levels, to protect the marine environment against the adverse effects of anthropogenic activities;

- determine the advantages, limitations and compatibility of these various mechanisms in terms of their practicality and effectiveness in protecting the environment on a sectoral and contaminant-specific basis and in providing a scientifically-defensible and holistic approach to pollution prevention;

- prepare a report synthesising current pollution control/prevention mechanisms, identifying the components and inter-relationships of those frameworks best suited to harmonised implementation and allowing for achievement of sustainable use and protection of the marine environment.

Impacts of anthropogenically mobilised sediments in the coastal environment: GESAMP approved terms of reference for a working group to carry out a study on the subject.[11]

3.9 REGIONAL MARINE POLLUTION EMERGENCY RESPONSE FOR THE MEDITERRANEAN SEA (REMPEC)

REMPEC, which is situated in Malta, is better known under the acronym ROCC (Regional Oil Combating Centre for the Mediterranean). The change in name followed a decision of the Contracting Parties to the Convention for the Protection of the Mediterranean Sea against Pollution to extend the objectives and the functions of the Centre to hazardous substances other than oil in accordance with the Protocol on cooperation in cases of emergency. The Sixth Ordinary Meeting of the Contracting Parties (Athens, October 1989) approved the change of name, objectives and functions of the Centre and recommended a programme for future activities.

Although most public concern about marine pollution has, in the past, concentrated on problems associated with oil, many of the chemicals carried by sea are far more dangerous to the marine environment. The number of different chemicals and other goods of this type is growing all the time as the world becomes more industrialised and industry itself becomes more complex. It has been estimated that up to 15% of all goods carried in conventional dry cargo ships are dangerous to some degree; if liquid substances carried in chemical carriers or tankers are included, then the total is around 50%.

It is with this situation in mind that the activities of REMPEC were widened to include harmful substances other than oil, thus bringing the functions of the Centre into line with the emergency Protocol. The Centre is therefore entering another phase which involves the development of a new area of action but is at the same time maintaining and reinforcing past action.

In view of these extended functions a regional Workshop on Combating Accidental Pollution of the Mediterranean Sea by Harmful Substances was convened in Malta between 23-26 May 1989. The Workshop provided the participants with an overview of the problems related to maritime transportation of harmful substances and accidental spillages of the substances. The participants also discussed and approved recommendations concerning actions which have to be taken at national and regional levels for the prevention and response to accidental pollution of the Mediterranean Sea by harmful substances.

Among REMPEC's functions is the development and maintenance of a regional system of information but particularly the establishment and exploitation of a partially computerised data base as well as the adaptation to the region of a computerised marine pollution emergency decision support system.

Within the framework of the objectives and functions of the Centre and taking into account the guidelines adopted by the Mediterranean Coastal States during their intergovernmental meetings, several activities have been or are being developed.

Publications
On the information side the Centre has produced several publications, including a *Catalogue of Spill Response Equipment and Products,* and an *Inventory of Companies Offering Services in Cases of Emergency in the Mediterranean.* The Centre has also compiled a *Guide for Combating Accidental Marine Pollution in the Mediterranean* and a *Concise Dictionary of Antipollution Terms.* All documents are produced in French and English, the official languages of the Centre.

Activities
REMPEC helps those countries which make a request with assistance in the preparation or adaptation of their national contingency plans. Yugoslavia has been a recent beneficiary. Greece will be the next one.

During the week of 6-9 March 1989 the Centre organised ALERTEX 89, an alertness and communication exercise with a view to test the level of preparedness of the Mediterranean Coastal States to communicate with the Centre and among themselves and to cooperate in case of a serious pollution accident. Practically all of the Mediterranean Coastal States and EEC took part in the exercise, which provided some valuable evidence on the use of the Standard Alert Format.

Training Courses
In line with its objectives REMPEC organises training courses regularly. A practical course on Containment and Recovery Techniques at Sea was held from 28 September to 4 October 1988 at the Scuola Ambiente in Ercolano, Italy. The MEDEXPOL 89 Course was on the use of Dispersants and other Products in Response to Oil Spills. It was held in Marseilles, France between 16-21 October 1989. The cost of both courses was shared between the Centre and the EEC. CASTALIA SpA (Societa Italiana per l'Ambiente) provided the logistic support for the course held in Italy while the Port Autonome de Marseille did the

same for MEDEXPOL 89.'

REMPEC also provides assistance to countries which so request in organising national training courses. A Course on Maritime Pollution, Prevention, Control and Response was held in Alexandria, Egypt. This was organized by the Arab Maritime Transport Academy with technical support from REMPEC. Forty-eight participants attended the course which was held from 10–19 February 1990 and was financially supported by the European Commission.

In May 1990 REMPEC organised MEDIPOL, a specialised training course on harmful substances. In October 1990, there will be a Regional Seminar on Financial Questions, Liabilities and Compensation for Consequences of Accidents Causing Pollution or other Harmful Substances.

On the top of its Agenda for future publications REMPEC has an inventory of Equipment and Products which could, under certain conditions, be put at the disposal of a state which so requests. REMPEC also intends to prepare Guidelines for the use of dispersants in the Mediterranean.[12]

3.10 INTERNATIONAL OIL POLLUTION COMPENSATION FUND (IOPC Fund)

The IOPC Fund is an intergovernmental organisation the function of which is to provide supplementary compensation to victims if, and to the extent that, the shipowner's liability is insufficient to provide adequate compensation for pollution damage caused as a result of an oil spill from a laden tanker. The IOPC Fund was set up in October 1978 under the International Convention on the Establishment of an International Fund for Compensation for Oil Pollution Damage, 1971 (Fund Convention). This Convention was adopted to supplement the International Convention on Civil Liability for Oil Pollution Damage, 1969 (Civil Liability Convention, CLC).

'Oh God, just look at my hair'

During the period covered by the Yearbook, five States — Canada, Cyprus, Qatar, the Seychelles and Vanuatu — became Parties to the Fund Convention, bringing the number of Member States to 43. It is expected that several more States will soon join the IOPC Fund.

Incidents and Settlement of Claims

During the period seven incidents occurred that gave rise to claims against the IOPC Fund, namely the *Kasuga Maru No 1, Fukkol Maru No 12, Tsubame Maru No 58, Tsubame Maru No 16, Kifuku Maru No 103* and *Dainichi Maru No 5* incidents, which took place in Japan, and the *Nancy Orr Gaucher* incident which occurred in Canada.

The most important of the new incidents was the *Kasuga Maru No 1* incident which occurred in December 1988 in Japan, when a Japanese coastal tanker (480 GRT) capsized and sank in stormy weather. The sunken tanker, lying at a depth of approximately 270 metres, was leaking oil for several months. Extensive fishing is carried out by local fishermen in the area around the site of the incident. Major operations for the purpose of preventing the surfacing oil from coming ashore were carried out by applying dispersants, mainly from helicopters. Claims for compensation totalling Y570 million (£2.5 million) were submitted to the IOPC Fund during the period July to September 1989. These claims were settled at an aggregate amount of Y442 million (£1.9 million), and the claims were paid in November and December of that year.

In previous yearbooks, information has been given concerning the *Tanio* incident (France, 1980), the most important case in which the IOPC Fund has so far been involved. In the period 1983–1985 the IOPC Fund made part payments to claimants totalling FFr221 million (£18.2 million). A final payment of FFr939,191 (£90,000) was made by the IOPC Fund to claimants in October 1988. As mentioned in the *1984 Yearbook*, the IOPC Fund took legal action against the shipowner and other parties in the Civil Court of Brest (France) in order to recover the money paid as compensation. The French Government and other claimants took similar action to recover the part of their claims which would remain uncompensated under the Civil Liability Convention and the Fund Convention. In December 1987 an out-of-court settlement was reached between the IOPC Fund and the French Government, on the one side, and all defendants in this case, on the other. As a result of this settlement the IOPC Fund recovered US $17.5 million (£9.5 million). When all claims arising out of this incident had been paid, there was a considerable balance on the claims fund established for this incident. In accordance with a decision taken by the IOPC Fund Assembly, an amount of £13.9 million was reimbursed on 1 February 1989 to the persons who had paid contributions to cover the claims arising out of this incident.

In respect of the *Patmos* incident, which occurred in March 1985 in the Straits of Messina, Italy, claims totalling £37 million were lodged against the owner of the *Patmos* and the IOPC Fund. Most of the claims have been settled out of court and paid. However, the IOPC Fund has become involved in complex court proceedings in Italy concerning some claims. One claim related to a salvage reward due by the cargo owner to the salvors in subrogation.

In its judgment, the Court of first instance made a general statement to the effect that salvage operations could not be considered as preventive measures (as defined in the Conventions) since the primary purpose of such operations was that of rescuing ship and cargo; this applied even if the operations had the further effect of preventing pollution. On the basis of this position of principle, the Court of first instance rejected, *inter alia*, the cargo owner's claims. In January 1988, an out-of-court settlement was reached in respect of that claim. Under the settlement no payment was made in respect of the salvage award. In the record of the court hearing at which the settlement was approved, it was stated that the cargo owner waived his claim in respect of remuneration for salvage. The main outstanding issue relates to a claim submitted by the Italian Government for compensation for damage to the marine environment which was rejected by the Court of first instance. The position taken by the IOPC Fund and by the Court of first instance in respect of this claim was set out in the *1986–1987 Yearbook* (page 27). This claim is being considered by the Court of appeal in Messina.

During the period covered by the present Yearbook, settlements were reached in respect of all claims against the IOPC Fund arising out of three major incidents: the *Oued Gueterini* (Algeria 1986). *Thuntank 5* (Sweden, 1986) and *Antonio Gramsci* (Finland, 1987) incidents.

The *Thuntank 5* case related to a tanker which ran aground in December 1988 in very bad weather on the east coast of Sweden 200 kilometres north of Stockholm. It was estimated that 150–200 tonnes of oil escaped. The oil affected various areas along a 150 kilometre stretch of coast, including a number of small islands. A claim submitted by the Swedish Government for clean-up operations was settled at SKr23 million (£2.3 million).

The tanker *Oued Gueterini* was unloading bitumen in the port of Algiers, when part of the cargo was spilled onto the deck of the vessel. Some bitumen escaped into the water in the port area and approximately 15 tonnes entered the seawater intake of a power station, necessitating the shutdown of the station for a short period of time. Claims were submitted by the owner of the power-station relating to damage to the equipment, cost of cleaning and replacing some equipment and loss of profit as a result of the closure of the station. These claims were settled during 1989 for a total amount of £285,000.

As for the *Antonio Gramsci* incident, which took place in the Gulf of Finland in February 1987, a claim submitted by the Finnish Government relating to clean-up operations was settled at FM9.8 million (£1.5 million).

The 1984 Protocols to the Civil Liability Convention and the Fund Convention

As mentioned in previous yearbooks, a Diplomic Conference, held in 1984 under the auspices of the International Maritime Organisation (IMO) adopted two Protocols to amend the Civil Liability Convention and the Fund Convention, respectively. So far the Protocol to the Civil Liability Convention has been ratified by Australia, Federal Republic of Germany, France, Peru, St Vincent and the Grenadines and South Africa, whereas only the Federal Republic of Germany and France have become Parties to the

Protocol to the Fund Convention. In the United Kingdom, a bill which would enable the Government to ratify the Protocols has been approved by Parliament, and it is expected that the United Kingdom will soon deposit its instruments of ratification. Several other States have begun preparing legislation enabling them to ratify the Protocols.[13]

3.11 OPERATION OF THE 1972 LONDON DUMPING CONVENTION (LDC)

The Convention for the Prevention of Marine Pollution by Dumping of Wastes and Other Matter, 1972 (the London Dumping Convention) entered into force in 1975. The Convention regulates the disposal at sea of all types of wastes throughout the world. As of 1 January 1990, 63 states had become Contracting Parties to the LDC.

The Twelfth Consultative Meeting of Contracting Parties to the London Dumping Convention (LDC) met at the Headquarters of the International Maritime Organisation (IWO) from 30 October to 3 November 1989. The Meeting was attended by 37 Contracting Parties, four states which are not yet Contracting Parties, eight intergovernmental and ten non-governmental international organisations. The following were the principle issues discussed and decisions taken.

Report of the Scientific Group on Dumping
The report of the Scientific Group on Dumping was accepted. Actions were taken with regard to the following:

- *Organotin compounds*: the Consultative Meeting, noting the regulatory steps taken and discussed within many other forums concerning the use of organotin compounds as antifouling paints, agreed that there was no compelling need at this stage to transfer these substances from the grey list of the Convention (Annex II) to its black list (Annex I). However, the Scientific Group was requested to continue the evaluation of the hazards to the marine environment of these compounds.

- *Notification of sea disposal activities:* the Consultative Meeting agreed to the recommendations of the Scientific Group that all Contracting Parties should be urged to report to the Secretariat on their disposal activities (dumping and incineration at sea) as well as on their monitoring of the effects of such activities. To facilitate the latter issue a revised definition of "monitoring" for the purpose of the London Dumping Convention was adopted by a resolution. Contracting Parties were also urged to submit information on their experiences in waste management.

- *Future work*: main issues for future consideration by the Scientific Group include the field verification of laboratory tests, the preparation of monitoring guidelines, the development of processes and procedures for the management of wastes dumped at sea, and the review of the incineration of noxious liquid substances at sea.

Ambiguities
The Consultative Meeting agreed that the ad hoc Group of Experts on the Annexes to the Convention should continue

its review of the operational procedures of the Convention. The goal was to eliminate certain inconsistencies and ambiguities from the existing provisions and overcome difficulties caused by terminology and generally to improve the regulation of dumping within an holistic, waste management context.

General Wastes

The Contracting Parties adopted the inclusion in Annex III to the Convention (section A) of a paragraph containing the following text:

"In issuing a permit for dumping, Contracting Parties should consider whether an adequate scientific basis exists concerning characteristics and composition of the matter to be dumped to assess the impact of the matter on marine life and on human health."

Disposal of Radioactive Waste at Sea

The Consultative Meeting agreed to convene another meeting of the Inter-Governmental Group of Experts on Radioactive Waste Disposal at Sea (IGPRAD) in the autumn of 1990. IGPRAD will address a number of critical issues relating to political, social and economic matters related to the sea disposal of radioactive wastes. In this connection, it was generally felt that the involved studies would most not likely be completed before 1993.

Disposal at Sea of Decommissioned Nuclear Powered Vessels and Disposal into a Subseabed Repository of Low-Level Radioactive Wastes

The Consultative Meeting confirmed that the requirements of the Convention applied to the disposal at sea of any vessel, whether military, nuclear powered or non-nuclear powered, commissioned or decommissioned. It also considered itself the appropriate forum to consider disposal of low-level radioactive wastes into a subsea-bed repository accessed from the sea. With respect to some other questions of what would constitute 'dumping at sea' under the Convention, it was decided to reconvene the Group of Legal Experts on Dumping with a view to clarifying some uncertainties.

With regard to accidents at sea involving releases of radioactive material, Contracting Parties were requested to provide all relevant information to the International Atomic Energy Agency (IAEA).

Matters Related to Incineration of Wastes and Other Matter at Sea

To assist with the implementation of resolution LDC 35(11), on the status of incineration at sea, the Consultative Meeting requested the Scientific Group to:

* provide advice to the Contracting Parties on how to conduct the re-evaluation;

* review clean technology and practical availability of land-based alternatives; and

* take into account all relevant information on incineration technology and associated environmental implications.

Other Matters

The Consultative Meeting expressed the view that the issue of liability was an extremely important one. It was agreed that its Group of legal Experts on Dumping should continue to examine the issue of liability and compensation.

The Meeting agreed that Contracting Parties should be invited to review the provisions of the LDC in the light of the Basle Convention and to submit their views *inter alia* on the need to recommend any additional measures within the LDC or the Basle Convention in order to enhance the effectiveness of either convention with respect to the environmentally sound disposal of wastes.

The Meeting welcomed the activities of the Secretariat in convening national and regional seminars on the control and prevention of pollution by waste disposal at sea and its support for the International Ocean Disposal Symposium.

A major focus at the Meeting was a special public session on the health of the marine environment, as well as on the progress and needs in relation to the basic principles of the LDC.

The Meeting was informed of the extensive cooperation and collaboration with other organisations and bodies involved with the prevention and control of marine pollution. These include the Joint Group of Experts on the Scientific Aspects of Marine Pollution (GESAMP), the Intergovernmental Oceanographic Commission (IOC), the United Nations Environment Programme (UNEP), the Oslo Commission, the Helsinki Commission, and the International Council for the Exploration of the Sea (ICES).

The Twelfth Consultative Meeting agreed to establish a Steering Group to examine a long-term strategy for the Convention. Its resolution LDC.38(12) on a long-term strategy for the Convention calls for an evaluation of the status of the Convention, its implementation and possible new directions.

The Thirteenth Consultative Meeting will be held from 29 October to 2 November 1990.[14]

3.12 ORGANISATION FOR ECONOMIC COOPERATION AND DEVELOPMENT (OECD)

The Organisation for Economic Cooperation and Development (OECD) was set up under a convention signed in Paris on 14 December 1960, which provides that OECD shall promote policies designed:

* to achieve the highest sustainable economic growth and employment and a rising standard of living in Member Countries, while maintaining financial stability and thus contribute to the development of the world economy;

* to contribute to sound economic expansion in Member as well as non-member countries in the process of economic development;

* to contribute to the expansion of world trade on a multilateral, non-discriminatory basis in accordance with international obligations.

In 1988, the OECD launched a new programme of activities in the area of accidents involving hazardous substances. A high level Meeting was organised under the chairmanship of Mr A Carignon, French Minister for the Environment. Subsequently, two legally binding Council Acts were adopted in the area of information to the public concerning hazardous installations and international information exchange on accidents causing transfrontier damage.

In 1989, the Council adopted a recommendation on the application of the polluter-pays principle to accidental pollution. The purpose of this new Act is to allocate as far as possible the costs of accident prevention and response to the operator of the fixed installation involved or potentially involved in an accident. In particular, the operator will be asked to pay the clean-up costs and environmental rehabilitation costs as in the case of oil spills at sea.

The financial costs of accidental pollution from fixed installations were estimated. It was found that the total damage actually paid to victims was very much smaller than the pollution prevention costs and than the total costs of other accidental risks borne by industry. A survey of pollution insurance and of compensation funds for accidental pollution showed that pollution insurance is now evolving and that full compensation will soon be provided in a few countries to all victims of accidental pollution whether the polluting event is sudden or gradual.

Hazardous Wastes

In May 1988, after extended discussion and negotiation among Member Countries, the OECD Council decided on a 'Core List' of wastes for which consensus had been reached that they require control when proposed for disposal following transfrontier movement [OECD Council Decision C(88)90 (Final)]. This Core List contains 17 generic waste types and 27 constituents. Comparison with national lists indicate that the core list covers at least 85% of all wastes considered to be hazardous by OECD Member Countries.

In addition to wastes covered by the Core List, all other wastes which are considered to be or are legally defined as hazardous wastes in the country from which these wastes are exported or in the country into which these wastes are imported are also subject to control under terms of Council Decision C(88)90 (Final). This Decision also includes a definition of the terms 'wastes' and 'disposal' for purposes of control of transfrontier movements and provides for a uniform classification system referred to as the International Waste Identification Code (IWIC). The use of the IWIC allows virtually all wastes deemed to be hazardous by most countries to be described satisfactorily in terms of potential hazard, activity generating the wastes, physical form, generic description and constituents, as well as indicating the disposal or recycling operation for which the wastes are intended.

The Core List of hazardous wastes, as well as the definitions of 'wastes' and 'disposal', have been incorporated in the Convention on the Control of Transboundary Movements of Hazardous Wastes and their Disposal, which was adopted in Basle on 22 March 1989 (see 3.3.12).[15]

3.13 INDIAN OCEAN MARINE AFFAIRS COOPERATION CONFERENCE (IOMAC)

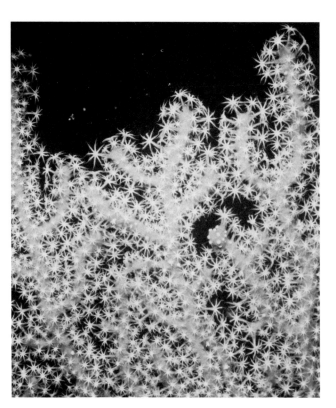

Melithaeid sea-fan, Indonesia. Many sea-fans collected in minutes for the curio trade take decades to grow to maturity. They are also sensitive to increased sediment in water, due, for example to logging activities and unplanned coastal development. Photo: D George

Since its establishment, IOMAC has significantly progressed in building and strengthening marine affairs capabilities in the wider Indian Ocean Region in accordance with the Declaration, Programme of Cooperation and Plan of Action adopted by the First IOMAC Conference in Colombo, Sri Lanka in 1987.

IOMAC is an intergovernmental regional organisation in an early stage of development which has been characterised by a deliberate avoidance of what apparently has come to be regarded by participating states as premature formalisation. IOMAC operates through the following organs:

- the *Conference* — plenary meetings of all states and organisations participating in IOMAC, convened once every three years, with the Second IOMAC Conference scheduled for September 1990 in Tanzania;

- the *Standing Committee* consisting of representatives of 17 Indian Ocean states, but open to all IOMAC participants, meeting annually and coordinating the national institutions (focal points) for marine affairs development;

- the small *Secretariat* in Colombo, which functions with the guidance of the Standing Committee and the Secretary-General, Ambassador Hiran WA Jayewardene of Sri Lanka;

- functional *Working Groups* established, when necessary, for specific tasks, such as the IOMAC Technical Group on Offshore Prospecting for Mineral Resources or a Technical Cooperation Group (TCG).

Participation in IOMAC is open to:

- the regional Indian Ocean states (Members), including both 38 coastal and 12 Asian and African land-locked states;

- the major maritime users (MMU) — states from outside the Indian Ocean region determined principally on the basis of the global tonnage in shipping traversing the area (observers);

- international organisations (observers), which include the UN Office for Ocean Affairs and the Law of the Sea, ESCAP, ESCWA, IMO, FAO, UNEP, WMO, UNESCO–IOC, UNOSAD, UNDP, UNRFNRE, World Bank, UNU, AALCC, SACEP, Cooperation Council of the Arab Gulf States, IHB, IWC, ESA, and several non-governmental organisations (ICOD, NILOS, IOI).

The IOMAC Programme of Cooperation and Plan of Action articulate a broad range of activities (short, medium and long-term) to be carried out at national, subregional and regional levels in application of the 1982 UN Law of the Sea Convention in the fields of marine environment, mineral resources, fisheries, ocean science and services, maritime transport and communications, marine affairs information system, technical assistance and training.

Recent Activities
In the period June 1988–February 1990 the activities of IOMAC included:

- July 1988: the First Prospecting for Mineral Resources in the Indian Ocean held in Karachi, Pakistan, which formulated the Offshore Prospecting Programme (OPP);

- November 1988: the 3rd Meeting of IOMAC Standing Committee held in Colombo, Sri Lanka;

- October–December 1988: the Second IOMAC/IOI Regional Training Programme in Marine Affairs held in Kuala Lumpur, Malaysia, with cosponsorship of NILOS;

- January 1989: the Meeting of Legal and Fisheries Experts and Second Meeting of IOMAC Statute held in Jakarta, Indonesia, devoted to the establishment of a new Indian Ocean Tuna Commission (within or outside the FAO framework) and to the drafting of IOMAC Statute with a view to completion of a final draft by the 1990 Second IOMAC Conference;

- July 1989 : United Nations/IOMAC Workshop on Oceanographic/Marine Space Information Systems held in Karachi, Pakistan, which provided for the establishment of the IOMAC Marine Space Application Programme (MSAP);

- July 1989: the Fourth Meeting of IOMAC Standing Committee held in Colombo, Sri Lanka;

- October–December 1989: the Third IOMAC/IOI Regional Training Programme in Marine Affairs held in Egypt;

- February 1990: Marine Affairs Information Workshop held in Jakarta, Indonesia.

At its Third and Forth Meetings, in 1988 and 1989, the Standing Committee considered the specific issue of dumping hazardous wastes and toxic materials, noting that in recent times Indian Ocean States and other developing countries have encountered disposal of such substances in their adjacent offshore areas. The Committee emphasised the necessity of taking stricter preventive and control measures with regard to hazardous waste disposal and envisaged the convening of a special workshop to facilitate the relevant action.

The IOMAC Standing Committee also took note of the concern expressed by the Indian Ocean states with regard to pollution of the sea by oil, especially non-compliance with the existing standards and rules, and recommended the Secretary-General to identify an effective mechanism to monitor and develop regional capabilities of coastal states for prevention of such environmental hazards with assistance from the competent international agencies. In addition, the meeting discussed the potential adverse impact on the marine environment of the proposed deep sea-bed mining in the Indian Ocean and recommended that a team of experts from within and outside the region should begin consideration of this subject. This initiative seems particularly important in view of the fact that environmental aspects have been undertaken by PrepCom only in 1989.

The IOMAC Secretariat has sought the assistance of the UN with respect to streamlining and harmonising the delivery of UN system support for IOMAC and the Indian Ocean Region generally. Since bilateral aid agencies and non-governmental institutions do not have a mechanism for coordination of their marine-related aid programmes, IOMAC could fulfil the useful role of harmonising those activities in the context of an overall plan of assistance to the Indian Ocean states.[16]

3.14 PERMANENT COMMISSION OF THE SOUTH-EAST PACIFIC (CPPS)

The Permanent Commission of the South-East Pacific (CPPS) is an organisation established by Agreement on 18 August 1952, signed and adopted by the governments of Colombia, Chile, Ecuador and Peru. As a marine organisation it has as its main objectives the coordination of the marine policy adopted by the Ministers of Foreign Affairs of the South Pacific System, in the legal and scientific field, and the formation and diffusion of the knowledge, economy and protection of the environment. It consists of a secretary-general and three deputy secretary-generals for Legal, Economic and Scientific Affairs. CPPS also coordinates various regional activities as part of international projects in cooperation with their respective international organisations.

In the pollution field, work continues through the Action Plan for the Protection of Marine Environment and Coastal Areas of the South-East Pacific, adopted in November 1981 by the Plenipotentiaries Representatives of Colombia, Chile, Ecuador, Panama and Peru, with the purpose of evaluating and monitoring continuously regional marine pollution. By general agreement, the Convention for the Protection of the Marine Environment and Coastal Area of the South-East Pacific and its three supplementary legal agreements have been ratified by Colombia, Chile, Ecuador, Panama and Peru. These agreements are now effective, together with two new Protocols recently signed by the above-mentioned governments: the Protocol for the Protection of the South-East Pacific against the Radioactive Contamination, and the Protocol for the Conservation and Administration of Protected Marine Areas in accordance with the legal framework of regional marine contamination. A total of 46 institutions and laboratories coordinated by five national focus points are involved in a study of marine contaminants of regional interest in more than 25 geographical areas from Charco Azul in Panama down to the Magellan Strait in the southern- most part of Chile.

Recent Activities

- International Seminar of the Investigation and Monitoring of Marine Pollution in the South-East Pacific, Cali, Colombia, 6–8 September 1989.

- Fourth Experts Meeting (Consultative Group) of the Action Plan for the Protection of the Marine Environment in Coastal Areas of the South-East Pacific, Cali Colombia 13-16 September 1989.

- Fourth Intergovernmental Meeting of the Action Plan, and First Meeting of the High Contracting Parties to the Lima Agreement and its supplementary instruments, Paipa, Boyacá, Colombia, 20–22 September 1989.

- Experts Meeting of the Regional Contingency Plan against Oil Spill Pollution in the South-East Pacific. Notification and Alert Plan, Viña del Mar, Chile, 22–24 November 1989.[17]

3.15 INTERNATIONAL WHALING COMMISSION

The International Whaling Commission (IWC) was set up under the International Convention for the Regulation of Whaling signed in Washington on 2 December 1946. This Convention was concluded in order to provide for the proper conservation of whale stocks and thus make possible the orderly development of the whaling industry.

The main duty of the IWC is to keep under review and revise as necessary the measures laid down in the Convention governing the conduct of whaling. These measures provide for the complete protection of certain species of whales; designate specified ocean areas as whale sanctuaries; set the maximum catches on whales which may be taken in any one season; prescribe open and closed seasons and areas for whaling; fix size limits above and below which certain species of whale may not be killed; prohibit the capture of suckling calves and female whales accompanied by calves; and require the compilation of catch reports and other statistics and biological research.

The Commission encourages, coordinates and funds research on whales, in particular through its sponsorship of a second 'International Decade of Cetacean Research', currently concentrating on sighting surveys of the Antarctic minke whale stock. It also promotes studies into related matters such as the humaneness of the killing operations, and the management of aboriginal subsistence whaling which is currently permitted from Denmark (Greenland), St Vincent and The Grenadines, the USSR (Siberia) and the

Japanese continue to get special permit for scientific whaling; see 3.15.

USA (Alaska). The Commission also takes an interest in marine pollution and degradation of the marine environment generally insofar as these matters may have an impact on the health, behaviour and breeding of stocks of whales.

Membership of the Commission is open to any country in the world and at present comprises 37 member governments, each represented by a commissioner appointed by the government. The commissioner may be accompanied by experts and advisers, and the Scientific Committee is made up of scientists nominated by the member governments and invited experts. Meetings of the Commission are also attended by observers from non-member governments, other intergovernmental organisations, and international non-governmental organisations by invitation.

The Scientific Committee is presently engaged in an in-depth evaluation of the status of whale stocks in the light of management objectives and procedures as part of this process of a comprehensive assessment.

Since the introduction of the pause in commercial whaling, some member governments have implemented major research programmes which include the sampling of whales caught under special permits which the Convention allows them to grant. Some other members have expressed concern that this procedure could be abused in order to circumvent the so-called moratorium and intense debate is taking place on the scientific, legal and political aspects of this problem, particularly since it raises questions of sovereign states' jurisdiction in relation to the operation of an international convention.

Recent Activities

At the 40th Annual Meeting held in Auckland, New Zealand, 30 May - 3 June 1988, the Commission discussed research permit proposals from Iceland and Norway, and adopted Resolutions which expressed the view that they did not fulfil all the Commission's guidelines and effectively recommended that those governments refrain from issuing such permits. In the event, these governments, and Japan, all carried out their planned research programmes. At this meeting the IWC also set catch limits for continuing aboriginal subsistence whaling in the areas where it is already permitted.

Further Resolutions directed to Iceland, Japan and Norway were adopted at the 41st Annual Meeting held in San

Further Resolutions directed to Iceland, Japan and Norway were adopted at the 41st Annual Meeting held in San Diego, California, USA, 12 June 1989. These requested the governments concerned to reconsider their scientific research permits, which were then issued after modification in each case. The Commission noted the remarkable progress made to improve killing methods in the Alaskan bowhead fishery with the cooperation of Norwegian specialists. It also extended by three years the period for which the Indian Ocean is designated as a whale sanctuary, following its original 10 year designation in 1979. There was also much discussion of the socio- economic implications of zero catch limits, but a request for an interim catch quota for the Japanese small-type coastal whaling operation was not agreed.

The 42nd Annual meeting of the International Whaling Commission was held in the Netherlands, 2–6 July 1990. The First of the comprehensive assessments by the Scientific Committee of North Pacific grey whales, Southern Hemisphere and North Atlantic minke whales was considered, together with progress on developing an alternative management procedure to replace the existing scheme for commercial whaling.

Further work on the comprehensive assessment of whale stocks will continue in the following year, concentrating on North Atlantic fin whales and North Pacific minke whales, as well as a revised management procedure. These matters will then be considered by the Commission at its 43rd Annual Meeting to be held in Reykjavik, Iceland in May 1991.[18]

3.16 COMMONWEALTH SECRETARIAT

The modern Commonwealth, which celebrated its 40th anniversary last year, is an association of 49 (4 developed and 45 developing) countries in equal partnership devoted to cooperation in the interests of freedom and development. The association is based mainly on principles laid down by Heads of Commonwealth Governments.

The Commonwealth Secretariat, established by the Heads of Commonwealth Governments in 1965, coordinates the work of the association, organises intergovernmental consultations, services, Commonwealth meetings and committees, conducts programmes of cooperation and acts as a clearing house of information. It organises regular meetings of Heads of government, Ministers and senior officials, and is active in international political and economic affairs and in areas of functional cooperation such as education, health, law, science, youth and the role of women in development. The Secretariat includes a technical assistance arm — the Commonwealth Fund for Technical cooperation (CFTC).

Commonwealth leaders at their Vancouver Summit in October 1987 welcomed the report "Our Common Future" of the World Commission on Environment and Development and agreed that environmentally sustainable development should be central to national and international policy. They considered the threats posed particularly to low lying island country of the Maldives and set up an expert group to examine the implications of natural disasters resulting from possible climate change. The group consisted of nine Commonwealth experts with Dr Holgate, Director General of the International Union for Conserva-

tion, of Nature and Natural Resources (IUCN) as Chairman. It held three meetings in May 1988, January and July 1989 and produced its report *Climate Change: Meeting the Challenge* which it hoped would contribute to the discussions in the wider international community.

Commonwealth leaders meeting in Kuala Lumpur in October 1989 considered the Report and agreed to the establishment of a group of experts on the environment who could monitor and evaluate climate change taking into account global developments. They also issued the Langkawi Declaration on the Environment which commits Commonwealth Governments to a wide ranging 16 point action programme. The Programme supports efforts to prevent marine pollution, including curbing ocean dumping of toxic wastes. Following the Report on Climate Change and the Langkawi Declaration, a Conference on Small States was convened in November 1989 by the Government of the Maldives and ended with the Male Declaration on Global Warming and Sea Level Rise. The Small States intend to work, collaborate and seek international cooperation to protect the low-lying small coastal and island states from the dangers posed by climate change, global warming and sea-level rise.

Commonwealth leaders at Vancouver also welcomed the Secretariat's Study Conservation for Sustainable Development which focuses on the scope for Commonwealth action to counteract soil erosion, desertification and drought in Africa and asked the Secretariat to continue its special programme to build sustainable agriculture. The Secretariat is also assisting African institutions to build environmentally sound conservation strategies and land-use plans and is promoting the integration of conservation into farming. In implementing the programme developed in its Study on the Scope for Commonwealth Action on Soil Erosion, the Secretariat held six conferences and workshops between 1988 and mid–1989.

The Secretariat's Economic Affairs Division serviced the work of the expert group on climate change. The Food Production and Rural Development Division is responsible for the work on sustainable agriculture. A number of other divisions and programmes have also been concerned with environmental issues including marine pollution.

The Industrial Development Unit has helped several countries improve pollution controls in industry. Illustrative of the work done in this area is the designing of systems for effluent disposal and water purification in mining areas in one member country.

The Technical Assistance Group (TAG)

The TAG deals with pollution in two areas of activity, mining and petroleum and fisheries. TAG advises governments on the negotiation of agreements and the drafting of legislation in relation to mining and petroleum, and access to fisheries.

TAG has recently offered assistance to a government in drafting regulations dealing with the use of mercury in mining operations. In such practices, the mercury frequently enters rivers in the mining area, thus polluting the waters.

The Commonwealth Science Council (CSC)

Activities of this Council related to pollution of the sea have been developed under the Environmental Planning Programme over the past six years. The programme's main thrust has been in the problems of coastal zone management, especially in the tropics. Activities have included surveys of marine biological resources, training of coastal zone managers and the involvement of local communities in awareness raising. Assistance is being provided to the governments of the South Pacific and to institutions concerned with research and management of natural resources to achieve sustained development of the coastal area. The Caribbean is also receiving similar technical support. Here, tourism and population pressures have resulted in many island states of the Eastern Caribbean facing conflicting pressures on use of the environment. CSC's projects have contributed to identification of issues in management and research needs, as well as the creation and improvement of university-level training activities in the region. Most recently a manual was published: the *Workbook of Practical Exercises in Coastal Zone Management for Tropical Islands.* A new activity in the Environmental Planning Programme was launched in August 1989 at a Pan-Commonwealth meeting on Chemistry and the Environment. This project will be developed during 1990 and will relate to toxic waste and other aspects.[19]

3.17 CARIBBEAN COMMUNITY (CARICOM)

The First CARICOM Ministerial Conference on the Environment was held in the Port of Spain, Trinidad and Tobago from 31 May–2 June 1989. The objectives were as follows:

- to achieve increased appreciation of the significance of the issues and needs relevant to management and protection of the Caribbean environment, and of the relationship between environment and development;

- to identify matters for priority attention in the region in relation to the environment; and

- to identify approaches that would allow for better regional coordination and monitoring of activities, agencies and resources.

Delegations from 12 Members States of the Community, and from 4 observer countries were present. Observer regional and international institutions also attended.

The Meeting agreed the following:

- that, at the regional level, Ministers with responsibility for Environmental Matters should meet at appropriate intervals for the purpose of conducting policy and programme review and establishing the goals and guidelines for action; to this end it supports the proposal by the Prime Minister of the Republic of Trinidad and Tobago that a Standing Committee of Ministers responsible for the Environment be established;

- approve, also at the regional level, the establishment of a consultative forum of agencies whose activities in the region are relevant to the development of Caribbean environmental programmes and projects. The function of this forum will be to pursue the identification of, and the allocation of, responsibility for action on programmes, projects and studies relating to the priority problems and strategic approaches set out above. The

CARICOM Secretariat will have the responsibility for convening this forum;

- reiterate the Community's commitment to the effective development of the Caribbean Environmental Health Institute as a regional technical, advisory and project implementing facility in the environmental field.

The CARICOM Secretariat was mandated to arrange consultations and negotiations with donor agencies for support to Caribbean environmental programmes and projects on the basis of the policies and guidelines laid down at ministerial level and the results of the work of the consultative forum. In this connection, appreciation to those bilateral and multilateral agencies which have been actively supporting these programmes was expressed.[20]

3.18 AFRICAN MINISTERIAL CONFERENCE ON THE ENVIRONMENT (AMCEN)

At the request of its African members, the Governing Council of UNEP in May 1983 invited the governments of Africa to convene, in cooperation with regional organisations and institutions, an African Environmental Conference to discuss national environmental priorities and identify common problems worthy of a regional programme of action to deal with serious environmental problems in Africa.

Following a careful preparatory process involving extensive consultations with African Governments, experts from African countries and several UN Agencies involved in African development, UNEP, in cooperation with OAU and the ECA, organised the First African Ministerial Conference on the Environment in Cairo in December 1985. In its resolution at the end of its first session, the Conference, among other things, decided to strengthen cooperation between African governments in economic, technical and scientific activities, with the prime objective of halting and reversing the degradation of the African environment in order to satisfy the food and energy needs of the peoples of the continent. The Conference further recommended that regional cooperation should focus on the essential resources of water, soils, plant cover and forest, fauna, energy and seas, by means of a progressive reorientation of economic development strategies.

To implement the above objectives the Conference decided to institutionalise itself and to establish a number of organisational and operational machineries at the regional, sub-regional and national level, and UNEP was invited, in cooperation with the OAU and ECA, to assume the functions of the secretariat for the Conference.

Seas

With regard to the AMCEN Committee on the Seas, its basic functions are to provide:

- a consultative forum for coordination of the four UNEP-sponsored Regional Seas Action Plans for the protection and development of the marine and coastal environment in which all African Coastal States (with the exception of South Africa) participate;

- mechanisms for improved information exchange and for enhancing participation and commitment of individual countries to the Regional Seas Action Plans and to AMCEN's programme; and

- mechanisms for raising funds needed to support the Regional Seas Action Plans or specific additional projects.

UNEP's support networks of national institutions (universities, government laboratories etc) have been established to aid each environmental area and to develop measures for its control. UNEP coordinates the incorporation with FAO, IOC and WHO.

The weak institutional infrastructure in all of the African States was the most serious impediment for the effective functioning of the networks. Therefore, UNEP's assistance (amounting to about US$6 million in cash since 1976) was mainly concentrating on:

- purchase of pollution monitoring equipment;

- maintenance services for this equipment;

- training of local personnel in use of the equipment; and

- training of local administrators and policy-makers.

At present functional networks exist in the Mediterranean and West and Central African Regions. The networks in Eastern Africa and the Red Sea regions are in the early formative stages.[21]

3.19 SOUTH PACIFIC FORUM FISHERIES AGENCY (FFA)

The South Pacific FFA is based in Honiara, the capital of the Solomon Islands. It was established in 1979 following a directive by the South Pacific Forum which saw the need for increased regional cooperation in fisheries in the South Pacific. The need for the establishment of a regional fisheries organisation arose from the dramatic changes to the management, use and ownership of the ocean's resources proposed by the Third United Nations Law of the Sea Conference in the 1970s. Pacific Island countries were quick to appreciate the significance of the changes proposed under UNCLOS III to areas of national fisheries jurisdiction and access to living marine resources.

One of the major factors contributing to the Agency's past success has been the commitment to it shown by the 16 politically independent and self-governing Pacific Island countries which make up its membership. Current members are Australia, Cook Islands, Federated States of Micronesia, Fiji, Kiribati, Marshall Islands, Nauru, New Zealand, Niue, Palau, Papua New Guinea, Solomon Islands, Tonga, Tuvalu, Vanuatu, and Western Samoa. These countries, together with resource experts from around the globe recently participated in a Conference which, in addition to celebrating ten years of activity of FFA, developed a Corporate Plan which will be used to assist the Agency in responding to issues that are likely to be important to the development and management of South Pacific fisheries in the coming decade.

Throughout the ten years that the Agency has been in operation, it has set international standards at national and regional levels in the dealings of coastal states with distant

water fishing nations. The major achievement of the decade in this activity area was the successful negotiation of a Treaty on Fisheries between certain Pacific Island states and the United States of America. The Agency is also active in providing a broad range of assistance with the development and implementation of national fishing initiatives in member countries. For the majority of coastal states served by the Agency, fish are one of the few natural resources offering economic development potential.

During the last two years, the issues of driftnets and South Pacific albacore have attracted a considerable amount of the Agency's work effort. In late 1988, it became apparent that the rapid increase in driftnet fishing effort, from approximately twenty boats in previous years to an estimated 130–160 boats for the 1988/89 season was of considerable regional concern. South Pacific island nations aspire to develop their own fleets of vessels to harvest the substantial tuna resources of the region which are mainly taken by foreign fleets at present. Mindful of the long term need to manage the region's shared marine resources, Pacific Island countries actively promote the rational development and management of these resources.

In addition to threatening the sustainability of the South Pacific albacore resource, there were other concerns associated with the non-target catch of marine mammals, reptiles and birds killed or injured in the nets. Environmental concerns relating to the fate of discarded nets, known as ghost nets, which continue to fish for indefinite periods after loss figured highly in discussions.

The FFA was instrumental in coordinating initiatives by South Pacific countries and territories to draft a Convention banning driftnets from the region. In July 1989 a meeting of the Heads of Government of South Pacific Forum countries in Kiribati called for a cessation of this fishing practice in the South Pacific in a statement known as the Tarawa Declaration. This led, in November 1989, to the signing of the Driftnet Convention in Wellington, New Zealand which calls for an immediate ban on driftnet fishing or any activity associated with it in the South Pacific. The South Pacific initiatives were taken up in the General Assembly of the United Nations and in December 1989 a Resolution on large-scale Pelagic Driftnet Fishing and its Impacts on the Living Marine resources of the World's Oceans and Seas was adopted by consensus. The Resolution calls for a cessation of driftnet fishing activities in the South Pacific by 1

3.20 INTERNATIONAL COUNCIL FOR THE EXPLORATION OF THE SEA (ICES)

The International Council for the Exploration of the Sea (ICES) is an intergovernmental organisation established to coordinate marine scientific research in the North Atlantic and adjacent seas, and to provide advice to its member governments and cooperating regulatory commissions on fisheries and marine pollution issues. Founded in 1902, ICES is the world's oldest intergovernmental organisation concerned with marine sciences, and is headquartered in Copenhagen.

During the period June 1988 to February 1990, ICES has been involved in the following activities relevant to marine pollution:

Monitoring Contaminants in the Marine Environment

A focus of ICES activities in recent years has been the development of methodologies for the assessment of monitoring data. In this connection, a statistical procedure developed to assess temporal trends in contaminant levels in fish and shellfish has been extensively applied in analysis of data from the Cooperative ICES Monitoring Studies Programme (CMP) and the Joint Monitoring Programme (IMP) of the Oslo and Paris Commissions. The first report on the analyses of these data sets, concerning comtaminants in fish muscle tissue, was published in the ICES Cooperative Research Report series (No. 162) in January 1989.

During the meeting of the ICES Marine Chemistry Working Group in February 1990, the final evaluation of data submitted for the ICES Baseline Study of Trace Metals in Coastal and Shelf Sea Waters (1985-87) was completed. The report on this study is due to be published in the latter part of 1990 and the study will provide an important contribution to the formulation of ICES advice to other organisations that are planning to coordinate similar types of studies.

Analytical Quality Assurance and Intercalibration Activities

Quality assurance of analytical data continues to be an important issue for ICES, and in this context a number of ICES-coordinated analytical intercomparison exercises have been organised.

In Phase I of the ICES/OSPARCOM/IOC Intercomparison Exercise on Analyses of Chlorobiphenyls (individual PCBs), 57 laboratories from 18 countries submitted results, most of the laboratories demonstrating acceptable performance. The full exercise is planned to consist of four phases, the second of which, involving the analysis of extracts of marine sediments and seal blubber, will be undertaken in 1990.

Similarly, the first stage of the Fourth Intercomparison Programme on Analyses of Hydrocarbons, involving the determination of 10 selected polycyclic aromatic hydrocarbons (PAHs), was completed in 1989, with 17 laboratories from 10 countries submitting results. Plans have now been developed for the second stage of the exercise, which will involve analysis of extracts of marine sediments.

A third major intercalibration activity during 1989, in the field of chemical oceanography, was an Intercalibration Exercise on Analysis of Nutrients in Sea Water, in which 66 laboratories from all 18 ICES member countries participated. The exercise will provide a first basis for quality assurance of nutrient data from a number of new monitoring activities, such as the nutrient component of the monitoring programmes being coordinated by the Paris Commission and the North Sea Task Force, as well as a planned ICES project involving nutrient measurements.

Finally, the first stage of yet another multi-stage intercomparison was completed in 1989; the First Phase of an Intercomparison Exercise on Trace Metals in Suspended Particulate Matter (SPM), with 29 laboratories submitting results by February 1990.

Contaminants

Three years of preparatory work resulted in January 1990 in the final detailed plans for the ICES/IOC Workshop on Biological Effects Techniques, which took place 12–30 March 1990 in Bremerhaven, FRG. The workshop incorporated biological techniques in the fields of fish gross pathology, cellular pathology and biochemistry; bioassays; benthos sediment bioassays and benthic studies. The final plans include the coordination of vessel operations involving six research vessels from FRG and the Netherlands and the activities of 70 participating scientists. The main part of the work consists of investigations of two known contaminant gradients, a north-west transect across the German Bight, extending into the Elbe Estuary plume and around the zone of influence of a Dutch oil platform.

The results of the workshop will be reviewed in the first instance in order to provide advice on possible biological effects techniques for inclusion in the routine monitoring programmes, specifically the new North Sea Task Force monitoring programme. A concluding workshop is scheduled to take place in Copenhagen in September 1991 to present the results and recommendations from the initial workshop with a view to developing protocols for the use of successful techniques and the interpretation of their results.

North Sea Task Force

One result of the Second International Conference on the Protection of the North Sea (London, 1987) was identification of the need for an enhancement of scientific knowledge and understanding. In response to this, ICES and the Oslo and Paris Commissions were charged with the formation of a North Sea Task Force (NSTF) to oversee the further development of harmonised methods for monitoring, modelling, and assessing environmental conditions at the national and international levels.

In relation to this decision, the NSTF was established and had its inaugural meeting in December 1988. One of the remits for the Task Force is the preparation of a Quality Status Report of the North Sea, to be completed in 1993. Among other tasks completed during 1989, the Task Force prepared papers on three special topics which were submitted to the Third North Sea Conference held in The

Hague in March 1990. These papers concerned: algal blooms; the epidemic death of seals in the spring of 1988; and sensitive issues on which progress has been made since the Second North Sea Conference in 1987. ICES provided information on the first two of these topics on the basis of the 1989 report of the Advisory Committee on Marine Pollution.

Forthcoming Meetings and Conferences
During the period May 1990 to June 1992, ICES will hold a fairly large number of internal scientific meetings, with participation generally restricted to experts from member country research institutes. The 78th and 79th Statutory Meetings will be held in Copenhagen from 4–12 October 1990 and in La Rochelle, France from 26 September–4 October 1990, respectively. Several scientific symposia (open to the public) relating wholly or in part to marine pollution are scheduled:

1) Hydrobiological Variability in the ICES Area, 1980-1989 to be held in Mariehamn, Finland from 5–7 June 1991;
2) Patchiness in the Baltic to be held in Mariehamn, Finland from 3–4 June 1991; and
3) Measurement of Phytoplankton Primary Production: From Molecular Base up to Global Space to be held in April 1992.[23]

3.21 MINISTERIAL CONFERENCE ON THE PROTECTION OF THE ARCTIC ENVIRONMENT

The Government of Finland has for many years been concerned about the state of the Arctic environment which is extremely fragile and vulnerable. In recent years the Arctic has witnessed a sharp increase in the rate of natural resources development. Pollution from the North Atlantic and the Northern Pacific, land-based pollution from rivers, air pollution, navigation, oil drilling and other economic activities have created a serious threat to the Arctic environment. One major oil tanker accident alone or a blow-out at an oil drilling platform may drastically change the environmental situation in the Arctic Ocean which still is among the least polluted of the world's oceans.

Against this background the Government of Finland took the initiative in proposing a ministerial conference of the eight Arctic countries — the Nordic countries, Canada, the Soviet Union and the United States — on the protection of the Arctic environment by sending a letter to the governments of seven other Arctic countries in January 1989. All the responses to the initiative were positive in principle. Therefore, the Finnish government invited representatives of those governments to attend a consultative meeting at Rovaniemi, Finland, from 20–26 September 1989.

This Meeting was an historic one in the sense that for the first time all the Arctic countries were meeting at governmental level. The meeting also turned out to be a success.

After adoption of the agenda two Working Groups were established, the first to review the state of the environment in the Arctic and the need for further action, the second to consider existing international legal instruments for the protection of the Arctic environment and the organisation of future co-operation.

A common concern about the Arctic environment was expressed. The Arctic environment already shows signs of serious deterioration. Economic activities in the Arctic region and long-range transportation of pollutants have contributed to this alarming situation. The pollution is already causing changes in some parts of the Arctic ecosystem and there is particular concern over threats to the health of indigenous peoples from toxic substances in the Arctic food chain.

The importance of the interaction between the environment in the Arctic and other geographical areas was stressed. The Polar basin seems to function as the final depository of a number of air and seaborne pollutants. Air pollution also contributes to deterioration of the forests and the state of the environment more generally.

All these changes call for urgent action to combat a worsening of the Arctic environment. It was underlined that the Arctic environment is extremely fragile and vulnerable and therefore in need of special measures. While there are a number of legal instruments, such as the London Dumping Convention, applicable to the Arctic ecosystem, none has been elaborated for the specific purpose of protecting the Arctic environment and no delegation suggested that the existing system of legal measures was adequate.

The process will continue and the 2nd Consultative Meeting will take place in the Spring of 1991, in Canada, and the Ministerial Conference on Protection of the Arctic Environment will be held in the autumn 1991 in Finland.[24]

3.22 ANTARCTIC TREATY SYSTEM (ATS)

The conservation of the Antarctic region has been at the core of the discussions taking place within the Antarctic Treaty System (ATS), for the last two years. The adoption of the Convention on the Regulation of Antarctic Mineral Activities (CRAMRA), in Wellington, New Zealand, in June 1988, followed by the opening of the one year signature period from 25 November 1988 until 1989, has been one of the central elements of the Antarctic scenario up to the present time.

This Convention has been signed by Argentina, Brazil, China, Chile, Czechoslovakia, Denmark, Finland, Japan, German Democratic Republic, New Zealand, Norway, Poland, South Africa, Sweden, USSR, Uruguay, United States, United Kingdom. The outcome of this Convention and the chances of obtaining the necessary ratifications in accordance with the final clauses and the Final Act of the IV Special Consultative Meeting, appear linked to a larger discussion about the alternatives for the protection of the Antarctic environment and how to ensure the effectiveness of the moratorium on minerals prospecting, exploration and development already in existence.

The CRAMRA embodies strong environmental principles, for its Article IV includes a provision that "no Antarctic mineral resource activity shall take place until it is judged, based upon assessment of its possible impacts, that the activity in question would not cause significant effects in global or regional climate or weather patterns" (Art. 4, 3). For a later stage, the Convention envisages the adoption by consensus of a liability protocol, which must set further rules and procedures regarding the means by which claims against operators are to be assessed and adjudicated. This protocol may prescribe appropriate limits on liability.

The 15th Antarctic Treaty Consultative Meeting (ATCM) held in Paris, 9–20 October 1989, was the occasion to study this subject, as well as to deal with other environmental issues that came under the mandate of the Antarctic Treaty Consultative Parties in accordance with Recommendations adopted in previous meetings.

During the preparatory meeting of the 15th ATCM (Paris, 9–13 May 1989) Chile proposed the adoption of Comprehensive Measures for the protection of the Antarctic environment and its associated and dependent ecosystems. In accordance with this proposal, a set of rules of a mandatory nature that will regulate all activity in Antarctica, should be adopted.

On the other hand, Australia made public on 22 May 1989 that it will not favour the CRAMRA and will support the negotiation of a comprehensive environmental protection convention, within the framework of the Antarctic Treaty. France joined this position, stating its willingness to cooperate with Australia to negotiate an agreement to declare Antarctica a wilderness reserve.

These proposals, together with papers introduced by the United States and New Zealand, were not discussed in detail. While France/Australia's proposal is focused on the establishment of centralised institutions invested with strong powers to govern all Antarctic activities, the other documents are inspired by the idea of setting clear and enforceable obligations, accompanied by monitoring schemes and liability principles. One of the points that the French-Australian document raises is the relationship between a new convention and the other components of the Antarctic Treaty System (ATS) that might be superseded by a new instrument.

The relationship between the examination of more stringent measures to protect the Antarctic environment and the negotiation of the liability protocol provided by CRAMRA was apparent at the Paris meeting. From a political point of view, any progress on the first issue should be accompanied by the consideration of the second one. This factor becomes relevant if we take into account that the decision-making process within the Antarctic Treaty System is based upon the rule of consensus.

The 15th ATCM adopted two Recommendations related to these questions in order to convene two meetings in 1990, both due to be held in November in Chile, in order to: (1) explore and discuss all proposals relating to the protection of the Antarctic environment and dependent and associated ecosystems; and (2) explore and discuss all proposals relating to Article 8 (7) of the CRAMRA. This Article refers to the question of liability resulting from damages derived from mineral prospecting, exploration and development activities. In accordance with its wording, the protocol should be concluded before any exploration permit is issued.

Concerning the first item, the ATCM recalls the designation of the Antarctic as a special conservation area and recommends to undertake as a priority objective the further elaboration, maintenance and effective implementation of a comprehensive system for the protection of the Antarctic

environment and its dependent and associated ecosystems aimed at ensuring that human activity does not have adverse impacts on the Antarctic environment or its dependent or associated ecosystems or compromise the scientific, aesthetic or wilderness values of the Antarctic. The principles annexed to the final Report of the 15th ATCM reflect those included in Article 4 of the CRAMRA. The Resolution adopted by the 44th UN General Assembly (1989) included the concept of the Antarctic being declared a world park or a wilderness reserve. The Antarctic Treaty Parties did not take part in the voting.

Specific problems related to the conservation of the Antarctic Treaty area have also been subject to recommendation during the 15th ATCM. These are:

Marine Pollution

The ATCM recommended that States take measures within their competence to ensure compliance with the relevant provisions of six conventions, by their vessels engaged in or supporting Antarctic operations: the LDC, MARPOL 73/78, the 1978 International Convention of Standards of Training, Certification and Watchkeeping for Seafarers with Annex, the 1976 International Convention on Load Lines, and the 1972 Convention on the International Regulations for Preventing Collisions at Sea. It was also recommended that Antarctic Treaty Parties which are parties of MARPOL 73/78 take action to secure formal designation of the waters south of $60^{o}S$ as a special area under Annexes I and V of that Convention. The 1988 Convention on the Control of Transboundary Movements of Hazardous Wastes and their Disposal (see 3.3.12) is also mentioned as a relevant instrument to be considered in the future.

Waste Disposal

A revised Code of Conduct on waste disposal was approved by the 15th ATCM, following the guidelines suggested by a Scientific Committee on Antarctic Research (SCAR) panel of experts. This Code requires that governments ensure compliance by their national programmes and by their private operators subject to their jurisdiction. This item considered the natural linkage between waste disposal and scientific activity, reaffirming the desirability of relying increasingly on returning wastes to the country of origin.

Protected Areas System

The 15th ATCM approved two new categories of protected areas. These areas are:

- Special Reserved Areas to protect areas of outstanding geologic, glaciologic, geomorphic, aesthetic, scenic and wilderness value;

- Multiple-Use Planning Areas, to provide for cooperative planning and coordinated management of activities in areas where multiple activities may interfere with one another or cause cumulative environmental impact. The Special Protected Areas designated on the basis of the Agreed Measures for the Protection of Antarctic Fauna and Flora were also dealt with at the 15th ATCM.

Environmental Impact Assessment

The ATCM examined the implementation of Recommendation XIV-2 on environmental impact assessment, applicable to scientific and logistic activities. Related to the point of international scientific cooperation, the question of environmental monitoring and the establishment of an Antarctic Scientific and Environmental Data System, were subject to Recommendations.

A Recommendation calling for development of cooperative research programmes on ozone depletion was adopted along with four others dealing with the productivity of scientific programmes. A SCAR conference will be convened in 1991 to deal with these subjects at greater length.

Other Meetings

In addition to these actions undertaken by the ATS, there have been other meetings which have dealt with specific problems related to the Antarctic natural resources. These include the 20th Meeting of the SCAR, September 1988, in Hobart, Australia; the second review of the 1972 Convention for the Conservation of Antarctic Seals, September 1988, London; the 7th and 8th Meetings of the Commission established by the 1980 Convention on the Conservation of Antarctic Marine Living Resources (CCAMLR), November 1988 and 1989, respectively, Hobart, Australia.

The Commission of CCAMLR will hold the 9th meeting in October–November 1990, in Hobart, Australia and the 10th meeting in Santiago, Chile, in October–November 1991. SCAR 21st and 22nd meetings were due to be held 15–27 July 1990, in Sao Paulo, Brazil and for August 1992 in Bariloche, Argentina. The next ATCM will take place in Bonn, FRG, in 1991; a Special ATCM has been convened 12–30 November 1990, in Santiago, Chile.

3.23 SUMMIT OF THE SEVEN

The Heads of State or Government of the United States (George Bush), France (Francois Mitterrand), the FRG (Helmut Kohl), the UK (Margaret Thatcher), Italy (Ciriaco De Mita), Japan (Sosuke Uno) and Canada (Brian Mulroney) and European Commission President (Jaques Delors) held their 15th Summit from 14–16 July 1989 in Paris, with Francois Mitterrand in the chair. Each Head of State or Government was assisted by his or her Foreign and Finance Ministers. Mr Mitterrand was also presiding in his capacity as President-in-Office of the European Council. The Summit adopted an economic declaration and a series of political declarations. Below are some of the main points relating to the marine environment.

33. Decisive action is urgently needed to understand and protect the earth's ecological balance. We will work together to achieve the common goals of preserving a healthy and balanced global environment.

34. We urge all countries to give further impetus to scientific research on environmental issues, to develop necessary technologies and to make clear evaluations of the economic costs and benefits of environmental policies. In this connection, we ask all countries to combine their efforts in order to improve observation and monitoring on a global scale.

35. We believe that international cooperation also needs to be enhanced in the field of technology transfer in order to reduce pollution or provide alternative solutions.

36. We believe that industry has a crucial role in preventing pollution at source, in waste minimisation, in energy conservation, and in the design and marketing of cost-effective clean technologies. The agricultural sector must also contribute to tackling problems such as water pollution, soil erosion and desertification.

37. Environmental protection is integral to issues such as trade, development, energy, transport, agriculture and economic planning.

In order to achieve sustainable development, we shall ensure the compatibility of economic growth and development with the protection of the environment.

We encourage the World Bank and regional development banks to integrate environmental considerations into their activities. International organisations such as the OECD and the United Nations and its affiliated organisations, will be asked to develop further techniques of analysis which would help governments assess appropriate economic measures to promote the quality of the environment. We ask the OECD to examine how selected environmental indicators could be developed. We expect the 1992 UN Conference on Environment and Development to give additional momentum to the protection of the global environment.

39. The depletion of the stratospheric ozone layer is alarming and calls for prompt action. We welcome the Helsinki conclusions related, among other issues, to the complete abandonment of the production and consumption of chloro-fluorocarbons covered by the Montreal Protocol as soon as possible and not later than the end of the century. Specific attention must also be given to those ozone-depleting substances not covered by the Montreal Protocol.

40. We strongly advocate common efforts to limit emissions of carbon dioxide and other greenhouse gases.

We need to strengthen the worldwide network of observatories for greenhouse gases and support the World Meteorological Organisation initiative to establish a global climatological reference network to detect climate changes.

41. We agree that increasing energy efficiency could make a substantial contribution to these goals. We urge international organisations concerned to encourage measures, including economic measures, to improve energy conservation and, more broadly, efficiency in the use of energy of all kinds and to promote relevant techniques and technologies.

45. The increasing complexity of the issues related to the protection of the atmosphere calls for innovative solutions. New instruments may be contemplated. We believe that the conclusion of a framework or umbrella convention on climate change to set out general principles or guidelines is urgently required to mobilise and rationalise the efforts made by the international community.

46. We condemn indiscriminate use of oceans as dumping grounds for polluting waste. There is a particular problem with the deterioration of coastal waters. To ensure the sustainable management of the marine environment, we recognise the importance of international cooperation.

We express our concern that national, regional and global capabilities to contain and alleviate the consequences of maritime oil spills be improved. We urge all countries to make better use of the latest monitoring and clean-up technologies. We ask all countries to adhere to and implement fully the international conventions for the prevention of oil pollution of the oceans. We also ask the International Maritime Organisation to put forward proposals for further preventive action.

47. We are committed to ensuring full implementation of existing rules for the environment. In this respect, we note with interest the initiative of the Italian government to host in 1990 a forum on international law for the environment with scholars, scientific experts and officials, to consider the need for a digest of existing rules and to give in-depth consideration to the legal aspects of environment at the international level.[26]

3.24 GLOBAL LEGISLATORS FOR A BALANCED ENVIRONMENT (GLOBE)

Tin cans litter the beach instead of shells. Plastics—less harmful to man than animals—take centuries longer to degrade. Photo: CEE

The Congressional Institute for the Future, a non-profit organisation which educates federal policy-makers about critical emerging issues, has undertaken a new environmental project. Recognising the global scope of current and future environmental challenges and the expanding market forces in the European Community, the Institute has fostered the creation of Global Legislators for a Balanced Environment (GLOBE), a small working group of legislators from the United States Congress and the European Parliament.

GLOBE consists of 28 parliamentarians who recognise that the environmental problems we are facing today require multilateral cooperation for solutions. Many of the potential solutions exist in harnessing market forces and using technology provided by the marketplace. In recogni-

tion of the volume of the 1992 European Market and the existing United States market, Members or the European Parliament and Members of Congress have banded together to educate themselves, share information and fight environmental degradation.

In its efforts to address environmental issues, GLOBE will reach out to the Japanese Diet in its second year and the Supreme Soviet in the future. With these two additional parliamentary groups GLOBE will be active in the four largest markets in the world.

Issues

GLOBE will address only those environmental issues which cannot be solved unilaterally. These global challenges originate in individual countries but the effects are felt worldwide and solutions will require joint action. GLOBE will review possible ways to pre-empt future environmental problems also. Issues to be addressed include:

- global warming/climate change ozone depletion
- ozone depletion
- waste management
- oceans and coastal resources

GLOBE will address those issues by searching for new ideas and examining the following:

- renewable energy resources and energy efficiency initiatives, alternative technologies for chlorofluorocarbons, pollution control tradeable permits, new and alternative storage techniques, international and domestic conservation, using market forces to stop depletion of natural resources, market efficiency mechanisms.

Actions

GLOBE members envision action through:

- six workshops, two per year leading up to 1992, where members can exchange information both formally and informally on environmental problems and innovative solutions;
- comparative analysis on legislation created by the United States and the European Parliament to address existing problems;
- periodic delegation briefings by environmental expert advisory groups.

The GLOBE project is propitious for two reasons. The European Parliament is rapidly expanding the breadth of its scope. As the only body in the European Community elected by its people its responsibilities can only grow. GLOBE is in on the ground floor, assuring a strong, positive relationship with the United States on issues of mutual concern. By 1992, when the Community is at full force, the United States and the European Parliament will be firm global partners. Secondly, environmental issues are a major public concern as is evidenced by newspaper headlines, public opinion polls, talk shows, conferences and policy discussions. By forging a relationship among those parliamentarians who have notable environmental records, the leaders in the field can take unified action to diminish and divert global degradation.

Mechanisms and Topics
Through intense biannual two-day workshops, quarterly reports, delegation briefings by experts and continuous communication, the United States and the European Parliament, in the first phase, will tackle environmental crises and emerge as partners in the drive to clean up the environment.

GLOBE's success will be measured in part by new initiatives which develop from GLOBE meetings. A group of environmental experts will serve as GLOBE's advisory board. After 1992 GLOBE members will seek funding from their parliamentary bodies.

3.25 UN CONFERENCE ON ENVIRONMENT AND DEVELOPMENT, 1992, BRAZIL

On 22 December 1989, the General Assembly of the United Nations adopted a resolution 44/228 on the United Nations Conference on Environment and Development. The Assembly decided to convene the Conference at the highest possible level of participation, spanning two weeks' duration to coincide with World Environment Day, 5 June in 1992. The Government of Brazil offered to host the Conference.

One of the aims of the Conference is to elaborate strategies and measures to halt and reverse the effects of environmental degradation and to promote sustainable and environmentally sound development in all countries. The Assembly affirmed that the protection and enhancement of the environment are major issues that affect the well-being of peoples and economic development throughout the world, and that the promotion of economic growth in developing countries is essential to address problems of environmental degradation. At the same time it was noted that, at present, the largest part of current emissions of pollutants into the environment originates in developed countries, and therefore they have the main responsibility for combating such pollution. The Assembly listed the following environmental issues as being among those of major concern in maintaining the quality of the Earth's environment:

- protection of the atmosphere by combating climate change, depletion of the ozone layer and air pollution;
- protection of the quality and supply of freshwater;
- protection of the oceans and all seas, including enclosed and semi-enclosed seas;
- protection and management of land resources;
- conservation of biological diversity;
- environmentally sound management of biotechnology;

- environmentally sound management of wastes;
- improvement of the living and working environment of the poor;
- protection of human health conditions and improvement of the quality of life;

The Assembly emphasised the need for strengthening international cooperation, particularly between developed and developing countries.

Included in the objectives of the Conference are:

- to examine the state of the environment and changes that have occurred since the 1972 UN Conference on the Human Environment;
- to identify strategies to deal with major environmental issues;
- to promote the further development of international environmental law;
- to promote open and timely exchange of information on national environmental policies, situations and accidents;
- to promote international cooperation.

Notes and References
1. Report from OALOS.
2. Report by Prof. PK Mukherjee, IMLI.
3. Report from IMO statement.
4. Compiled from Siren reports.
5. Report from UNEP office, Jamaica.
6. Compiled from Basle Secretariat statement.
7. Report from UNDP.
8. Report from WHO.
9. Report from FAO.
10. Report from IOC.
11. Report from GESAMP.
12. Report from REMPEC.
13. Report from IOPC Fund.
14. Compiled from LDC report.
15. Report from OECD.
16. Report from IOMAC Secretariat.
17. Report from CPPS.
18. Report from IWC.
19. Report from Commonwealth Secretariat.
20. Report from UNEP office, Jamaica.
21. Report from AMCEN.
22. Report from FFA.
23. Report from ICES.
24. From Statement, Ministry of Foreign Affairs, Finland.
25. Report by Maria Tereas Infante, University of Chile.
26. Compiled from Summit Declaration.
27. Report from GLOBE.

4 Activities of Intergovernmental Agencies and Meetings in Europe

Foreword

by Carlo Ripa di Meana
European Commissioner for Environment,
Nuclear Safety and Civil Protection

Alongside other organisations operating in Europe, the European Communities have continued to play an important role in dealing with various aspects of protection of the environment. One of the priority issues has been transport and disposal of toxic waste.

The problem of waste management represents one of the most acute and difficult issues which states, local administrations and societies have to deal with. The evolution of waste management during the past decades has shown that the problems involved have now reached such dimensions that the management of waste cannot be confined to a regional or national framework.

In the past, a solution to waste elimination was generally found at a local level; a solution to the current diversity of wastes can be found only in a larger framework where economic considerations play a fundamental role. In this context, the European Community has adopted a certain number of framework measures, leaving to Member States a large degree of manoeuvrability.

Given the strong interdependence between waste management and various industrial and commercial activities, the absence of a Community concept of waste management might affect the protection of the environment. It might also influence the completion of the internal market, creating competition distortion and unjustified shifting of investments.

Community action in the field of waste management must therefore be based on precise principles and guided by global and strategic thinking in the medium and long term, as well as by the determination of general priorities which are scheduled to be implemented in the period up to the year 2000.

In the light of these considerations, the Community last year adopted a strategy paper which reflects the ideas of the future development of EC waste management policy. The strategy has been developed around five major policy axes:

- preventive action is of paramount importance and will have to be actively pursued by continuing and strengthening existing practices for promoting clean technologies. Moreover, during 1990, the Commission will present to the Council a proposal for legislation at a Community level to introduce ecological labelling.

- waste which is unavoidably produced should, as far as possible, be re-used or recycled. The present policies promoting re-use or recycling will, *inter alia*, be completed by new legislation on plastic and metallic containers and by research into waste exchange systems.

- insofar as some waste nevertheless continues to be produced, it should be disposed of in an environmentally sound way. Therefore, strict standards will be developed for landfill, in particular concerning site selection, site development, site operation, pre-treatment of waste to be dumped, the type of waste accepted and post-closure supervision.

Standards of municipal waste incinerators have already been agreed on by the EC Council; attention will now be given to industrial waste incineration.

The dumping and incineration of waste at sea must, also in accordance with international obligations, be phased out. The Commission made a proposal to that effect in 1985.

- transport conditions must be optimalised. The Commission published a report on this subject in 1987 and is pursuing its work as set out in that report.

- remedial action must not be neglected. Work will be initiated on a scheme for remedying irretrievable damage, particularly with regard to abandoned landfills, in co-operation with Member States.

However, over and above these five policy axes, perhaps the most important part of the strategy paper is the chapter devoted to the question of the movement of waste on completion of the internal market. This is a difficult and delicate question not only because of increasing public awareness of, and opposition to, the movement of waste, but also because of emerging divergent national policies. Understanding the problem involved in the movement of waste is critical if environmental requirements are to be built into the completed Single Market.

The starting point of any discussion of the problem is the fact that the Court of Justice in Luxembourg has decided that waste falls within the Treaty definition of 'goods' and is thus entitled to freedom of movement throughout the Community. On the other hand, excessive transport of waste clearly presents a potential danger for the environment. Unrestricted freedom of movement might lead to certain regions of the Community having no waste disposal capacity, while other regions (often those which are highly sensitive from an environmental point of view) would be overburdened with waste from outside.

The principle therefore adopted in the strategy paper is the following: *waste should be disposed of in the nearest suitable centres, making use of the most appropriate technologies to guarantee a high level of protection for the environment and public health.*

In my opinion, this principle will allow the development and maintenance of an adequate waste disposal infrastructure in every region of the Community, while at the same time allowing movement of waste over national frontiers where this is desirable from the viewpoint of national disposal of waste.

Why should national frontiers have any relevance in questions of waste disposal? For example, it may be more logical and efficient for Hamburg's waste to be disposed of in Denmark if the only suitable disposal plant in the Federal Republic is situated in, say, Munich. Why oblige densely populated regions to create incineration plants or regions with high groundwater levels to create landfills when other nearby regions have better conditions for such installations?

Anyhow, if preventive action is to feature in the forefront of our policies — as it does — the fact remains that a society without waste, or even without waste disposal, is purely utopian.

All the data available underline the fact that, despite every effort at waste reduction, the overall waste quantities still increase each year.

Therefore, we must create optimal conditions for a high-level waste disposal infrastructure throughout Europe, responding to the specific needs of the relevant waste streams.

Such a policy can be brought about by the criteria set out above.

There is one important exception: we do not intend to apply the proximity criterion to waste that is contractually destined for re-use, recycling or regeneration. Valorisation of waste merits strong support, and should reduce the need to find space for the disposal of waste. The strategy document represents an important step forward for the Community in the area of waste disposal. It represents a secure and stable base for environmental protection in the run-up to 1993.

Carlo Ripa di Meana

Carlo Ripa di Meana
European Commissioner for Environment,
Nuclear Safety and Civil Protection

4.1 EUROPEAN COMMUNITIES

4.1.1 Council of Ministers of the Environment

In the course of 1988–89, the Council concentrated its efforts on the measures necessary to allow the entry into force of the Common Market in 1992 relating to environment, and on problems relating to atmospheric pollution and climate change.

On the first issue, the most spectacular results were achieved in the field of biotechnology with the adoption of a Directive on the Use of Modified Organisms and adoption of a common position of the Council, on a directive to allow assurance that voluntary dissemination of these organisms cannot harm either man or his environment. Adoption of a directive on emissions from industries manufacturing titanium dioxide, which had remained for a long time on the desk of the Council, will allow for a drastic limit to be imposed on waste from an industry responsible for very large emissions of waste into the sea, whilst providing these industries with a legal framework for investment of depollution measures which are particularly expensive.

Regulation of emissions from cars involves two major concerns. Strict reinforcement of norms relating to cars with small cylinders received support of the Council, thus opening the way to reinforcement of all norms relating to emissions from cars. This success was achieved due to the persistence and diplomacy of then Commissioner Stanley Clinton Davis, and one can but wish his successor Mr Ripa di Meana the same success, in the days to follow, in adopting a complete and up-to-date regulation which will set standards for our manufacturers in the years to come.

As far as atmospheric pollution is concerned, one should first quote adoption, under FRG presidency, of the directive on the great combustion installations, which represents the principal source of pollution from fixed sources. Regulation of incinerators of household refuse is another positive result of the work of the Council.

In the field of climate change, the Council decided to suppress the use of chorofluorocarbons by the end of the century, thus allowing the European Community to play an active role in negotiations of the revision of the Montreal Protocol. A Council Resolution was adopted in June 1989; this Resolution invites the Commission to elaborate and implement an ambitious work programme, destined to serve as the basis for future decisions which the Council will have to take in this matter. The Commission should in the course of the year, submit its first concrete proposals in this matter. Equally, protection of tropical forests was given much attention by the Council.

The political will of the Council, but also that of the Commission and the European Parliament is essential for the Community environmental policy to achieve the progress expected by all Europeans. However, it is crucial that the Commission and Member States provide the instruments which are necessary to implement these objectives if concrete results, which will be adhered to by all, are to be realised. In this context, the French Presidency concentrated its efforts on the proposal for the creation of an European Environmental Agency — an idea launched by the President of the Commission, M Delors. Acceptance by all Member States of this ideal led to a unanimous adoption by the Council in November 1989 of a draft regulation. This agency will be independent and will concentrate its initial activities on ascertaining the state of environment in the Community and problems relating to environmental data. The Agency should allow the Commission to undertake non- administrative duties and provide it with information which is essential in order to elaborate its policies. The setting up of a network of national centres should facilitate integration of competent systems within Member States into the Community system.

The Council did not allow the large agenda of priority items to give it reason to neglect other environmental issues. Only a few of the most important issues relating to marine pollution are referred to here. The Commission has not found itself able in the two years covered by the Yearbook to present numerous proposals relating to protection of fresh and marine waters. However, the Council requested that it embark upon examination of an important directive on the protection of waters from nitrates; all delegations, in particular those who had to bear the consequences of algal blooms, agreed on the importance of the directive. Unfortunately, in view of the implications of any such directive on agriculture, the Council has not yet been able to reach an agreement.

By contract, as a result of a French initiative, in June 1989 the Council adopted a resolution on the major technical and natural risks, requesting the Commission to undertake certain work. An important element of this declaration concerns maritime transport and risks linked with transport of chemical products.

The work carried out in the last few years by the Council has been comprehensive and enabled the Community environmental policy to progress, contributing also to an improvement in the quality of life of Europeans. During preparations towards the entry into force of the great market in 1992, the Council work should allow that, in conformity with stipulations in the Single European Act, the single market will improve rather than degrade the environment. At the international level, the recent events in Eastern Europe also influenced the work of the Council. The principle of the opening of the Agency to East European countries was accepted, and the Community played an important part in the work of the Conference on Security and Cooperation in Europe (CSCE) which met last autumn in Sofia. More generally, the Community, represented by the presidency of the Commission, plays an increasingly active role in international bodies by applying negotiating rules established by the Council[1].

4.1.2 European Parliament

Marine pollution is a frequent concern of the European Parliament's Committee on the Environment, Public Health and Consumer Protection. By way of example, since 1987 Parliament has adopted no less than 15 resolutions on the dangers threatening the North Sea. International responses are crucial when it comes to addressing and devising policy responses to marine pollution.

The European Community (EC) must play a particularly important part in this process. Indeed, some now regard the EC, in environmental matters, as a federal system. I would not go quite so far yet, for what we actually have is an imperfect federal system because of an inappropriate balance of powers between its institutions. The role of the EC can by no means be underestimated, however.

Attempts by European states to tackle the problems of marine pollution reveal a curious contradiction. On the one hand most European countries assess the technical and scientific expertise to overcome almost every problem in other fields — from medicine to space travel to computer science — while on the other hand they continue to use archaic methods to dispose of their waste, spreading it on land, in the air and finally in the sea.

One example is the disposal of industrial waste and, in particular, the dumping of dilute acid. At the Second International Conference on the Protection of the North Sea, it was decided that the dumping of industrial waste in the North Sea would be stopped by the end of 1989. One EC Member State (Germany) stopped dumping dilute acid by the deadline. It even invested in a land-based dilute acid recycling plant. No less than three other Member States (UK, France and Spain) will continue to discharge these substances until 1992 or 1993. At a technical level then, we already have methods for the disposal of waste which, if consistently employed, could contribute to the reduction of waste and therefore of marine pollution, the problem is to translate this technical capability into environmental policy and, ultimately, legislation.

International coordination of policy responses to environmental problems are the single most effective way to tackle the threats now faced by the marine environment. Marine pollution, like air and other forms of pollution, respects not a single national frontier. Indeed, we can speak of there always having been, long before the idea of 1992 was born, a single market in emissions! The role of the European Community is therefore clear. Community legislation should increasingly be developed to prevent the pollution of Europe's seas.[2]

4.1.3   Commission of the European Communities (CEC)

A unique event took place in Frankfurt in June 1988. This was the seminar attended by European environment Ministers to work out Community Water Policy for the Nineties. The occasion was unique because it afforded the Ministers the opportunity for a frank and open debate without the responsibility of defending their entrenched national positions. At the conclusion of the seminar, the Ministers agreed to "expand and intensify the Community policy and legislation on the protection and management of Community water resources". They identified six main areas of work for the Commission:

- ecological quality of surface waters
- waste water treatment

- dangerous substances
- diffuse sources
- water resources
- integration.

Ecological Quality of Surface Water

It was agreed that there should be a general improvement in the ecological quality of Community waters. However, Ministers recognised that improvements could not be achieved everywhere in the short term. In 1990 the Commission will prepare a directive with the objective of fulfilling these aims. The task will not be easy since the ecology of Community waters varies dramatically from the temperate northern latitudes to the drier southern regions. Furthermore, the methods of measuring ecological quality depends very much on the type of water that is being studied. Nevertheless, the Commission will prepare a measure which will provide the framework within which ecological improvements can be achieved.

Waste Water Treatment

Many of the worst pollution problems in the Community are linked to the lack or inadequacy of sewage treatment. There are more discharges of municipal waste water than from any other source, ranging from small villages to large conurbations, and many environmental problems are caused by insufficient treatment. At present there are no general requirements at Community level to treat either industrial or sewage effluents before they are discharged into the aquatic environment although certain directives, eg the Dangerous Substances Directive, apply to the discharges of specific substances.

The Commission has now published a proposed Directive for the treatment of municipal waste water. As a general principle, the Directive proposes a minimum of secondary (biological) treatment. However, in so-called 'sensitive areas', additional treatment may be necessary in order to meet specific environmental needs, eg. a reduction in discharges of nutrients. For discharges to coastal waters, the Directive envisages that primary treatment of sewage may be sufficient provided that the hydrographic conditions are favourable and provided also that comprehensive studies have shown that the discharged waste water will not result in an adverse effect on the environment.

Dangerous Substances

Progress on dangerous substances has accelerated since the adoption of Directive 86/280/EEC. This is a framework Directive covering all the general provisions for dangerous substances and enabling new annexes to be added for fixing discharge limits and quality values for individual substances.

By Directive 88/347/EEC, limit values and quality objectives were laid down for seven new List I (black list) substances: aldrin, dieldrin, endrin, isodrin, hexachlorobenzene, hexachlorabutadiene and chloroform.

A further means of speeding up progress agreed at the Frankfurt seminar was that the identity of additional List I substances should be agreed at Council by unanimity,

A cuttlefish (Sepia officinalis) *photographed near Plymouth, UK. It feeds on shrimps and small crustacea on sand/gravel beds. Many believe its numbers have declined over the last 30 years, but reasons for this are unknown.*
Photo: D George

leaving the actual limit values, quality objectives, measurement methods and dates for compliance to be determined by by qualified majority in the Council in accordance with Article 130 S, second indent of the Single European Act. The first amendment to the Dangerous Substances Directive using this new procedure has been proposed by the Commission and concerns limit values and quality objectives for 16 biocides.

Diffuse Sources
Many of the dangerous substances which are detected in Community Waters are not directly discharged from point sources but originate from a number of diffuse sources. Ministers agreed at the Frankfurt seminar that more attention should be paid to this problem and in particular to the environmental effects of intensive agricultural practices.

As a first step, the Commission has proposed measures to limit the quantity of nitrates reaching surface and underground waters in the Community. The measures are needed because of the build-up of nitrate in drinking water supplies where preventive action is the only satisfactory long-term solution. Eutrophication is the second major problem caused by nitrates (and phosphates). This is when the level of nutrients in the water stimulates the growth of plant and algal life, which absorb oxygen when they decay and cause extensive damage to fish, plants and to other organisms. Algal blooms have caused serious economic damage to fisheries in the Baltic and the North Sea and to tourism in the Adriatic in the last two years.

The Commission's proposal includes measures to control the handling and application of animal manure, the application of chemical fertiliser and certain other land management practices. The proposed Directive will operate on the principle of 'subsidiarity', ie the framework for the rules will be established at Community level leaving the detailed application to be determined nationally or regionally taking into account varying environmental conditions.

The Commission is also considering a proposal to deal with phosphates and measures for the control and use of pesticides.

Water Resources
The problem of water resources is one which has not hitherto been dealt with at Community level. Ministers felt that water resource problems could not be divorced from those of water quality and should be addressed as part of an overall policy for water. In some Member States the shortage of water dominates all other considerations. The Commission is examining this aspect of water policy to see how it can be given greater emphasis in a Community framework.

Integration
Ministers at the Frankfurt seminar felt that there was more scope for integrating water policy with other aspects of Community environmental policy. A number of Directives or proposals have attempted to deal with the cross-media aspect of certain pollutants. For example, Directive 87/217/EEC on the prevention and reduction of environmental pollution by asbestos deals with all forms of asbestos emissions for particular industrial sites.

The Community 'Task Force'
As reported in the *1987/88 Yearbook* (p.38), the Commission has established a Community 'Task Force', a network of government and non-government experts or liaison officers able to provide rapid assistance and advice to authorities confronted by serious pollution incidents involving oil or other harmful substances. In the last two years, the Task Force has been called into action to assist the local authorities on four occasions: the *Marao* incident off Southern Portugal, the *Kharg 5* off the coast of Morocco, the *Aragon* at the invitation of the Spanish authorities, and the oil pollution affecting Porto Santo Island of the Madeira archipelago. The *Kharg 5* incident was the first in which the Community Task Force rendered its assistance to a non-Member State.

The Commission is continuing its programme of national and international training courses in order that those responsible are better equipped to deal with major pollution accidents at sea when they occur. National courses have been held in Greece, Italy, Spain and France with Community support and financial assistance.

Special Action Programmes
The Commission is also committed to improving the environment in a very practical way by funding projects designed to solve specific and acute environmental problems. The exceptional vulnerability of the Mediterranean Sea and its highly sensitive ecosystem was recognised in 1985 when the Mediterranean Special Action Programme (MEDSAP) was launched. The Community contributed 1.5 million ECUs to projects in 1988, 5 million ECUs in 1989

and 9 million will be spent in 1990.

A similar programme to clean-up the coastal zones and coastal waters of the Community's northern seas was launched in 1989 with funds of 2 million ECUs. A further 2 million ECUs will be available for 1990 to assist projects connected with nutrient abatement measures, advanced wastewater treatment and pollution prevention technologies ('clean technology'), integrated ecological management plans, alternatives to biocide use and projects designed to solve localised marine problems.[3]

4.1.4   European Court

It is the task of the Court of Justice of the European Communities (hereinafter ECJ) "to ensure that in the interpretation and application" of the treaties (as amended) establishing the three Communities and of implementing legislation "the law is observed" (Art, 164 EEC; Art, 31 ECSC; Art, 136 EAEC). In principle, the Court is free to depart from its previous decisions but in practice will rarely to do so.

In the period under review no case bearing directly on marine pollution has come before the Court. There are, however, three decisions of the Court that are indirectly relevant to the protection of the marine environment.

In Case 309/86, Commission v. Italy (not yet reported) the Commission brought an action for a declaration that Italy had failed to adopt within the period prescribed, the measures needed to comply with Directives intended to secure the approximation of the laws of the Member States relating to methods of testing the biogradability of non-ionic surfactants and anionic surfactants respectively. Italian law did meet in part the objects that the Directive sought to achieve. The Court, in its judgment of 2 March 1988, affirmed the principle enunciated in previous rulings that Community Directives must be effectively transposed into national law to ensure their application in full. The overriding aim of both Directives is to reduce pollution "*du milieu naturel*" by prescribing common standards of biogradability and testing, thereby precluding Member States from invoking Art. 36 EEC to prohibit the use and marketing of detergents that satisfy these standards.

In Case 322/86 Commission v. Italy (not yet reported) the Commission sought a declaration that Italy had failed to adopt within the prescribed period, the provisions necessary to implement Directive 78/659 EEC concerning the quality of fresh waters needing protection or improvement to support fish life. The Directive required, *inter alia*, that Member States designate (i) waters requiring protection or improvement, and (ii) salmonid and cyprinoid waters. Within five years all waters so designated had to be subject to programmes to reduce pollution in accordance with physio-chemical level as prescribed in the Directive. A complaint was lodged by the World Wide Fund for Nature that pollution in Italian rivers and lakes was the result of the failure to transpose the Directive into Italian law.

The court, in its judgment of 12 July 1988, once more rejected as insufficient for the effective transposition of Directives into national law, the contention of Italy that its existing legislation conferred on public authorities powers of management and protection of waters sufficient to secure implementation of the Directive.

In Case 187, Saarland and Others v. Minister for Industry and Others (1989) 54 CMLR 529. This arose from a reference by the Tribunal administratif de Strasbourg requesting a preliminary ruling on the interpretation of Art. 37 EURATOM. The proceedings were part of a long-running campaign by the Saarland and several environmental groups to challenge the validity, *inter alia*, of French ministerial orders concerning four nuclear power installations under construction at Catteno, Northern France, close to the borders with Luxembourg and the Federal Republic of Germany. One of the orders (dated February 1986) challenged, specified fixed annual limits to the radioactive content of liquids discharged into the River Moselle. One of the grounds of challenge was the failure of the French authorities to comply with Art. 37 EURATOM. This reads:

> Each Member state shall provide the Commission with such general data relating to any plan for the disposal of radioactive waste in whatever form as will make it possible to determine whether the implementation of such plan is liable to result in the radioactive contamination of the water, soil or airspace of another Member State.

The Commission shall deliver its opinion within six months, after consultations the group of experts referred to in Article 31.

On 29 April 1986, the day after proceedings were instituted, the French Government sent the Commission the 'general data' required under Art, 37. Ministerial authority for the start up of the installations was given in July and August. The Commission's opinion was given on 22 October 1986; three days later a nuclear reaction was set in motion in the first of the installations. The ECJ ruled that the context and purpose of Art.37 conferred great importance on the Commission's opinion which, despite its non-binding character, had to be put to '*un examen approfondi*' by the Member State. Where, as here, a Member State has made disposal of radioactive waste subject to authorisation, full effect can be given to the opinion only if it is brought to the notice of that state before a final decision authorising disposal is taken, Accordingly, Art. 37 must be interpreted as meaning that the 'general data' of a plan for the disposal of radioactive waste has to be provided to the Commission before definite authorisation for such disposal is given by the Member State concerned.

Since the great majority of nuclear installations planned for the future will be sited along the coasts and rivers of Member States the ruling can be seen as a modest extension of Community control over liquid radioactive discharges to the marine environment.[4]

4.1.5   Recent EEC Legislation

Council Directive 88/347/EEC of 16 June 1988 (Official Journal of the European Communities (hereafter OJ) 1988 L158/35)
This amends Annex II to Directive 86/280/EEC on limit

values and quality objectives for discharges of certain dangerous substances included in List I of the Annex to Directive 76/464/EEC by adding to the list of products for which emission values and quality objectives for the aquatic environment have to be observed, aldrin, dieldrin, endrin, isodrin, hexachlorobenzene, hexachlorobutadiene and chloroform. In September 1988 the Commission proposed a number of further additions to the list: dichloroethane, trichloroethylene, perchloroethylene and trichlorobenzine (OJ 1988 C253/4).

Council Directive 88/609/EEC of 24 November 1988 on the Limitation of Emissions of Certain Pollutants into the Air from Large Combustion Plants (OJ 1988 L336/I)
Requires a reduction in overall annual emissions from existing large combustion plants (50 MW or more) of sulphur dioxide from 1980 levels in three stages (by 1993, 1998 and 2003 respectively), the percentage reductions specified varying from Member State to Member State. Similarly, emissions of nitrogen oxides must be reduced in two stages (by 1993 and 1998), again by varying percentages. New plants are subject to emission limit values in respect of sulphur dioxide, nitrogen oxides and dust.

A ship trails its load of sewage sludge at sea. This practice has now been stopped by most North Sea states. The UK has pledged to phase out sludge dumping in the North Sea by 1998 (see 4.2). Photo: Greenpeace

Council Directive 89/369/EEC of 8 June 1989 on the Prevention of Air Pollution from New Municipal Waste Incineration Plants (L163/32)
Sets emission limit values for dust, heavy metals, hydrochloric acid, hydrofluoric acid and sulphur dioxide from new plants incinerating domestic, commercial, trade and similar refuse. National authorities are to lay down emission values for other pollutants when they consider this to be appropriate because of the composition of the waste to be incinerated and the characteristics of the incineration plant. The Directive has to be implemented by 1 December 1990.

Council Directive 89/429/EEC of 21 June 1989 on the Reduction of Air Pollution, from Existing Municipal Waste Incineration Plants (OJ 1989 L203/50)
Provides that existing municipal waste incineration plants must meet the same emission limit values as new plants by 1 December 1996 in the case of plants with a nominal capacity equal to or more than 6 tonnes of waste per hour and by 1 December 2000 in the case of other plants.

Council Directive 89428/EEC of 21 June 1989 on Procedures for Harmonising the Programmes for the Reduction and Eventual Elimination of Pollution caused by Waste from the Titanium Dioxide Industry (OJ 1989 L201/56)
This Directive provides that after the end of 1989 waste from the titanium dioxide industry produced by the sulphate or chlorine processes shall not be dumped in inland or marine waters, and solid, strong acid and treatment wastes from the sulphate process and solid waste and strong acid waste from the chloride process shall not be discharged into inland or marine waters. In addition, the directive sets emission limits for other kinds of waste which must be reached by the end of 1989 or 1992, depending on the type of waste: alternatively, instead of complying with emission standards, a Member State may use quality objectives provided that these achieve the same effect. The Directive also sets atmospheric emission limits for the titanium dioxide industry.

Proposed EEC Legislation on Marine Pollution

Draft Directive on the Protection of Natural and Semi-Natural Habitats and of Wild Fauna and Flora (COM (88) 381)
Although not predominantly concerned with marine pollution, this draft directive (published by the Commission in August 1988) does nevertheless have some bearing on the matter. The draft directive requires Member States to conserve the habitats of wild fauna and flora in their territories, including maritime areas under their sovereignty or jurisdiction. Conservation in this context includes the obligation to avoid the pollution of habitats.

Draft Directive of the Disposal of Polychlorinated Biphenyls and Polychlorinated Terphenyls (COM (88) 559)

The Commission published this draft directive, which would replace an existing directive on the topic (Directive 76/403/EEC), in October 1988. The main interest of the Directive as far as marine pollution is concerned is Article 3, which provides that Member States shall take the necessary measures to prohibit the uncontrolled disposal of PCBs and end the incineration of PCBs on incinerator ships by 1995.

Draft Directive concerning the Protection of Fresh, Coastal and Marine Waters against Pollution caused by Nitrates from Diffuse Sources (COM (88) 708)

This draft directive, which was published by the Commission in December 1988, lays down limits for spreading slurry and manure; requires Member States to draw up programmes for the use of the nitrogen fertilisers that reflect the real requirement of crops allowing for nitrogen content of the soil; proposes a code of good farming practice; and sets a limit value for the nitrogen content of sewage treatment effluent. These provisions would apply to areas where the surface freshwater or groundwater intended for the abstraction of drinking water contains or could contain 50 mg/l nitrate and to areas of land which drain into fresh or marine waters where eutrophication is already occurring or liable to occur.

Draft Directive concerning Minimum Requirements for Vessels entering or leaving Community Ports carrying Packages of Dangerous or Polluting Goods (COM (89) 7)

The purpose of this draft directive, which was put forward by the Commission in May 1989, is to require vessels entering or leaving Community ports carrying packages of dangerous goods to observe a number of minimum standards designed to improve shipping safety, safeguard human health and protect the marine environment. These standards include communicating certain information to port authorities at least 24 hours before arrival; establishing as soon as possible radiotelephone communication with coastal radio stations; making use of the services provided by radar stations and any local vessel traffic service; using available pilotage services; and notifying the authorities of a Member State of any deficiencies or incidents which may adversely affect the manoeuvrability of the vessel or constitute a hazard to the marine environment, or of any leak or discharge of dangerous goods, before entering that Member State's territorial sea.

Draft Directive concerning Municipal Waste Water Treatment (COM (89) 518)

This draft directive, which was proposed by the Commission in November 1989, lays down minimum requirements for the treatment of municipal waste water. Receiving waters are to be classified into three types: in general secondary (biological) treatment will be required as a minimum level of treatment. The draft directive also seeks to control the discharge of industrial waste waters which are of a nature similar to municipal waste water and which do not enter municipal waste water treatment plants

before discharge to the environment. The draft directive requires the disposal of sewage sludge to sea by dumping from ships, by discharge from pipelines, or by other means to be eliminated by the end of 1998. Before that date Member States shall ensure that the total amount of dry matter in sludge disposed of at sea does not increase and that the amounts of toxic, persistent or bioaccumulable materials contained therein are progressively reduced.[5]

4.2 THIRD INTERNATIONAL CONFERENCE ON THE PROTECTION OF THE NORTH SEA

The Third International Conference on the North Sea was held in The Hague on 7–8 March 1990. It was attended by ministers from Belgium, Denmark, FRG, France, the Netherlands, Norway, Sweden, Switzerland and the UK. To protect further the North Sea environment the participants decided to adopt a comprehensive set of common actions.

Inputs of Hazardous Substances
It was agreed that:

* all substances that are persistent, toxic and liable to bio-accumulate should be covered by reduction measures as required in the London Declaration;

* for each of the substances listed to achieve a 50% reduction of inputs via rivers between 1985 and 1995, and of atmospheric emissions by 1995 or 1999 at the latest;

* for dioxins, mercury, cadmium and lead, reductions of 70% or more should be achieved between 1985 and 1999;

* a substantial reduction in the quantities of pesticides reaching the North Sea, and by 31 December 1992 to control strictly the use of pesticides.

Phasing out of PCBs
The Conference agreed:

* to take measures to phase out and destroy in an environmentally safe manner all identifiable PCBs as soon as possible with the aim of complete destruction of capacitors and transformers by 1995 or by the end of 1999 at the latest.

Input of Nutrients
The Conference agreed:

* that for the North Sea catchment area as a minimum level of treatment, urban areas with more than 5,000 people and industries with a comparable waste water load should be connected to sewage treatment plants with secondary (biological) or equally effective treatments;

* that further measures are required in order to achieve a reduction of 50% for inputs of nutrients between 1985 and 1995 into areas where this may cause pollution;

* to establish common assessment and reporting procedures for calculating the reduction of nutrients.

Dumping and Incineration at Sea

The Conference noted that almost all North Sea countries have stopped the dumping of sewage sludge at sea. The UK has given a firm undertaking to stop dumping sludge as soon as possible, at the very latest by 1998. The Paris Convention has been invited to undertake before 1992 a review of alternative methods of sludge disposal.

Almost all North Sea States have stopped dumping at sea of industrial waste. The UK has given a firm undertaking to end industrial waste dumping as soon as possible, and no later than the end of 1992.

Further action needs to be taken to improve the quality of dredged materials disposed of in the North Sea by reducing inputs of contaminants to rivers and estuaries.

The Conference agreed to phase out marine incineration by 31 December 1991 and to seek agreement on this date within the Oslo Commission by 31 December 1990.

Pollution from Ships

It was agreed to improve control and to deter all ships from contravening the requirements of MARPOL 73/78, and to improve legal instruments and rules aimed at the minimisation of intentional and accidental pollution.

Pollution from Offshore Installations

It was agreed to reduce further operational discharges from offshore installations and to take intitiatives towards the further improvement of safety and reduction of the risk of calamities involving offshore installations. PARCOM will be requested to assess the risks such accidents pose to the marine environment. The OSCOM will be invited to continue its work in developing guidelines with the aim of ensuring that offshore installations are disposed of in an environmentally satisfactory manner.

Discharges and Disposal of Radioactive Waste

It was agreed to continue to apply the Best Available Technology to reduce radioactive discharges.

Airborne Surveillance

It was agreed to use airborne surveillance as a tool for adequate control and surveillance at sea and as an aid to the enforcement of existing regulations.

Wadden Sea

The Conference gave high priority to the implementation of the measures agreed in this Declaration which are likely to have a special significance for the Wadden Sea.

Enhancement of Scientific Knowledge

To further enhance scientific knowledge of the North Sea ecosystem, the Conference invited the North Sea Task Force (see 4.5) to continue its research into exceptional algal blooms and on the epidemic death of seals, and to address several sensitive ecological issues in its 1993 Quality Status report.

Other

It was agreed to coordinate action, with the aim of increasing coastal state jurisdiction, in accordance with international law, including the possibility of establishing Exclusive Economic Zones in the areas of the North Sea where they do not exist and to request the Netherlands to initiate the coordination of this action and to submit the findings to the North Sea Ministers by the beginning of 1992.

The Conference will endeavour to obtain the early entry into force of the Salvage Convention and will actively contribute to the work of IMO in preparing the Convention on Hazardous and Noxious Substances.

It was agreed to give further protection to marine wildlife in the North Sea and to tackle important gaps in knowledge which remain.

It was agreed to consider the impact of fisheries (including fish farming) on the North Sea ecosystem and the impact of the marine environment on fisheries resources.

Future Cooperation

A Working Group Meeting at ministerial level will be held in 1993 to discuss the 1993 Quality Status Report on the North Sea, to evaluate the actions taken within IMO on Annex I and Annex II of MARPOL 73/78, and to discuss problems encountered with the implementation of the North Sea Conference Declaration with regard to nutrients and pesticides.

The Fourth International Conference on the Protection of the North Sea will be held in Denmark in 1995.[6]

4.3 OPERATION OF THE MEMORANDUM OF UNDERSTANDING ON PORT STATE CONTROL (MOU)

The Memorandum of Understanding on Port State Control is an international agreement between 14 European countries to coordinate and to harmonise their efforts in respect of inspections of foreign ships in their ports. The principal aim of the agreement is to eliminate the operation of substandard ships. The Memorandum came into operation on 1 July 1982 and each of the participating maritime authorities will maintain an effective system of port state control to ensure that foreign merchant ships visiting its ports comply with the standards laid down in the following international conventions and all amendments thereto in force:

- the International Convention on Loadlines, 1966;
- the International Convention for the Safety of Life at Sea, 1974, as amended by the Protocol of 1978;
- the International Convention for the Prevention of Pollution from Ships 1973, as modified by the Protocol of 1978 relating thereto;
- the International Convention on Standards of Training, Certification and Watchkeeping for Seafarers, 1978;
- the Convention on International Regulations for Preventing Collisions at Sea, 1972;
- the Merchant Shipping Convention, 1976 (ILO Convention no.147);

Each country will have to achieve an annual total of inspections corresponding to 25% of the estimated number of individual foreign merchant ships which entered the ports of its state during a 12 month period.

The Port State Control Committee is the executive body instituted by the MOU. It is composed of representatives from each of the 14 participating maritime authorities and of the Commission of the European Communities. The International Maritime Organisation (IMO) and the International Labour Organisation (ILO) participate as observers.

The secretariat, functioning within the Netherlands' Ministry of Transport and Public Works, is situated at Rijswijk (near The Hague), The Netherlands. The secretariat acts under the guidance of the Committee and prepares meetings, and provides such assistance as may be required to enable the Committee to carry out its functions.

The computer centre, used by the 14 partners of the MOU, is operated by the French maritime administration and is located at Saint-Malo, France. The results of each port state inspection carried out anywhere in the MOU region are entered into this computer and thus made available to the other partners to the Memorandum through on-line terminals.

Recent Activities

On 13–16 September 1988 the 12th meeting of the Contracting Parties to the Bonn Agreement was held in Copenhagen. The Secretariat of the Memorandum of Understanding was invited to participate as an observer at this meeting for the purpose of arriving at improved communication procedures between the Bonn Agreement and the Memorandum of Understanding, to ensure stringent PSC on ships suspected of discharge violations.

On 21–23 February 1989 a special Seminar for Surveyors was held in Santander, Spain, which was dedicated to the Control of Pollution by Noxious Liquid Substances in Bulk (MARPOL Annex II) .

The Port State Control Committee meetings were held at: Athens, Greece, 18–19 May 1988; Rome, Italy, 16–17 November 1988; Lisbon, Portugal, 11–12 May 1989; Stockholm, Sweden, 29–30 November 1988.

Other Seminars for Surveyors were held at: Oslo, Norway, 8–10 June 1988; Dublin, Ireland, 7–9 June 1989.[7]

4.4 OPERATION OF THE 1972 OSLO CONVENTION

The Convention for the Prevention of Marine Pollution by Dumping from Ships and Aircraft 1972 (Oslo Convention) was opened for signature in Oslo on 15 February 1972 and entered into force on 6 April 1974. The Convention applies to the North Sea and North-East Atlantic and makes it an offence to dump material in the sea without licence from the relevant authority. To date it has been ratified by 13 maritime states of Western Europe. Operation of the Convention is supervised by the intergovernmental Oslo Commission (OSCOM) with a small secretariat in London, which is shared with the Paris Commission (see 4.5).

The major work of the Commission is to regulate the dumping and incineration of wastes at sea, to draw up codes of practice for particular substances and activities, and to collect data on different types of waste. Substances with high potential for damaging the environment on account of their toxicity, persistence and bio-accumulation, eg cadmium, mercury, are placed on a 'black list' and may not be dumped. Substances with a lesser degree of these characteristics, eg leads, cyanides and fluorides, are placed on a 'grey list' and may be dumped only when special precautions are taken. Wastes not on these lists still require a specific or general permit before being dumped at sea. Exceptions are made for black list substances when only a trace is present in the waste to be dumped. This aspect of

Discarded netting collected on a beach in Yorkshire, UK. Old netting has the potential to cause marine life much suffering and death. Photo: TBG

the Commission's work, ie. deciding on what proportion constitutes a 'trace', involves highly technical discussions.

Activities 1988

The Fourteenth Meeting of the Oslo Commission was held in Lisbon from 22–24 June 1988. The Commission met jointly with the Paris Commission from 20–21 June 1988 (see 4.5 for report).

Amendments to the Oslo Convention

Annex on Rules on Incineration at Sea All Contracting Parties, except Belgium and Spain, had ratified the amending Protocol.

Amendment of Annexes I and II to the Convention All Contracting Parties, except four, have ratified the amending Protocol.

Implications of the Second International Conference of the Protection of the North Sea

The Oslo Commission is regarded as a useful forum in which to discuss matters relating to dumping and incineration at sea. However, a simple transfer of the North Sea Conference Declaration to the whole Oslo Convention area could present problems.

Dumping at Sea

The Swedish delegation presented a proposal for a Recommendation on the Reduction and Cessation of Dumping of Industrial Wastes and Sewage Sludges at Sea. The Draft Recommendation proposed the phasing out of sewage sludge dumping in the whole Convention area by 31 December 1989. The Commission agreed that the Ad Hoc Working Group on Dumping could be used to evaluate all industrial waste dumping permits. It decided to invite the Standing Advisory Committee for Scientific Advice (SACSA) to review the provisions of OSCOM Recommendation 86/1.

Incineration at Sea

Termination date The Danish Delegation presented its proposal on the phasing out of incineration at sea: no new incineration permits should be issued after 31 December 1988 and incineration should be prohibited after 31 December 1990.

The Commission decided to terminate incineration at sea by the Contracting Parties to the Oslo Convention and within the Oslo Convention area by 31 December 1994. In addition the OSCOM Decision 88/1 states:

* that the North Sea States shall take steps to minimise by not less than 65% the use of marine incineration by 1 January 1991;

* that Contracting Parties shall not export wastes intended for incineration to marine waters outside the Convention area, nor allow their disposal in other ways harmful to the environment;

* that it is preferable that the waste to be incinerated be loaded in a harbour of the country from which it originates instead of being exported to another country.

The inclusion in Decision 88/1 of the provisions of the London Conference Declaration which relate to the export of wastes and the loading of wastes in a harbour in the country of origin, were regarded as important provisions by most delegates.

Wastes containing PCBs

The Commission discussed a proposal to clarify whether a concentration limit of 50ppm PCBs/PCTs should be set to define whether or not waste presented for incineration at sea should be classified as problem waste. It was agreed that this was a technical question and the matter should be referred to SACSA.

Scenarios for the Handling of Organochloride Waste

The Commission noted that SACSA had discussed essentially two options to bridge the period between 1990 and 1994:

* to keep the amounts of waste to be incinerated at sea at the level of about 35,000 tonnes per annum;

* to ensure adequate storage facilities pending definite treatment of the waste on land for a maximum period of 4 years.

SACSA report

The Fifteenth Meeting of SACSA was held from 21–25 March 1988.

Disposal of Sewage Sludge The Commission agreed that sewage sludge disposal should remain on SACSA's agenda and that Contracting Parties should be invited to submit new information.

The Commission took note of SACSA's preparation of a report surveying the policies implemented by France, Belgium and the Netherlands to eliminate phosphogypsum wastes. It endorsed SACSA's arrangements for keeping the review document up-to-date, including the long-term effects of land storage and land-based disposal options.

SACSA's proposal to reconvene the Ad Hoc Working Group on Dumping to examine the case of industrial wastes was also considered. Contracting Parties which have abandoned dumping at sea were invited to elucidate their disposal options on land.

Monitoring of Dumping Grounds

Taking note of SACSA's preparation of an overview of the reports, the Commission agreed:

* that the overview was useful and should be updated for discussion at the next SACSA Meeting;

* that SACSA should consider at its next Meeting the appropriate frequency of monitoring.

Transfrontier Movement of Industrial Waste

The Commission adopted OSCOM Recommendation 88/1 concerning the export of wastes for disposal at sea.

Reports on Dumping and Incineration at Sea

The Commission took note of the following reports:

* on incineration permits;

- on the permits and approvals issued in 1986;
- on the amounts of wastes dumped at sea in 1986;
- on trends in the amount of wastes dumped at sea from 1976 to 1986.

The Commission noted that there had been an improvement in the 'quality' of industrial liquid and sludge wastes and of sewage sludge, resulting in marked reduction in most heavy metal inputs.[8]

Monitoring the Marine Environment
See 4.5.

North Sea Task Force
See 4.5.

Activities 1989
The Fifteenth Meeting of the Oslo Commission was held in Dublin from 12–14 June 1989. At that meeting the following issues were discussed:

Amendments to the Oslo Convention
A Protocol to amend the Oslo Convention by means of a new Annex IV containing mandatory rules on incineration had been ratified by 12 of the 13 Commission Members. The 13th Instrument of Ratification is to be deposited in the very near future at which stage the Protocol can enter into force.

Dumping in Internal Waters
The Commission adopted a Protocol for including dumping in internal waters within the scope of the Oslo Convention. The Commission invited the Norwegian Government, as Depository Government, to convene a diplomatic conference for amending the Convention in accordance with Article 25. Contracting Parties have submitted to the Secretariat their definition of internal waters.

Policy on Dumping at Sea
It was decided that the principles of reduction and cessation of dumping of hazardous materials, as set out in the 1987 North Sea Conference, should be applied by coastal states. The Oslo Commission accordingly adopted a decision that Dumping of industrial wastes in the North Sea shall cease by 31 December 1989 and in other parts of Convention waters, by 31 December 1995, except for those industrial wastes for which there are no practicable alternatives on land and which cause no harm to the marine environment.

Removal and Disposal of Offshore Installations
The Commission decided to develop guidelines for the management of removal and disposal of offshore platforms, but until these became available the provisions for bulky wastes (Annex II to the Oslo Convention) would apply to the disposal of platforms at sea. The Netherlands and FRG are preparing a list of substantive items to be included in the guidelines.

Future Work Programme
An Ad Hoc Working Group of Legal Experts will meet in 1990 to consider whether Article 15(1)(c) of the Convention should be amended to take account of the rights and obligations of Coastal States under the provision of the United Nations Convention on the Law of the Sea (UNCLOS).

The special case of abandoning or toppling of offshore platforms at site will be reviewed by a legal expert group in 1990.

The Standing Advisory Committee for Scientific Advice (SACSA) will review in 1990 the guidelines adopted in 1980 on the methods of monitoring dumping grounds for sewage sludge and dredge spoil, together with the guidelines adopted in 1986 for the disposal of dredged material

Following the Commission's decision in 1988 on the termination of incineration at sea by 1994, the first indication of progress towards this aim are becoming visible. SACSA will assess statistics for 1988 and 1989 at its 1990 meeting. It will also attempt to make annual incineration forecasts.

Disposal of pipes, metal shavings and other material resulting from offshore exploration and exploitation activities will be reviewed by SACSA in 1992 when it was hoped that the Contracting Parties concerned would be able to report on the actual situation on the seabed.

4.5 OPERATION OF THE 1974 PARIS CONVENTION

The Convention for the Prevention of Marine Pollution from Land-Based Sources 1974 (Paris Convention) was opened for signature in Paris on 4 June 1974 and entered into force on 6 May 1978. The Convention aims at controlling land-based discharges into the North-East Atlantic and dependent seas (excluding the Baltic and Mediterranean). The Contracting Parties comprise the European Communities and 13 individual states.

The Convention aims at complementing the Oslo Convention (see 4.4) and there is a joint monitoring programme for sites at which both discharges and dumping occur. The task of supervising implementation of the Convention is carried out by the Paris Commission (PARCOM) which shares its secretariat with the Oslo Commission.

Activities 1988
The Tenth Meeting of the Paris Commission was held in Lisbon, 15–17 June 1988. In addition, the Commission met jointly with the Oslo Commission (Tenth Joint Meeting) from 20–21 June 1988 to discuss matters of mutual interest.

CVC
The Paris and Oslo Commissions have established a group comprising the Chairmen and Vice-Chairmen of both the Commissions with the task, among others, of taking initiatives and putting forward proposals to the Commissions which could promote their efficient operation. The Group

(CVC) met twice during the Commissions' inter-sessional period and their recommendations were examined at the Commissions' Tenth Joint Meeting. It was agreed that 1991/1992 was a suitable target date for holding a meeting at Ministerial level in the OSPARCOM framework.

Third North Sea Conference
The CVC raised the following points to be taken into account by the Netherlands in planning the Third Conference:

- the special status of those members of the Oslo and Paris Commissions which are not North Sea riparian States;
- the preparations for the North Sea Conference and the ongoing work of the Oslo and Paris Commissions;
- the need to consider whether the functions of the Ministerial Conferences on the North Sea should be transferred to OSPARCOM.

Review of Waste Disposal Policies
The Commissions agreed that the next review of waste disposal policies should take place in the two years 1990/1991.

Joint OSPARCOM/ICES Liaison Committee
The meeting agreed that it would be of general benefit to both organisations to establish a Joint OSPARCOM/ICES Liaison Committee. This would serve as a steering group to provide general guidance and comments to the ICES governing bodies and to the Commissions on the implementation of that part of the work of the two organisations where there is a common interest.

Implications of the Second International Conference on the Protection of the North Sea

The Paris Commission adopted PARCOM Recommendation 88/3 of 17 June 1988 as a first approach on the use of the best available technology, whereby Contracting Parties agree that measures adopted under Article 4 of the Convention should be applied in such a way as to prevent any increase in pollution in Convention waters, in other sea areas, or in other parts of the environment, or any risk to the health of industrial workers or the general population. The FRG reserved its position on this Recommendation.

In considering the additional work called for by the North Sea Conference on the input of pollutants via the atmosphere, the Commission agreed:

- to invite the ATMOS Working Group to identify the steps which need to be taken to move from the present pilot phase to a comprehensive, long-term monitoring programme for measuring atmospheric inputs to the sea, and to establish a detailed timetable for this progression;
- to agree that the ATMOS Working Group should prepare, for adoption by the Commission at its Eleventh Meeting in 1989, emission standards for four major industrial sectors.

Discharges of radioactive substances

The Paris Commission unanimously adopted PARCOM Recommendation 88/5, by which the Contracting Parties agree to respect the relevant recommendations of the competent international organisations and to apply the best available technology to minimise and, as appropriate, eliminate any pollution caused by radioactive discharges from all nuclear industries, including reprocessing plants, into the marine environment.

Mercury Pollution

The Commission agreed to publish the synthesis document in its next Annual Report. The conclusions of the synthesis report, including proposals for a possible PARCOM Decision in 1989, will be considered further at TWG in 1989.

Tributyl-tin Compounds

Organotin compounds should be placed on the waiting list of hazardous substances with a view to being placed in Part I of Annex A to the Paris Convention. The Paris Commission should draw the attention of the IMO to restricting the use of TBT compounds on seagoing vessels. The Commission also agreed to adopt a further PARCOM Recommendation aimed at reducing the amounts of organotin compounds which may reach the aquatic environment as a result of docking activities.

Working Group on the Atmospheric input of Pollutants to Convention Waters

The Commission agreed that mercury, cadmium and nitrogen compounds should be regarded as top priority. However, the Atmospheric Inputs Working Group and the Technical Working Group had been unable to reach an agreement on limit values for emissions.

Pollution by PCBs

The Commission examined the measures taken by the Contracting Parties to implement the 1984 Decision on PCBs and PCTs. These are now widely prohibited by most Contracting Parties. Some Contracting Parties intend to phase out the use of PCBs in some important applications by 1995; others consider that a time limit of 20 years would be necessary.

The Commission requested TWG to prepare an information note for consideration by PARCOM on products marketed as PCB substitutes.

Report of the Working Group on Oil Pollution (GOP)

It was agreed that, in future, oil spills from exploration and exploitation installations will be reported when exceeding one tonne. The Commission endorsed TWG's conclusion on the 40 mg/l standard.

Nutrients

The FRG proposed a Recommendation on the reduction of nutrient inputs to the Paris Convention area. After discussion, the Commission adopted PARCOM Recommendation 88/2 on the reduction of nutrients.

Monitoring

Contracting Parties were required to report every second year on a mandatory basis the results of national programmes to assess the quality of fish for human consumption. Information about national guidance values for fish quality and concentrations found was only supplied by the Netherlands.

The Commission noted that the procedures agreed in 1986 for the purpose of assessing hazards to human health are not being followed by most delegations. Consequently, the Commissions agreed that countries which have information about national guidance values should submit the information to the Joint Monitoring Group (JMG) in 1989.

Concerning the monitoring of contaminants in biota, the Commissions took note that the ICES Working Group on Monitoring Strategies had been requested to provide its advice to JMG.

Establishment of the North Sea Task Force
The Second International Conference for the Protection of the North Sea endorsed the need for further development of assessment of environmental conditions at national and international level. The Ministers requested ICES and the Commissions to consider together the optimal means to achieve these ends, including the possible benefits of a joint Working Group or 'Task Force'. ICES and OSPARCOM were also requested to prepare and publish, taking the 1987 Quality Status Report (QSR) as a basis, further reports on the quality of the North Sea at regular intervals, commencing in 1991.

The Commissions agreed that the North Sea should be the subject of the first regional assessment to be undertaken jointly by ICES and OSPARCOM, as requested by Ministers at the Second North Sea Conference.

It was agreed that 1993 would be the target date for the production of the next QSR.

The Commissions adopted the CVC proposals for the establishment of a mechanism to implement the provisions of the North Sea Conference Declaration on the enhancement of scientific knowledge and understanding.

Ratification of the Protocol amending the Paris Convention
It was reported that, since the last meeting, the Protocol amending the Paris Convention to include within its scope pollution of the maritime area from the atmosphere, had been ratified by the Governments of the United Kingdom, Iceland, Spain and Norway.

The Protocol will only enter into force after all States party to the Convention have deposited their instruments of ratification with the French Government. The five states — Portugal, Belgium, Denmark, Ireland and the Federal Republic of Germany – whose ratification procedures are still outstanding reported on their progress.[9]

Activities 1989
The Eleventh Meeting of the Paris Commission was held in Dublin from 19–22 June 1989. This year, special attention was paid to refineries, platforms, nuclear installations, titanium dioxide and chloralkali industries. In general, all data refer to the year 1987, and will be published together with the Commission's Annual Report later in 1990.

Important work is going on to assess discharge levels and to define suitable control technologies for a number of industrial sectors. At present, the work is in the phase of information collection and collation, and the progress report shows that work is going on as planned. In order to develop the theoretical basis for this work, a definition of the so-called precautionary principle has been worked out and was adopted by the Commission. This means that this principle is formally recognised as one of the key elements in the future work of the Paris Commission. One of the cornerstones in the application of the precautionary principle is the concept of Best Available Technology (BAT). At this year's meeting the Commission was able to take further steps as regards harmonisation in practical terms, of the application of BAT in Member Countries.

Recognising the increasing importance of the diffused sources of inputs of pollutants into the sea, the Commission agreed upon a programme for control of certain mercury containing products and appliances, in particular concerning hospital thermometers and dentistry. There also was general support for a substantial reduction of the mercury content of batteries, but due to an expected EEC regulation, the formal decision will only be able to take force later this autumn.

In view of the importance of atmospheric transport as one route of inputs of pollutants into the seas, the Commission agreed to establish a comprehensive network for the monitoring of, among other elements, certain pesticides and heavy metals as well as nitrogen compounds. Progress was also made as regards programmes and measures to reduce the inputs of pollutants from the refinery industry and for substances causing eutrophication.

As regards refineries, all countries but two have agreed upon principles and measures substantially to reduce the discharges of oil from this industry. Concerning the eutrophying substances, causing algal blooms, a detailed list of possible measures for, among others, the agricultural sector, municipal waste waters and industry has been agreed upon. The programme is designed to substantially reduce the input (of the order of 50%) of such substances into areas effected by eutrophication, before the year 1995.[10]

4.6 OPERATION OF THE 1974 HELSINKI CONVENTION

Under the 1973 MARPOL Convention the Baltic Sea was one of the five waters designated as 'special areas'. On 24 March 1974 the seven states bordering the Baltic Sea signed the Convention on the Protection of the Marine Environment of the Baltic Sea Area (Helsinki Convention) which entered into force in May 1980. The Convention works through an intergovernmental agency, the Helsinki Commission (HELCOM) which makes recommendations to the governments of the states concerned.

4.7 THE COUNCIL OF EUROPE

The Council of Europe, created in 1949, is Europe's oldest and geographically most extensive political institution. The 21 Member States of the Council of Europe are: Austria, Belgium, Cyprus, Denmark, France, Federal Republic of

Germany, Greece, Iceland, Ireland, Italy, Liechtenstein, Luxembourg, Malta, Netherlands, Norway, Portugal, Spain, Sweden, Switzerland, Turkey and the United Kingdom. Its headquarters are established in Strasbourg.

The aim of the Council of Europe is to achieve greater unity between its members. This is pursued by discussion of questions of common concern and by agreements and common action.

The Council of Europe includes the first international Parliamentary Assembly in history which holds three sessions a year. It is through the Assembly that European public opinion on major issues is expressed.

In the environmental field, the European Committee for the Conservation of Nature and Natural Resources has, since 1962, been the steering organ of the Intergovernmental Work Programme. Among its achievements are the creation of the European Diploma, the Water and Soil Charters and numerous studies on the landscape, fauna and flora.

Documentation and Information Centre for the Environment and Nature

The role of the Council's Documentation and Information Centre for the Environment and Nature is to keep Europeans, both in Member States and elsewhere, informed about the state of the natural environment and steps taken to improve it. This information effort is conducted mainly through a 10-language monthly *Newsletter* and an illus-trated quarterly entitled *Naturopa*, which are circulated to persons in positions of responsibility (politicians, administrators, scientists, teachers and journalists). In addition, the Centre deals with a large range of enquiries, organises campaigns on topical subjects, such as soil conservation, freshwater management, the preservation of wetlands and the conservation of wildlife and natural habitats.

Notes and References

1. Report by Minister Brice Lalonde.
2. Report by Ken Collins, MEP.
3. Report by LJ Brinkhorst, Director General DGXI, CEC.
4. Report by John Woodliffe, Faculty of Law, Leicester University.
5. Compiled by Robin Churchill, Cardiff Law School, UWIST.
6. Compiled from the Final Declaration of the Third International Conference on the Protection of the North Sea.
7. Report from Secretariat of the MOU.
8. Compiled from Oslo Commission Thirteenth Annual Report.
9. Compiled from Paris Commission Tenth Annual Report.
10. Compiled from PARCOM press release.

5. Activities of Governments, Government Departments and Statutory Bodies

Foreword

by Klaus Töpfer
Federal Minister for the Environment
Federal Republic of Germany

It is only by international solidarity that we can guarantee the protection of the seas. National efforts are important, but they should be made part of an international concept.

The Federal Republic of Germany has made the North and Baltic Seas focal points of its endeavours in the field of environmental protection. The First International Conference on the protection of the North Sea, held in Bremen in 1984, originated from an initiative of the FRG. Two further conferences have since taken place — in 1987 in London, and in March 1990 in The Hague (see 4.2) — with considerable progress being made towards protecting the marine environment.

The FRG is also striving for rapid progress within the framework of the Helsinki Commission, whose task it is to protect the Baltic Sea.

These activities stem from the recognition that our seas cannot simply be used as waste dumping sites. We must take action not only in places where it has been proven that the marine environment is already damaged. Rather we must take precautionary action and protect our seas for their own sake. Thus the anachronism whereby waste is disposed of at sea must everywhere become a thing of the past. A stop must be put to this practice as soon as possible wherever it continues to be carried out. This demand is the logical conclusion of the precautionary principle, the maxim under which I work.

This principle is being recognised by all the North Sea States. It means that avoidance measures ought not only to be taken in cases where there is conclusive scientific evidence of damage caused by inputs of pollutants, but, rather as soon as observations suggest that this may be the case.

This principle holds true for all input routes.

I have already mentioned my demand that all dumping of waste of whatever kind be stopped. This must include, of course, the incineration of hazardous waste at sea.

A further, direct, source of marine pollution is presented by shipping. The emphasis here must be laid both on better monitoring to ensure the provisions of existing agreements, such as the MARPOL Convention 73/78, are adhered to, and on further measures to tighten up legislation in force at the moment. Thus, for instance, the regulations in the MARPOL Convention for discharges of residues of oil and chemicals must be made more stringent, and be brought up to comply with the limit values already set in special areas. If not, then we need more special areas. The stringent regulations required in a special area according to MARPOL 73/78 Annex I (oil) and II (chemicals) should apply especially to the North Sea, an area at particular risk since it has the highest shipping density in the world. Internationally, too, we must not neglect taking measures to reduce the risk of accidents involving tankers. This must include considering the introduction of double bottom tankers and worldwide cooperation in combating oil catastrophes. The negotiations taking place at this moment in the IMO are the best way towards this goal, I hope that it will be possible to sign a Convention to this effect before the end of 1990, and that this will be accepted by as many countries as possible, including even those countries which have not ratified MARPOL 73/78.

Action also needs to be taken to reduce marine pollution from drilling platforms. In particular it should be examined whether it might not be possible to conclude regional conventions — such as the 1989 Kuwait Protocol — to deal fully with this problem.

Another important factor in protecting the marine environment is dealing with land-based pollution.

There is a particular obligation to reduce pollutant inputs via rivers and from the atmosphere. Thus the states bordering the North and Baltic Seas have committed themselves to reducing discharges of nutrients and other pollutants via rivers by around 50% by 1995. Furthermore, at the Third International Conference on the Protection of the North Sea held in The Hague in March 1990, this decision was extended to include atmospheric inputs from 17 different substances, insofar as such a reduction is possible using the best available technology.

In addition, a further decision was taken in The Hague to reduce by around 70% amounts of four particularly hazardous substances — dioxines, mercury, cadmium and lead — again insofar as this is possible using the best available technology.

The spirit of the precautionary principle also lies behind these decisions. The best available technology is to be used even if it cannot be fully proven scientifically that the inputs of specific pollutants will have detrimental effects.

A particular challenge is posed to the FRG by measures to clean up the River Elbe, a major source of North Sea pollution. While international cooperation has led to considerable success in dealing with the problems of the River Rhine, there is still a great deal of work to be done to cope with the difficulties surrounding the Elbe. The main inputs stem from river catchment areas in East Germany and Czechoslovakia, and most of these (95% of mercury inputs) flow with the River Elbe across the intra-German border. According to GDR sources, discharges from this country alone account for around 20% of the entire mercury input into the North Sea. This shows the special need to set up a commission for the protection of the Elbe, for which negotiations have already been taken up with the governments of the GDR and Czechoslovakia, the European Community also being included. I am optimistic that it will be possible to sign an agreement for a Commission for the Protection of the Elbe as early as this coming summer (1990). A decisive factor must be that the future Commission should begin its work as speedily as possible and that it should be commissioned to give priority to important tasks such as developing programmes for the drastic reduction of inputs of waste water.

In summary, may I stress that the protection of the marine environment is based on two pillars:

- strict national legislation on the environment, and

- international cooperation for the purposes of harmonisation.

All countries in the world must make known the measures they have at their disposal, so that proper in-depth negotiations can begin on the necessary international harmonisation. This part of the ACOPS annual report is a valuable contribution to this task.

Dr Klaus Töpfer
Federal Minister for the Environment,
Nature Conservation and Nuclear Safety of the
Federal Republic of Germany

5.1 LEGISLATIVE OUTLINES

When compiling this chapter, we invited individuals from a selection of countries throughout the world to send us information on their national legislation. The reports that appear here are not necessarily intended to be a comprehensive review of legislative instruments from these countries, but rather give an indication of the recent legal developments in the field of marine pollution. In section 5.1, Legislative Outlines, reports of legislation that relate to marine pollution have been included from four countries. Section 5.2, Legislative Update, contains updates on legislation of countries that we have already featured in previous Yearbooks. Hence, for example, the information under 5.2.2, The Netherlands, is legislation on maritime matters that has been passed since The Netherlands legislation appeared in the *1987/88 ACOPS Yearbook* (p 52). For an indication of measures that would have appeared previously, please refer to the appropriate section of Chapter 7, under the heading of National Legislation.

5.1.1   Jamaica

Harbours Act
Section 19 of the Harbours Act makes it an offence for any person who throws or deposits or permits to be thrown or deposited any oil or mixture of oil residues in any channel leading into or out of any harbour or any place within the limits of the harbour other than a place specially designated for such purposes.

Beach Control Act
Under Section 7 of the Beach Control Act the Minister may make an order declaring any part of the foreshore or floor of the sea as a protected area. In addition within such a protected area the Minister may prohibit various activities such as the disposal of rubbish or other waste matter.

Wildlife Protection Act
Under Section 10 of the Wildlife Protection Act every person who places any poison lime or noxious material in any water with intent to take, kill or injure fish commits an offence under the Act. Section 11 of the Wildlife Protection Act also makes it an offence to place trade effluents or industrial waste from any factory in any harbour, stream or estuary containing fish.[1]

5.1.2   Uruguay

Laws Derived from Global Treaties

The UN Convention on the Law of the Sea
Uruguay has not yet ratified this Convention, but the Message of the Executive Power is already in the Senate and it is the firm intention of the new Government that will assume office on 1 March 1991, to ratify the Convention.

UNEP's Regional Seas Programme
This excellent programme provides some of the most relevant successes of UNEP. In spite of the fact that Uruguay has solicited the inclusion of the South Atlantic area in the RSP, the project could not be carried out because of the opposition of the Brazilian Government.

French grunts, pictured in the Turks and Caicos Islands in the Caribbean. Fished locally, they rely on good coral reef in which to feed and shelter. Photo: D George

General Resolutions and Recommendations of the UN System

As a full member of the UN, Uruguay participates in the field of marine protection. Very little is self-executing, but Uruguay is bound to respect and implement all of the requirements such as the "World Chart of Nature" and the report "Our Common Future".

Norms Derived from Regional Treaties

Uruguay belongs to two systems which refer to maritime pollution: the Organisation of American States (OAS), and the River Plata Basin Organisation.

The River Plata Basin Organisation

A sub-regional Organisation mostly concerned with "the harmonious and balanced development (of the countries concerned) and the best use of the great natural resources of the region, making sure its preservation for the future generations through a natural utilisation of the same resources" through the infra structural development of the Basin. This is one of the biggest Basins in the world, with 3.2 million km and in view of the random industrial development is a potential source of largescale land-based marine pollution.

The geo-political position of Uruguay, in the drain of the Basin, makes it, together with Buenos Aires, through the River Plata or Estuary, and then the continental platform, the receiving body of an unlimited amount of chemical pollutants and products of erosion caused by the irrational agriculture developed all along the main rivers.

The River Plata Basin Organisation was established by a meeting of the Ministers of Foreign Affairs of the five countries concerned, held in Brazilia, (Brazil), on 23 April 1969. The Treaty establishes as a main body the Conference of the Ministers of Foreign Affairs of the Member States, and as a permanent body the Intergovernmental Coordination Committee, consisting of representatives of the countries.

As the environmental problems of the Basin are vast the organisation was due to give them priority. Unfortunately, this has not been possible because of objections from a Member State. Notwithstanding, sooner or later the River Plata Basin Organisation will be called to take urgent action on the matter. The unchecked problems brought about by the deforestation in the upper part of the Basin, erosion, chemical land-based pollution from agro-toxins used without controls, acid rain and all kinds of environmental transgressions are creating an unbearable situation in the downstream countries, destroying their hydrical balance, polluting their waters and finally polluting the South Atlantic Regional Sea. At present, the River Plata Estuary is one of the most polluted estuaries in the world.

Laws Derived from Bilateral Treaties

This is the sector in which Uruguay has taken some more positive steps, engaging a process of real — if modest — progress in implementing a legal environmental protection system established by the delimitation treaties with Argentina.

The Treaty of Limits on the Uruguay River

This Treaty was signed on 7 April 1961, a complementary Protocol having also been signed on 16 October 1968. Later on a Statute of the Uruguay River was elaborated and approved on 26 February 1975. This Statute establishes an Administrative Commission of the Uruguay River, with its headquarters in Paysandú (CARU).

The philosophy of both instruments (concerning the Uruguay and the Plata Rivers) is quite modern. They do not limit themselves to the establishment of a borderline but consider both waterways as shared natural resources.

The Treaty of Limits on the River Plata

This was signed on 19 November 1973. To overcome historical difficulties the idea of establishing a rigid borderline was abandoned, and through a Statute two bands of exclusive jurisdiction adjacent to each country were established. So, it was not necessary to establish a borderline in the river.

The Convention also created an Administrative Commission and its headquarters were established in Martin Garcia Island. The Treaty came into force on 12 February 1974.

These two treaties established a coherent system for the protection of the fluvial-maritime environment, regulating not only the use and the exploitation of the water, but also of the natural resources, such as the minerals of the bed and subsoil and its fauna. The resources protected by these treaties include water, mineral resources, fauna and flora.

Cooperation Agreement between Argentina and Uruguay to Prevent and Combat Pollution of the Aquatic Environment

This agreement, signed on 16 September 1987, is a sort of regulation of treaties and statutes, aiming at maintaining and widening the cooperation between both countries to prevent pollution incidents and to combat their consequences in the aquatic environment. As the above mentioned instruments establish the obligation to protect and preserve the environment, it is necessary to harmonise the national policies and to establish contingency plans and compatible procedures to strengthen the joint action capacity. A coordination deed was elaborated between the Maritime Prefecture of both countries on 19 October 1989.

In December of the same year a Working Group on Pollution was created to improve the coordination and procedures. Some concrete actions have been taken, for instance, a joint action in Montevideo Bay subsequent to an oil spill. The process is still in action. Finally, last year a draft Contingency Plan was prepared by a inter-institutional Working Group, but is still at the stage of an internal draft.

Conclusions

Uruguay is provided with a comprehensive legal framework on environmental maritime law, particularly in the bilateral field. It is now necessary to elaborate a general National Environmental Policy to control land-based sources of marine pollution. To be really effective, this policy should be developed and applied at regional level. The action of Uruguay on its own will remain ineffective and even the Bilateral System with Argentina is not enough;. the River Plata Basin Organisation is the basic environmental unit that provides the potential for an effective policy.[2]

5.1.3 Federal Republic of Germany

Federal Water Act (Wasserhaushaltsgesetz)
Article 7A
With the 5th amendment to the Federal Water Act (Wasserhaushaltsgesetz — WHG), Article 7A, para. 1 have already provided the prerequisites that will allow the Federal Government to issue administrative regulations establishing hazardous substances on the basis of the best available technology. Thirty working groups have been set up to prepare administrative regulations for the various branches of industry.

The Federal Minister for the Environment's 10-Point Action Programme has established a priority list for the most important branches, according to which administrative regulations or drafts thereof were to be completed and issued by the end of 1989. Thus, administrative regulations for the limitation of hazardous substances on the basis of the best available technology were issued as follows:

18 May 1989:
chemical pulp production (notably AOX);
8 September 1989:
metal working and processing (heavy metals, AOX), power stations and waste incinerators with flue gas scrubbing facilities (heavy metals), leather production (heavy metals, AOX), landfill leachate (heavy metals, AOX), dental treatment (mercury);
19 December 1989:
non-ferrous metal production (heavy metals), waste water containing mineral oils (AOX, hazardous hydrocarbons), manufacture and processing of glass (heavy metals) as well as soda production (NH4; toxicity to fish), dry cleaning (AOX).

Furthermore, administrative regulations for the following areas are being prepared: chemical industry, fertiliser manufacture, textile industry, pulp and paper, hard fibreboard, refineries, ceramics industry, iron and steel, and coking plants. Administrative regulations will be issued in these areas within the next two years.

In individual cases, the Federal Länder may establish requirements in keeping with the best available technology before administrative regulations are issued on the basis of the work of the above-mentioned working groups. This is permitted by current water law and has already been practised in individual cases.

Waste Water Charges Act
In order to have an additional incentive to accelerate measures against water pollution, the Waste Water Charges Act was passed as early as 1976. According to the Waste Water Charges Act the waste water charge to be paid depends on the content of certain pollutants in the waste water and is calculated on the basis of the discharge values authorised in the official notices relating to water legislation, taking differently evaluated pollutants and groups of pollutants as its base.

The 10-Point Action Programme for the protection of the North and Baltic Seas includes a proposal to make phosphorus and nitrogen subject to the waste water charge. A bill to this effect will enter into force by 1991. This will provide an additional incentive for the necessary investments in water-protection measures in order to reduce phosphorus and nitrogen inputs as quickly as possible. the possibility of deducting these investments from the waste water charge for a limited period of time will act as an added incentive to accelerate the construction of new sewage treatment plants and the expansion of existing ones equipped for phosphorus and nitrogen elimination.

Washing and Cleansing Agents Act
The detergents industry has increasingly been developing phosphate-free detergents. At the beginning of 1988, two thirds of all detergents were already phosphate-free. the supply of these detergents and the phosphate reduction resulting from the Ordinance on Maximum Amounts of Phosphates in Washing and Cleansing Agents means that the sewage treatment plants and water bodies in the Federal Republic have been relieved of considerable amounts of phosphate. It was expected that washing and cleansing agents would contribute only 10 to 15% of the overall phosphorus input into the water bodies of the FRG by the end of 1988.

Amendment to the Chemicals Act (Chemikaliengesetz)
The Federal Government is also pursuing the principle of precaution outside the field of water protection proper in the interests of water protection

Thus, Article 17 of the Chemicals Act already provides the possibility to prohibit under certain circumstances the manufacture, marketing and use of specific hazardous substances and preparations or products containing such substances and preparations in order to protect man or the environment.

By amending the Chemicals Act the threshold for imposing such bans will be lowered to give greater consideration to the protection of man and the environment. It must be remembered, however, that measures for the limitation of substances can affect the free movement of goods within the EEC and therefore can only take effect if they comply with the relevant EEC regulations.

Amendment to the Federal-Emission Control Act (Bundes- Imissionsschutzgesetz BImSchG)
Subsequent to supplementing the Ordinance Concerning the Licensing Procedure Pursuant to the Federal Emission Control Act which was carried out in connection with the amendment to the Hazardous Incidents Ordinance (Störfall-Verordnung), work to amend the Federal Emission Control Act itself is now under way, with the main focus being placed on further improvements in the system of inspecting and supervising industrial installations. The Federal Minister of the Environment, Nature Conservation and Nuclear Safety is seeking to establish the prerequisites for technical

inspections of installations posing a potential threat to public safety. Such inspections will be performed by independent experts in a more comprehensive manner than in the past and the legal position of the Emission Control Officer will be strengthened by extending his power as required, in accordance with the respective on-site conditions. Other key areas to be focused on include, above all, the improvement of the regulations on clean-air planning, the introduction of the principle of precaution for installations not subject to licensing, increasing the efficiency of the instruments of free enterprise, as well as requiring operators — within the framework of the implementation of the Federal Government's soil protection concept — to ensure that installations that have been shut down to not pose any risk to the environment.

Act on Environmental Impact Assessment

The precautionary principle is taken into special account in connection with the translation of the EEC Directive on the assessment of the effects of certain public and private projects on the environment into national law. The draft of the Act on Environmental Impact Assessment was adopted by the Federal Parliament on 16 November 1989. The Act is expected to become effective in mid-1990.[3]

5.1.4 Australia

UN Convention on the Law of the Sea

Australia signed the United Nations Convention of the Law of the Sea in December 1982, but has not yet ratified it. There has been a trend over the past year, during discussions on UNCLOS, to give a higher profile to environmental issues and their implications for the Convention.

The Department's involvement in the Convention has primarily focused around negotiations on the environmental aspects of the development of a deep seabed mining code.

Environment Protection (Sea Dumping) Act 1981

This Act gives effect to the London Dumping Convention which was signed by Australia in 1973 and ratified in 1985. The Act is administered by the Department of the Environment and applies to any ships, aircraft or platforms in Australian waters and to Australian ships or aircraft in any part of the sea. Australian waters are defined, for the purposes of the Act, as seas within 200 nautical miles of the baseline from which the breadth of the territorial sea is measured. Where the baseline follows the sinuosities of the coast this is defined by lowest astronomical tide. The provisions of the Act relating to dumping or incineration do not apply within internal waters of Australian States or of the Northern Territory. The relevant definition of 'internal waters' is a narrow one and is based upon the state of the law at the date of Federation.

The Act was amended in January 1987 to prohibit entirely the dumping of radioactive waste or other radioactive matter at sea. The amendment defines such waste or material as that having an activity in excess of 35 becquerels

per gram. This value, which is based upon the activity of natural potassium, applies a prohibition to all those substances commonly regarded as radioactive in Australia.

The amendment gives continuing effect to the Government's policy of rigorous opposition to dumping radioactive waste at sea. It also accords with the provisions of the South Pacific Nuclear Free Zone Treaty, signed by the Prime Minister at Rarotonga in August 1985, and of the Convention for the Protection of the Natural Resources and Environment of the South Pacific Region (the SPREP Convention) which Australia signed in November 1987.

There is provision under Section 9 of the Act for the

Portfolio Minister, if he so wishes and is satisfied that under such an agreement there would be proper compliance with the requirements of the Convention, to make a declaration disapplying the Commonwealth Act in favour of corresponding State law with regard to dumping within coastal waters of the State. Such a declaration would not over-ride the amended provisions of the Commonwealth Act which prohibit the dumping of radioactive waste or other radioactive matter. Western Australia, South Australia and Queensland have expressed a wish to have such arrangements put in place. Negotiations over suitable memoranda of understanding, setting out the obligations and responsibilities of the parties, are in progress. A memorandum of understanding was signed with Tasmania and the Portfolio Minister made a declaration on 8 November 1988 disapplying the Commonwealth Act to Tasmanian coastal waters.

Discharges of an operational nature from ships, aircraft or platforms including such items as exhaust, cooling water, galley scraps and sewage and are not subject to the Environment Protection (Sea Dumping) Act but are regulated by the Protection of the Sea legislation under the responsibility of the Transport and Communications Portfolio.

Discharges from the land or from land-based structures such as pipelines or outfalls are not a Commonwealth responsibility, even when they take place into the marine environment. State authorities are responsible and to an increasing extent public expectations are coming into conflict with actual practice. There are moves underway for questions of uniform application of the relevant criteria and standards to be addressed.

South Pacific Environment Matters
The Department of the Environment is closely involved in the implementation of the Convention for the Protection of the Natural Resources and Environment of the South Pacific Region, known as the SPREP Convention. The Convention was developed through the South Pacific Regional Environment Program. It covers the protection, development and management of marine, terrestrial and coastal environments, including the preservation of rare or fragile ecosystems and depleted, threatened or endangered flora and fauna. The Department, in cooperation with other institutions, has focused on promoting cooperation between the island states of the South Pacific and those organisations in Australia which can offer environmental management expertise. Such cooperation has lead to investigating the possibility of establishing a proposal dealing with an international turtle programme in the South Pacific. A decision on whether this proposal will proceed was expected during 1989–90.

Torres Strait Treaty
The Torres Strait Treaty between Australia and Papua New Guinea was ratified in February 1985. The Treaty obliges the Contracting Parties to protect the marine environment and the traditional way of life of the inhabitants and establishes the Torres Strait Protected Zone for these purposes. Under Articles 13 and 14 of the Treaty, each Treaty Party is required to keep the other informed of developments which may have environmental implications for the Torres Strait Protected Zone.

In January 1989 the Department was represented at the First Meeting of the Torres Strait Environment Management Committee in Port Moresby.

The Department is also examining means of funding a proposal to conduct an environmental baseline study of the Torres Strait marine environment.

Timor Gap Agreement
An Agreement between Australia and Indonesia for a zone of joint cooperation in the Timor Gap was being developed during 1989. The purpose of the Agreement is to facilitate the exploration and development of resources in the area where previously there had been disagreement over which country's jurisdiction applied. The Agreement includes references to environmental principles, such as general obligations to protect the marine environment, including the application of environment impact assessment. It is expected that the details of the Agreement will be finalised in the second half of 1989.

Sea Installations Act 1987
The Act aims to:

- ensure that sea installations located in adjacent areas are operated with respect to the safety of the people using them and of the people, ships and aircraft near them;

- apply appropriate laws in relation to such sea installations; and

- ensure that such sea installations are operated in a man-

ner that is consistent with the protection of the environment.

In November 1988 the first sea installations permit under the act was issued to the John Brewer Floating Hotel off the coast of Townsville, Queensland.

Conventions on Wetlands
Australia is a signatory to the Convention on Wetlands of International Importance Especially as Waterfowl Habitat (the Ramsar Convention), the Japan-Australia Migratory Birds Agreement and the China-Australia Migratory Birds Agreement. The Department continued to receive numerous representations regarding the protection of wetlands and Australia's obligations under these international treaties and provided advice to the Portfolio Minister on a number of significant wetlands conservation issues.

Great Barrier Reef Marine Park Act 1975
The Great Barrier Reef Marine Park Act 1975 provides for increases in penalties and new enforcement and management provisions. A two-year moratorium on further floating hotels in the marine park was announced on 15 February 1989. During the moratorium, an assessment will be made on the operation of the existing John Brewer Reef floating hotel, and the floating hotel planned for Fitzroy Reef which was issued with a conditional permit by the Marine Park Authority. This examination will help determine whether earlier environmental assessments were accurate in their evaluation of the impacts the hotels would have on the reef environment.[4]

5.2 LEGISLATIVE UPDATE

5.2.1 United Kingdom

Merchant Shipping Act 1988
The parts of this Act, which received the Royal Assent on 3 May 1988, most directly relevant to marine pollution are Section 34 and Schedule 4, which implement the 1984 protocols to the International Convention on Civil Liability for Oil Pollution Damage, 1969 and the International Convention on the Establishment of an International Fund for Oil Pollution Damage, 1971. A number of other provisions of the Act also have some bearing on marine pollution. First, Part I of the Act allows orders to be made to tighten up the conditions under which ships can be registered in UK-dependent territories, to ensure that their registers cannot be used for sub-standard ships (as has happened in the past). Such orders have so far been made for Bermuda and the Isle of Man (SI 2251/1988). Secondly, Sections 30-32, which are a reaction to the *Herald of Free Enterprise* disaster, create new criminal offences of proceeding to sea with a ship which is unfit to go to sea without serious danger to human life, of operating a ship in an unsafe manner, and of doing any act on board a ship likely to cause loss or damage to that ship or its equipment or another ship. Thirdly, Section 33 creates a Marine Accident Investigation Branch separate from the Marine Directorate of the Department of

Transport. Lastly, Section 35 of the Act provides that regulations may be made to regulate the transfer of cargo, stores, bunker fuel or ballast between ships in the UK's territorial sea, in order *inter alia* to prevent pollution or hazards to the environment. No such regulations had been made at the time of writing.

Water Act 1989

The main purpose of this Act, which received the Royal Assent on 6 July 1989, is to privatise the water industry in England and Wales. Nevertheless, the Act also contains, in Chapter I of Part III, some important provisions on pollution. These provisions replace Part II of the Control of Pollution Act 1974 (although that Act continues to apply in Scotland, subject to amendments contained in Schedule 23 of the Water Act). The pollution controls of the new Act are largely based on the consent system for discharges contained in the 1974 Act, though with important modifications. First, implementation and enforcement of the system is by an independent body, the National Rivers Authority (see 6.3.6), which is established by the Act, and not by the water industry itself, as was formerly the case. Secondly, water quality objectives and targets for achieving them are for the first time given statutory form. Thirdly, the Act puts more emphasis on precautionary measures and on controlling diffuse as well as point sources of pollution: thus, for example, a new power is created of preventing and regulating the entry of nitrate into the waters to which the Act applies from agricultural activities by the establishment of nitrate sensitive areas. The Act's pollution controls apply to rivers, coastal and underground waters, and the inner three miles of the territorial sea (though orders may be made extending the controls to other parts of the territorial sea). At the time of writing some half-a-dozen statutory instruments had been issued to give further effect to the pollution control provisions of the Act. These include SI 1148, 1149, 1151, 1157, 1158 and 1160/1989.

Petroleum Production (Seaward Areas) Regulations 1988 (SI 1213/1988)

Like earlier Petroleum Production Regulations, these Regulations impose an obligation on those holding a licence to explore the UK's territorial sea or continental shelf for petroleum, or to produce petroleum therefrom, to prevent the escape of any petroleum into the sea.

Environmental Assessment (Salmon Farming in Marine Waters) Regulations 1988 (SI 1218/1988)
Harbour Works (Assessment of Environmental Effects) Regulations 1988 (SI 1336/1988)
Harbour Works (Assessment of Environmental Effects) (No 2) Regulations 1989 (SI 424/1989)

The first set of regulations implements EEC Directive 85/337 and relates to the assessment of the environmental effects of certain projects in respect of salmon farming in marine waters. The second and third sets of regulations implement the same directive in respect of harbour works.

The Merchant Shipping (Prevention of Pollution by Garbage) Order 1988 (SI 2252/1988)
Merchant Shipping (Prevention of Pollution by Garbage) Regulations 1988 (SI 2292/1988)
Merchant Shipping (Reception Facilities for Garbage) Regulations 1988 (SI 2293/1988)

These three statutory instruments implement Annex V of the MARPOL Convention. In particular, the third instrument imposes a duty on harbour authorities and terminal operators to provide the necessary reception facilities for garbage from ships.

Control of Pollution (Landed Ships' Waste) (Amendment) Regulations (SI 65/1989)

These regulations amend SI 402/1987 to apply to ships' garbage landed in Great Britain the same requirements that apply to tank washings landed there.

The Air Quality Standards Regulations 1989 (SI 317/1989)

These regulations implement various EEC directives (Nos 80/779, 82/884 and 85/203) setting air quality standards for sulphur dioxide, smoke, lead and nitrogen dioxide. The regulations place a duty on the Secretary of State to take any necessary measures to ensure that concentrations of the substances mentioned do not exceed the limits prescribed in the directives.

Control of Industrial Air Pollution (Regulation of Works) Regulations 1989 (SI 318/1989)
Health and Safety (Emissions into the Atmosphere) Regulations 1989 (SI 319/1989)

These regulations implement EEC Directive 88/609 on the Limitation of Emissions of Certain Pollutants into the Air from Large Combustion Plants (see 4.1.5).

Bills in Parliament

Environmental Protection Bill

This Bill, which was published in December 1989, is largely concerned with the terrestrial environment. Nevertheless, some of its provisions are relevant to marine pollution. Part I of the Bill establishes a new pollution control regime, Integrated Pollution Control. This regime will apply to prescribed industrial, commercial and other processes discharging into any environmental medium (including internal waters and the territorial sea). In relation to such processes the Secretary of State will be empowered to issue regulations setting limits on the concentrations or amounts of released substances and other requirements relating to the operation of prescribed processes; establishing environmental quality objectives and standards; and making national plans for controlling releases. Every prescribed process will require an authorisation before it is allowed to operate. Such authorisations will include conditions requiring the use of best available techniques not entailing excessive cost to prevent and minimise releases of substances, and to render harmless any substances which are released; and compliance with limits and plans, and achievement of quality standards and objectives, set by the Secretary of State.

More directly relevant to marine pollution is clause 111, which amends certain of the provisions of the Food and Environment Protection Act 1985 dealing with the dumping of waste at sea. The effect of the amendments is to extend the scope of the 1985 Act to include the dumping or incineration of waste within UK continental shelf limits by foreign vessels which have loaded the waste to be dumped or incinerated in foreign ports. Enforcement officers' powers are also to be increased to enable them to bring into port vessels suspected of illegal dumping, and the maximum fine on summary conviction is increased from £2,000 to £50,000.[5]

5.2.2   The Netherlands

Decision-making in the Netherlands in respect of North Sea policy takes place in four stages:

- preparation and execution of policy by government officials who meet in the framework of the Commission on North Sea Matters (ICONA), in which all ministries involved in North Sea decision-making (13 members) are represented;

- advice from independent organs, in particular North Sea Commission (constituted of representatives of organisations having some interest in the North Sea and independent experts). Occasionally other commissions give advice relating to North Sea matters;

- decision-making by ministers, in particular the Minister of Transport and Public works who has an overall coordinating responsibility for North Sea matters, and other ministers meeting in the MICONA (the equivalent of the aforementioned ICONA on ministerial level, consisting of the seven ministers involved), and ultimately the Council of Ministers;

- control by Parliament, particularly the Commission for the Harmonisation of North Sea Policy (see below).

Legislation 1988–89

The Shipping Traffic Act
On 1 September 1988 the Shipping Traffic Act entered into force. This Act replaces a number of old Acts, and is now the most important instrument for the regulation of maritime traffic (Stb. 1988, 352, 390).

Regulation of 7 October 1988 concerning the Transport of Dangerous Substances
The Regulation implements a number of EEC directives and provides for a number of obligations which have to be complied with before the transport of dangerous substances. The Regulation entered into force on 15 October 1988 (Stcrt.1988, 200).

MARPOL 1973/78
On 9 April 1988 the
Decree for the Prevention of Pollution by Noxious Liquid Substances in Bulk entered into force, implementing Annex II of MARPOL 1973/78 (Stb.1988, 122). This Decree is further implemented by the Decree of 29 April 1988, which contains lists of substances. The Decree closely follows the guidelines set out in Annex II to MARPOL 1973/78. Further, three regulations of 11 and 12 October 1988 provide for the implementation of MARPOL 1973/78 (Stcrt.1988, 39).

Another development relating to MARPOL 73/78 is the acceptance of Annex III and V (Trb.1988, 124). Annex V entered into force for the Netherlands on 31 December 1988 (Trb.1988, 199).

Following this acceptance on 31 December 1988, the **Decree Prevention of Pollution by Garbage from Ships** of 22 December 1988 entered into force. The Decree implements Annex V (garbage) of the MARPOL 1973/78 Convention and the Act on Prevention of Pollution by Ships 1986, which contains the general framework for the implementation of the MARPOL Convention. The provisions of the Decree are to a large extent congruent with the provisions of Annex V. The Decree is applicable to all Dutch ships and to foreign ships within the territorial waters of the Netherlands (Stb. 1988, 636).
On 16 February 1989, the
Decree on Reporting of Pollution Incidents by Ships of 23 December 1988 entered into force. The Decree implements the amended Protocol I of the MARPOL 73/78 Convention and the Act on Prevention of Pollution by Ships 1986. The provision of this decree are highly similar to the contents of Protocol I. A number of provisions contained in the Decree require further elaboration (stb.1988, 694).
A first Decree providing such further elaboration is the
Decree of 6 February 1989 implementing Arts. 3, 4 and 5 of the Decree reporting of incidents of pollution by ships. The Decree incorporates *inter alia* MEPC Resolution 25/26 which aims at the further elaboration of Protocol I and contains general guidelines for reports in case of pollution incidents (Stcrt.1989, 32).

Offshore Environment
The Decree of 14 December 1988 (Stert.1988, 246) concerning the discharge of drilling muds containing oil compounds from offshore platforms was withdrawn by the Decree of 18 January 1989, after it had appeared that the Decree of 14 December 1988 was adopted contrary the provisions on environmental impact assessment (Stcrt.1989, 17).

Voluntary Agreement on the Reduction of Phosphates
On 30 May 1989, a major step was made towards the implementation of the obligations accepted by the Netherlands in the North Sea Ministerial Declaration and the Rhine Action Plan. At that date the Government signed a *Voluntary Agreement on the Reduction of Phosphates* with a number of local authorities. The Agreement aims at the reduction of the input of phosphates in the Netherlands' surface waters. The Government considers the voluntary

agreement to be an adequate step towards the reduction of phosphates in the Rhine and the North Sea by 50% in 1995 (Stcrt. 1989, 102).

Pollution of Surface Waters

The Decree of 21 December 1989 amends the Decree implementation pollution of national waters (Stb.1985, 377) and increases the charges for polluters of national waters (both inland surface waters and marine waters). The increase aims to finance the intensified water quality policy, amongst others in pursuance of the Second North Sea Ministerial Declaration, the Rhine Action Plan and, on a national level, the Environmental Policy Programme[6] and the Third Government Report on Water Management[7]. The Decree entered into force on 1 January 1989 (Stb.1989, 597).

The Decrees of 21 December 1989 and 22 December 1989 aim to implement Directive 89/428/EEC concerning reduction of polluting waste from the titanium dioxide industry. The Decree of 21 December amends the Decree of 9 July 1975 (Stb.419), providing for a list of substances covered by the prohibition to dump these at sea, as contained in the Marine Pollution Act 1981. The Decree of 22 December implements the EEC Directive with respect to inland surface waters and provides for limit-values for sulphate and chlorides by discharges of the titanium dioxide industry. The Decree is based on Art. 1(a) of the Pollution of Surface Waters Act 1971. Both decrees entered into force on 31 December 1989 (Stb.1989, 956 and Stb.1989, 618).

International Agreements

In the 1988 survey of Minister of Foreign Affairs of treaties to which the Netherlands may become a party, but which have not yet been submitted to Parliament for approval, a number of treaties relating to the law of the sea and the marine environment were dealt with. The first category consists of treaties which are not expected to be submitted to Parliament in the near future, or with respect to which it has not been decided yet whether it is desirable to become a party. Within this category fall:

- the 1982 Convention on the Law of the Sea;

- the 1984 Protocol to the International Convention on Civil Liability for Oil Pollution Damage;

- the 1984 Protocol to the International Convention on the Establishment of an International Fund for Compensation for Oil Pollution Damage.

With respect to these two Protocols the position of the Netherlands depends on the attitude of states with rival ports.

In the category treaties with respect to which the Government has decided that it is not desirable to become a party, mention is made of the 1977 Convention on Civil Liability for Oil Pollution Damage Resulting from Exploration for and Exploitation of Sea-bed Mineral Resources. It is noted that the objectives of the Convention will be regulated in national legislation (K.II 1988-89, 20800, Ch.5

no.70).

Other Major Developments

National Environmental Policy Plan

On 25 May 1989 the National Environmental Policy Plan was presented to Parliament. Sustainable development, as defined and advocated in the Brundtland report, is the key concept in this report. The Plan consists of a comprehensive programme of environmental measures to be taken in the near future, aiming at tangible results by 2010. The Plan contains no new measures aimed at a reduction of marine pollution, but restates measures agreed on in an earlier stage, in particular within the framework of the North Sea Ministerial Conference (London, 1987) and the Rhine Action Programme. Relevant issues are, *inter alia*, a reduction of the input of nutrients, a reduction of the pollution of sediments, the introduction of an obligation for environmental impact assessment for activities in the Wadden Sea and the proposal to agree in cooperation with Belgium on an Action Programme for the Meuse and the Scheldt. Legislative action is being prepared (K.II 1988-1989, 21137, nos. 1-2).

Nature Conservation Policy Plan

Parallel to the National Environmental Policy Plan the Government also published a Nature Conservation Policy Plan. The emphasis of this policy is on terrestrial nature, but attention is also given to the needs and opportunities for nature conservation in the Netherlands' sector of the Continental Shelf of the North Sea. Pilot projects for the designation of marine protected areas in this sector of the North Sea are announced in this plan.

Third Government Report on Water Management

In August 1989, the Netherlands Government published its Third Government Report on Water Management. The main objective of water management in the Netherlands is to preserve a safe and habitable country and to develop sound water management systems based on sustainable use principles. The objectives for the North Sea and the Wadden Sea can be summarised as follows:

- healthy fish in a healthy sea;

- without external help a long and happy seal's life in the Wadden Sea;

- a source of resources and energy;

- an attraction for tourists.

According to the Report the future North Sea will be characterised by multiple use. Eutrophication is rare. The fish stocks are healthy and catches of a certain number of species are at a higher level than at present. The exploitation of oil, gas and sand deposits takes place under adequate environmental and safety conditions. The clean coast is a tourist attraction. Seals, porpoises and dolphins are regularly observed. Bird populations are stable and diverse. From the problem analysis in the Report, however, it is clear that the Netherlands' Government is far from fulfilling its objectives. Criticism has been voiced that the measures proposed by the Netherlands Government, especially in the fields of the inputs of dangerous substances and nutrients, are not sufficient to meet the objectives. As for the Rhine and the North Sea, the Report refers back to the Rhine Action Plan and the North Sea Action Plan as adopted in the international frameworks (KII 1988-89, 21250, nos. 1-2).

Exclusive Economic Zone

On 15 February 1989 the then minister for North Sea affairs sent a letter to the Netherlands Parliament containing the provisional point of view of the Netherlands' Government on the establishment of an Exclusive Economic Zone by the Netherlands. The essence of the letter contains a commitment by the Netherlands to bring in the common establishment of the exclusive economic zone in the North Sea into the negotiations in the framework of the Third North Sea Conference[8]. This has not resulted in the common establishment of EEZs in the North Sea. In the Hague Declaration it is said that the Netherlands will coordinate the action in respect of the EEZ[9].

Borcea incident

On 7 and 8 January 1988 the Romanian bulk carrier the *Borcea* lost 75 cubic metres of bunker oil (see *ACOPS Yearbook 1987-88*). This discharge caused the death of thousands of birds. On 13 June 1989 a special team published its report. The report was submitted to Parliament. The most important recommendations were:

- the Netherlands' Government should establish a special investigation team for environmental offences on the North Sea. This team should be available as soon as a disaster would occur;

- the Netherlands should take the initiative in the international frameworks to establish a reporting system for the North Sea and the mouths of rivers.

- the establishment of guidelines in respect of sampling, so that all samples are taken in such a way that they can be used as legal proof before the competent tribunals;

- the appointment of private laboratories for the purpose of analysing samples;

- the establishment of procedures which involve the Ministry of Foreign Affairs automatically so that diplomatic support can be obtained.

- The inclusion in the Netherlands' penal code of an offence which makes it punishable to pollute the environment with a penalty of two years in prison;

- to start a study into the possibility of application of the Netherlands' penal law to all environmental offences independently of the question of where the offence was committed if the offence would result in damage to the realm in Europe.

-

In March 1990 a Netherlands judge convicted the captain of the *Borcea* to three months imprisonment (Stcrt.1989, 113; K.II 1988-89, 20800 XII, nr. 75)[10].

Abbreviations

K.II: Kamerstukken Tweede Kamer (Parliamentary Papers of the Lower House, with year and number of the paper)

Trb: *Tractatenblad* (Official Journal of the Netherlands, with number)

Stb: Staatsblad (Official Journal of the Netherlands, with number)

Stcrt: *Staatscourant* (daily newspaper containing official notices and ministerial regulations).

5.2.3 Italy

During the years 1988 and 1989 the most relevant Italian enactments relating to the various aspects of the protection of the marine environment are the following.

National Legislation

Water Pollution

Decree of the Minister for the Environment no. 122 of 16 February 1988; GU no. 90 of 18 April 1988. Provisions for the granting of State contributions for projects against water pollution.

Rules for the Exploration and Exploitation of Deep Seabed Mineral Resources

D.P.R. no. 200 of 11 March 1988; GU no. 139 of 15 June 1988. This Decree contains regulations concerning the exploration of seabed mineral resources. It regulates the terms and the conditions for granting, extending, suspending and revoking permits for the exploration of deep-seabed mineral resources. In order to ensure the consistency of exploiting activities with an adequate protection of the marine environment, an environmental impact statement is to be enclosed within the applications for permits. The holder of a licence is requested, *inter alia*, to monitor the consequences of his activities on the marine environment and to draft an annual report. The Decree provides also for controls and inspections as well as for penalties against infractions.

Environmental Impact Assessment

Law no. 67 of 11 March 1988; GU Suppl. to no. 61 of 14 March 1988. This Law, relating to the 1988 State budget, provides *inter alia* for the allocation of 10 billion liras (US $8.1 million) for the drawing up of the project on the permanent crossing of the Strait of Messina inclusive of a prior environmental impact assessment.

Decree of the President of the Council of Ministers no. 377 of 10 August 1988; GU no. 204 of 31 August 1988. This Decree, issued in accordance to Art. 6 of Law no. 349 of 8 July 1986, concerning the establishment of the Ministry for the Environment, implements EEC Directive 85/337 of 27 June 1985. It is supplemented by technical rules on the drawing up of environmental impact studies Decree of the President of the Council of Ministers of 27 December

1988; GU no. 4 of 5 January 1989.

Bathing Waters
Law-decree no. 155 of 14 May 1988; GU no. 113 of 16 May 1988, converted. with modifications into Law no. 271 of 15 July 1988; GU no. 166 of 16 July 1988). This relates to the implementation in Italy of EEC Directive no. 76/160 on the quality of bathing waters.

Protection of Areas of Special Environmental Interest
Regional Law of Emilia Romagna no. 27 of 2 July 1988; GU S.S. no. 43 of 22 October 1988. Institution of the regional park of the delta of the River Po.

Decision of the Interministerial Committee for Economic Planning (CIPE) of 5 August 1988; GU Suppl. to no. 215 of 13 September 1988. Establishment of a new national marine park in the Gulf of Orosei, Sardinia.

Decree of the Minister for Merchant Marine no. 402 of 20 August 1988; GU no. 214 of 12 September 1988. Establishment for a period of three years of a biological protection zone in the marine area included between the perpendiculars to the coastline connecting Tor Paterno' to Villa Campello.

Decree of the Minister for Merchant Marine of 1 September 1988; GU no. 210 of 7 September 1988. Widening from 500 to 1,000 metres from the coast of the zone of biological protection around the Island of Montecristo, already established for the preservation of the monk seal (*Monachus monachus*) by the Ministerial Decree of 2 April 1981.

Decree of the Councillor for the Cultural Heritage of the Region of Sicily of 30 September 1988; GU no. 63 of 16 March 1989. Declaration of public interest of the basin of the old harbour Porto Grande of Siracusa, owing to its historical and naturalistic importance.

Decree of the Minister for Merchant Marine of 12 July 1989; GU no. 175 of 28 July 1989. The purpose of this Decree is to protect marine areas of archaeological, artistic and historical interest.

Decree of the Minister for the Environment of 21 July 1989; GU no. 177 of 31 July 1989. The Decree contains rules for the provisional bordering and measures for the protection of the National Park of the Tuscan Archipelago.

Law no. 305 of 28 August 1989; GU no. 205 of 2 September 1989. This Law, relating to the triennial programme for the protection of the environment, contains *inter alia* new provisions for the establishment of 'areas of high risk of environmental crisis'.

Decree of the Minister for Merchant Marine of 6 September 1989; GU no. 213 of 12, September 1989. Institution of a zone of biological protection around the Island of Pianosa. The aim of this Decree is to foster the reproduction and the growth of marine species of economic importance within an area up to 1,500 metres from the shore of the Island of Pianosa. According to Art. 2, the transit of certain vessels is prevented in the area.

Decree of the Minister of the Environment; GU no. 295 of 19 December 1989. Establishment of the marine natural reserve 'Islands Tremiti'. This reserve includes the coastal area surrounding the islands of S. Domino, S. Nicola, Ca-

prara and Pianosa up to the isobath of 70 metres. Different measures of protection are provided for within Zone A (integral reserve), Zone B (general reserve) and Zone C (partial reserve).

Waste
Ordinance of the Minister for the Coordination of Civil Protection no. 1471/FPC of 26 May 1988; GU no. 127 of 1 June 1988. Appointment of a commissary *ad acta* to carry out the operations of control and elimination of the industrial wastes stowed on the motor-ship *Zanoobia* moored at the harbour of Massa Carrara.

Ordinance of the Minister for the Coordination of Civil Protection no. 1500/FPC of 8 July 1988; GU no. 161 of 11 July 1988. Control and elimination of the noxious industrial wastes stowed on the motor-ship *Zanoobia*.

Ordinance of the Minister for the Coordination of Civil Protection no. 1508/FPC of 19 July 1988: GU no. 170 of 21 July 1988. First provisions to meet the situation consequent to the presence of toxic wastes of asserted Italian origin in Nigeria.

The crown-of-thorns starfish sometimes reaches 'plague' proportions, and lays bare large areas of coral by eating the living tissue. This reduces diversity of life on the reef which may then break up due to erosion. Some believe over-collection of Triton, a natural predator to the crown-of-thorns, for its attractive shell has contributed to the proliferation of this destructive starfish. In Australia this starfish is being closely researched and monitored (see 5.3.6).Photo: D George

Law-decree no. 397 of 9 September 1988; GU no. 213 of 10 September 1988, converted into Law no. 475 of 9 November 1988; GU no. 289 of 10 December 1988. Urgent provisions for the elimination of industrial wastes. This decree contains measures regulating *inter alia* the transfrontier movement of wastes produced in Italy, the restoration of areas polluted by wastes, the collecting, recycling and elimination of containers and emergency interventions.

Decree of the Minister for the Environment of 15 September 1988; GU no. 218 of 16 September 1988. Declaration of the state of emergency according to Art. 8 of Law-decree no. 397 of 9 September 1988, with regard to the elimination of toxic industrial wastes of Italian origin that the ships *Karin B* and *Deep Sea Carrier* carried from Nigeria (see 2.2.2).

Decree of the President of the Council of Ministers of 16 September 1988; GU no. 218 of 16 September 1988. Measures for the location of places and the individuation of methods to stock and control the elimination of toxic industrial wastes of Italian origin that *Karin B*, *Deep Sea Carrier* and other ships carried from Nigeria and Lebanon.

Ordinance of the Minister for the Coordination of Civil Protection no. 1557/FPC of 16 September 1988; GU no. 220 of 19 September 1988. Notification of errata-corrige in GU no. 236 of 7 October 1988. Exceptional measures providing for preliminary operations aimed at the stocking and the subsequently definitive elimination of noxious and toxic wastes that the ship *Karin B* carried from Nigeria to Leghorn.

Ordinance of the Minister for the Coordination of Civil Protection no. 1558/FPC of 16 September 1988; GU no. 220 of 19 September 1988. Notification of errata-corrige in GU no. 237 of 8 October 1988. Exceptional measures concerning the transport of noxious and toxic substances unloaded by the ship *Karin B*, from the harbour of Leghorn to the area appointed for a temporarily controlled stocking. The ordinance also contains rules providing for the elimination of the substances as well as for the restoration of the area affected by stocking.

Ordinance of the Minister for the Coordination of Civil Protection no. 1561/FPC of 21 September 1988; GU no. 231 of 1 October 1988. Further exceptional measures concerning the definitive elimination of noxious and toxic substances carried by the ship *Karin B*.

Ordinance of the Minister for the Coordination of Civil Protection no. 1563/FPC of 26 September 1988; GU no. 229 of 29 September 1988. Notification of errata-corrige in GU no. 236 of 7 October 1988. Measures for the financing of the expenses required for the elimination of the industrial toxic wastes that the *Karin B*, *Deep Sea Carrier* and other ships carried from Nigeria and Lebanon.

Decree of the President of the Council of Ministers of 28 September 1988; GU no. 229 of 29 September 1988. Provisions concerning the mooring, the cataloguing and the provisional controlled stocking of the industrial wastes carried, by the ship *Deep Sea Carrier.*

Ordinance of the Ministers for the Coordination of Civil Protection and for the Environment of 8 October 1988; GU no. 242 of 13 October 1988. Exceptional measures on the preliminary activities for the unloading, stocking and disposal of the waste shipped from Koko, Nigeria, to Ravenna.

Decree of the Minister for the Environment no. 457 of 22 October 1988; GU no. 256 of 31 October 1988. Rules on exportation and importation of waste.

Law-Decree no. 527 of 14 December 1988; GU no. 292 of 14 December 1988, converted into Law no. 45 of 10 February 1989; GU no. 35 of 11 February 1989. Urgent measures on disposal of industrial waste.

Decree of the Minister for the Environment of 26 January 1989; GU no. 33 of 9 February 1989. Declaration of an emergency situation relating to the disposal of waste of Italian origin coming from Lebanon and Nigeria.

Decree of the President of the Council of Ministers of 27 January 1989; GU no. 32 of 8 February 1989. Measures for the provisional stocking in the port of La Spezia of industrial waste unloaded from the ship *Jolly Rosso* coming from Lebanon.

Ordinance of the Minister for the Coordination of Civil Protection of 6 February 1989; GU no. 32 of 8 February 1989. Exceptional measures concerning waste unloaded from the ship *Jolly Rosso.*

Ordinance of the Minister for the Coordination of Civil Protection of 28 February 1989; GU no. 58 of 10 March 1989. Measures for the verification of the fitness of the port of Taranto for the docking of the ship *Deep Sea Carrier*, carrying waste from Nigeria.

Ordinance of the Minister for the Coordination of Civil Protection No. 1682/FPC of 8 April 1989; GU no. 91 of 19 April 989. Amendments to Art. 3 of Ordinance no. 1649/FPC of 6 February 1989. Containing exceptional measures concerning waste unloaded from the ship *Jolly Rosso.*

Decree of the Minister for the Environment of 26 April 1988; GU 128 of 3 June 1989, integrated by Decree of the Minister for the Environment of 28 June 1989; GU no. 159 of 10 July 1989. Discipline of the guarantee (garanzia fidejussoria) for transfrontier shipping of waste.

Decree of the Minister for the Environment of 26 April 1989 GU no. 135 of 12 June 1989. Institution of the national register of special wastes.

D.L. no. 25 of 30 June 1989; GU no. 152 of 1 July 1989. It provides, *inter alia*, for the extension of certain terms set forth under Law-decree no. 397 of 9 September 1988 containing urgent measures for the disposal of industrial waste.

Decree of the Minister for the Coordination of Civil Protection of 8 July 1989; GU no. 161 of 12 July 1989. Provisions for the mooring, listing and provisional stocking of the industrial waste carried by the ship *Deep sea Carrier.*

Ordinance of the Minister for the Coordination of Civil Protection no. 1764/FPC of 8 July 1989; GU no. 161 of 12 July 1989 and errata-corrige in GU no. 172 of 25 July 1989. Exceptional measures concerning operations that are to be carried out in the harbour of Leghorn and in the Region Tuscany for the disposal, the analysis, the provisional controlled stocking and the definitive elimination of the industrial waste coming from Nigeria and carried by the ship *Deep Sea Carrier.*

Ordinance of the Minister for the Coordination of Civil Protection no. 1778/FPC of 11 August 1989; GU no. 196 of 23 August 1989. Further measures aimed at the definitive stocking of the industrial waste carried by the ship *Jolly*

Rosso. This Ordinance has been amended by Ordinance of the Minister for the Coordination of Civil Protection no. 1821/FPC of 9 November 1989 (GU no. 271 of 20 November 1989).

Ordinance of the Minister for the Coordination of Civil Protection no. 1790/FPC of 3 September 1989; GU no. 225 of 26 September 1989. Further measures concerning interventions aimed at the definitive elimination of industrial wastes carried by the ship *Jolly Rosso.*

Eutrophication

Decision of 18 October 1988 of the Permanent Inter-regional Conference for the Protection of the Basin of the Po River; GU no. 265 of 11 November 1988. This Decision contains measures against eutrophication of the Adriatic Sea.

Ordinance of the Minister for the Coordination of Civil Protection no. 1697/FPC of 19 April 1989; GU no. 97 of 27 April 1989. This Ordinance contains urgent measures for the collection and the recycling of algae that have proliferated in the lagoon of Venice.

D.L. no. 227 of 13 June 1989; GU no. 138 of 15 June 1989, converted into Law no. 283 of 4 August 1989; GU no. 185 of 9 August 1989. The coordinate text is published in GU no. 219 of 19 September 1989. This Law contains urgent measures against phenomena of eutrophication of the coastal waters of the Adriatic Sea. To this aim, it regulates the collection and the elimination of algae and organic materials, the adaptation of coastal depuration plants, the reduction of effluents into the sea and the monitoring of eutrophication in the Adriatic Sea.

Decree of the Minister for the Environment no. 295 of 22 June 1989; GU no. 194 of 21 August 1989 and errata-corrige in GU no. 213 of 12 September 1989. This Decree contains rules for the financing of interventions aimed at the restraint of phenomena of eutrophication provided for under Art. 10 of Law-decree no. 667 of 25 November 1985, converted into law with modifications by Law no. 7 of 24 January 1986.

Decree of the President of the Council of Ministers of 27 July 1989; GU no. 257 of 3 November 1989. Putting into effect a monitoring system and elaboration of environmental data about the Adriatic Sea.

Ordinance of the Minister for the Environment of 11 August 1989; GU no. 193 of 19 August 1989. Financing of the interventions aimed at restoring and restraining the negative effects of mucilage to be carried out by the Regions Friuli-Venezia Giulia, Veneto, Emilia Romagna, Marche, Abruzzo, Molise, Puglia from 13 June 1989.

Ordinance of the Minister for the Environment of 11 August 1989; GU no. 193 of 19 August 1989. This Ordinance, amended by Ordinance of the Minister for the Environment of 18 August 1989 (GU no. 194 of 21 August 1989) provides for urgent intervention against mucilage along the Adriatic coast.

Ordinance of the Minister for the Environment of 11 August 1989; GU no. 193 of 19 August 1989 and Ordinance of the Ministry for the Environment of 22 September 1989; GU no. 250 of 25 October 1989. These Ordinances contain provisions concerning experimental interventions for the restraint and the mitigation of the effect of mucilage.

Law no. 424 of 30 December 1989; GU no. 6 of 9 January 1990. This Law regulates the allocation of State contributions in order to support the recovery of tourist and fishing activities in the areas affected by the exceptional phenomena of eutrophication and mucilage that occurred in 1989.

Institution of the Coast-Guard

Decree of the Minister for Merchant Marine of 8 June 1989; GU no. 146 of 24 June 1989. The activities of the Coast-Guard consist, *inter alia,* in the prevention and the abatement of pollution affecting maritime areas under the Italian State jurisdiction.

Other Legislation

Safety of Life at Sea

Decree of the Minister for Merchant Marine of 2 April 1988; GU no. 217 of 15 September 1988. Qualifications and examination programme to obtain the Certificate of Proficiency in survival craft in conformity with the 1974 SOLAS Convention and successive amendments and with the 1978 STCW Convention.

Rules for the Protection of the Soil

Law no. 183 of 18 May 1989; GU no. 120 of 25 May 1989. According to Art. 3 of this Law, the Public Administration undertakes to program and plan interventions aimed, *inter alia*, to protect and govern rivers, watercourses and wetlands and to defend coasts from the invasion and the erosion of the maritime waters.

Mediterranean Action Plan

Law no. 345 of 20 October 1989; GU no. 248 of 23 October 1989. Refinancing of the Mediterranean Action Plan (1988-1989).

Relating to MARPOL 73/78

The Decree of the Minister for Merchant Marine no. 289 of 8 March 1988 (GU no. 173 of 25 July 1988 contains measures for the implementation of some technical assessments provided for by rule 8 of Annex II of the 73/78 MARPOL Convention.

The Decree of the Minister for Merchant Marine of 3 December 1988 (GU no. 15 of 19 January 1989) contains provisions on the statement of fitness to Annex I of MARPOL 73/78 Convention.

International Agreements

1987 Kuwait-Italy *Agreement* on the access to and the treatment of ships in the port of the two States, entered into force for Italy on 18 September 1988 (Gu Suppl. to no. 2 of 30 January 1989)

1980 Convention on the Conservation of Antarctic Marine Living Resources implemented by Law no: 17 of 2 January 1989 (GU Suppl. to no. 23 of 28 January 1989), entered into force for Italy on 28 April 1989 (GU no. 93 of 21 April 1989).

1979 Hamburg Convention on Maritime Search and Rescue implemented by Law no. 147 of 3 April 1989 (GU Suppl. to no. 97 of 27 April 1989), entered into force for Italy on 2 July 1989 (GU no. 184 of 8 August 1989).

Jurisprudence

On 30 March 1989, the decision rendered by the Tribunal of Messina on 30 July 1986 on the *Patmos* case (see *ACOPS Yearbook 1987/88*, p. 56) was reversed by the Court of Appeal of Messina. This Court accepted *inter alia* the claim for environmental damage lodged by the Italian Ministry for Merchant Marine against the decision of the court of first instance. The judgment contains some interesting considerations about the qualification and the assessment of environmental damage as well as the open acknowledgement of the right or a state to claim reparation for the injuries suffered by its territorial waters and marine resources as a consequence of pollution.

The Court declares that the claim for environmental damage to the marine environment maintained by the Italian Government is consistent with the definition of 'pollution damage' laid down in the 1969 Brussels Convention on Civil Liability for Oil Pollution Damage. In the opinion of the Court, this definition covers an injury caused to the coastline and to other Coastal States' related interests of environmental character as the conservation of marine living resources, flora and fauna. In addition, in contrast with the decision of the Court of first instance, the Court of Appeal maintains that the ecological damage has an economical character, even if the injury affects immaterial goods devoid of pecuniary value according to market prices. The diminution of the economic value consists in the reduced enjoyment of the environment.

Finally, in the opinion of the Court, the State, as representing the national community, is entitled to claim reparation for environmental damage. This claim is grounded neither on the costs incurred by the State in order to restore the injury caused by pollution, nor on any other alleged economic loss, but on its function of protecting the interests of the national community in preserving the ecological and biological balance of the territory, the territorial sea included. The Court concludes that the Minister for Merchant Marine is entitled to claim reparation in respect of damage to the marine environment under Art V. and VI of the 1969 Brussels Convention. The concrete determination of damages is remitted to a definitive judgment.[11]

Abbreviations

D.L.   Decreto Legge (Decree Law)
D.P.R. Decreto del Presidente della Repubblica (Decree of the President of the Republic)
GU   *Gazzetta Ufficiale della Repubblica Italiana* (Official Journal of the Italian Republic)
S.S.   Special Series

Suppl. Supplement

5.3 SELECTED GOVERNMENTAL AND STATUTORY BODIES

5.3.1 United Kingdom

Overseas Development Administration (ODA)

The Overseas Development Administration (ODA) formulates and carries out British policies for the management of the aid programme. The basic purpose of the aid programme is to promote sustainable economic and social progress and alleviate poverty in the least developed countries. In 1988 over 130 countries received support under the aid programme.

In the current financial year (1989/90) the overseas aid budget stands at £1,500 million. About 60% of this is expected to be spent bilaterally and is focused on the poorest, Commonwealth and other countries with which Britain has close ties. The balance of approximately 40% of aid is channelled through multilateral organisations like the World Bank and the European Community programmes.

The ODA has long held the view that sustainable development must be linked to concern for the environment. If it is not, the ability of the natural environment to sustain and renew the many resources on which mankind depends will continue to be under threat. All ODA programmes and projects therefore take account of the need to use natural resources wisely as part of good agricultural and industrial practices, and it is the responsibility of all of those concerned with the aid programme to ensure that projects supported under it are environmentally sound.

Amongst its many activities the ODA has collaborated over a number of years with the University of Diponegoro in Indonesia on the sustainable management of its coastal fisheries. It has also agreed funding with the Government of Indonesia for research into improvements to the quality of the Kali Sunter river and its catchment area in and around Jakarta.

ODA has also agreed with the Government of Morocco to fund collaborative research for the development of a predictive model of oil pollution movement to be used as an emergency tool for coastal management and protection.

As part of the aid programme the ODA gives over 6,500 training awards a year to enable students from developing countries to to study in the United Kingdom. It also supports many more training awards in third countries. There is a need to promote in developing countries a better understanding of environmental issues like the degradation and pollution of the marine environment and ODA therefore stands ready to support such training if developing country governments give it a high priority within their programmes.

In China the ODA is assisting with a study of the feasibility of discharging wastewater, both industrial and municipal, into coastal waters at Jiao Zhou Bay, in Shandong/Jaingsu provinces. The study will verify standards of methods, and devise and introduce biological ways of monitoring pollutant impact on the marine environment.[12]

Nature Conservancy Council (NCC)

The Nature Conservancy Council is the statutory body responsible for advising the UK Government on nature conservation in Great Britain. Its work includes the selection, establishment and management of National Nature Reserves (NNR); the selection and management of Marine Nature Reserves (MNR); the identification and notification of Sites of Special Scientific Interest (SSSI); the provision of advice and the dissemination of knowledge about nature conservation; the support and conduct of research relevant to these functions. NCC also has a duty in the discharge of these functions to take account of actual or possible ecological change.

NCC's responsibilities for nature conservation are statutorily defined as meaning the conservation of flora, fauna or earth science features in Great Britain. Although NNRs and SSSIs mainly include the shore down to the mean low water mark. MNRs can be established out to the limits of the territorial seas and the advisory function extends beyond that into the Exclusive Economic Zone (EEZ). NCC also has an advisory role to Government on international conservation matters, for example those international conventions and directives which have a bearing on coastal and marine conservation. In particular:

- the UK is a party to the Ramsar Convention under which Contracting States are required to designate and protect wetlands of international importance and to formulate and implement their planning so as to promote the wise use of wetlands in their territory. At present the definition of wetlands under Ramsar include marine areas where the water depth does not exceed 6m.

- as a member of the European Community, the UK is bound by the EC Council Directive on the Conservation of Wild Birds. Member States are required to take measures to preserve a sufficient diversity of habitats or all species of wild birds in order to maintain their populations at an ecologically and scientifically sound level. Areas that are particularly sensitive for certain groups of birds are declared Special Protection Areas (SPA) and appropriately protected.

A number of coastal areas are now identified and protected as Ramsar sites or SPAs (15 & 20 sites respectively) and further proposed sites are being considered by Government.

NCC's advisory work in the marine environment includes close liaison with:

- the Department of Energy on oil and gas exploration and development licences as well as the implications of the introduction of alternative energy production such as tidal barrages;

- oil companies on the nature conservation aspects of their oil pollution contingency plans;

- the Marine Directorate of the Department of Transport (Marine Pollution Control Unit), the Ministry of Defence (Navy) and voluntary nature conservation bodies on oil pollution;

- the Department of the Environment on aggregate dredging and on the Ministerial North Sea Conferences;

- the Ministry of Agriculture, Fisheries and Food on dumping at sea and the introduction of alien species for shellfish farming;

- the Crown Estate Commissioners, Planning Authorities, and River Purification Boards on the potential pollution of Scottish sea lochs from the intensive farming of caged salmon;

- the River Purification Boards and National Rivers Authority on the discharge of sewage and other potential pollutants to estuaries and coastal waters.

The of NCC's advice depends on the application of technical knowledge gained by research, notably surveys, to describe the character, distribution and abundance of features, measurements of changes in time and location, and studies of causal and other relationships. In relation to the coasts and marine environment, NCC is building up a comprehensive information base of the physical features and the associated biological communities and species. Some of the current major projects which are contributed to the nature conservation information base are described below.

Coastwatch

This project has the objective of completing a survey of all the coastal and intertidal habitats and selected human activities around the whole British coast by direct observation. At October 1988, about 60% of the 19,000km of the British mainland coast and islands had been surveyed, with adequ-

This small community on underwater boulders in a Scottish loch includes sea squirts, sea firs, feather star, sea urchin and tube-worms. Sea lochs are currently being surveyed as part of the NCC Marine Nature Conservation Review of British marine ecosystems (see 5.3.1). Photo: NCC

ate resourcing Coastwatch should be completed in 1991. This basic habitat information is used in part to focus the more detailed coastal and marine site surveys which are described in this section (see also 6.2.4).

Seasearch

Seasearch is the subtidal equivalent to Coastwatch in so far as the objective is to map the basic benthic habitats and record any activities that appear to be damaging these habitats. However, as the direct observations require the use of divers, which limits the extent of the work quite considerably, the intention is to select the areas surveyed in relation to the programme of more detailed site surveys which follow, rather than attempt to cover all British waters.

The Marine Nature Conservation Review

In 1986, NCC began work on a major descriptive account of the marine flora and fauna of Britain's intertidal zone and the shallow subtidal area. This *Marine Nature Conservation Review* (MNCR) will draw together the extensive existing knowledge of marine habitats, communities and species and new surveys have been initiated to supplement and validate existing information. The objectives are to evaluate the conservation importance of marine habitats and the associated communities and species, with a view to applying protective measures as appropriate. The Review will also provide information to support more general measures to minimise the effects of pollution or industrial developments in the wider marine environment. Commencing in 1990, the results of the Review will be published sequentially as the assessment of each coastal sector and major habitat type is completed.

Seabird surveys

NCC's Seabirds at Sea research programme was started in 1979 and was initially designed to find out which areas of the North Sea were important to seabirds, particularly in relation to the potential threat of spilled oil from the exploration and transport of oil in the North Sea basin. The programme continued with the eventual aim of covering the whole of Britain's coastal water. As seabirds are particularly dependent on the offshore marine environment for much of their lives, they may be particularly vulnerable to disturbance by industrial activities in their feeding areas including effect on their food organisms.

Work has also now begun on a full-scale survey of breeding seabirds for the whole coast of Britain. This will update the results of the 1969/70 'Operation Seafarer' surveys and will identify areas where nearshore or onshore activities would potentially affect important seabird breeding colonies.

The Estuaries Review

The Estuaries Review is a new initiative by NCC with the purpose of preparing an overall nature conservation strategy for British estuaries. This Review is based on the collation and assessment of existing information, depending to a great deal on the results of other reviews and major survey projects described in this section. The basis of this Review will be a scientific document which will describe the features of nature conservation importance of British

estuaries and the types of activities and changes apparently affecting estuarine ecosystems as well as identifying activities and developments which are consistent with maintaining or enhancing natural estuarine ecosystems. The focus on the opportunities for maintaining and enhancing estuaries will help fulfil the UK's commitment to the 'wise-use' of wetlands as a signatory to the Ramsar Convention. Completion of the draft scientific document is scheduled for early 1990, when it will be circulated as a consultation document.

The North Sea Coastal Directory

In preparation for the Third Ministerial Conference on the North Sea, NCC has been commissioned by the Department of the Environment to produce a directory of the information held on Britain's North Sea coastal, estuarine, nearshore an offshore environment including biological, geological and geomorphological features. An assessment will also be made of the various activities that are apparently degrading the environment including dredging and spoil dumping. This review will draw on information already held by NCC and will be completed in draft in late 1990. A similar study is now underway for the Irish Sea.

Irish Sea Study Group

The main objective of the group is to gain an understanding of the impact of man's activities on the sustainable yield and long term environmental health of the Irish Sea. To accomplish this four specialist groups have been established to consider nature conservation, waste disposal, exploitation of living marine resources and planning, development and management. NCC are undertaking the coordination and collation of the nature conservation report which will provide a description and evaluation of the natural resources of the Irish Sea, as assessment of current and likely threats to those resources and a series of recommendations for the future management of the Irish Sea. The reports of all the specialist groups are to be completed in April 1990 and presented to a conference in the Isle of Man in October 1990.

Coastal Oil Spill Contingency Maps and Schedules

NCC has been commissioned by the Department of Transport's Marine Pollution Control Unit (MPCU) and British Petroleum International, to revise the coastal oil pollution contingency maps and schedules which are used by MPCU, local authorities, oil companies and NCC as a basis for deciding how to respond to oil spilled in coastal waters or beached. The maps and schedules record the information of sensitive habitats, communities, species and statutorily protected sites and other nature reserves for the whole British coast and advise on the oil spill treatment most appropriate with regard to its location and the habitat affected. This review depends on the information already held by NCC and is due for completion in 1990.

NCC is presently involved in the consultation over the Environmental Protection Bill and has published its detailed response to Government on all aspects of the proposed new legislation. In respect of the future role of Her Majesty's Inspectorate of Pollution and Her Majesty's Industrial Inspectorate for Scotland NCC urges that adequate resources will be made available for monitoring, enforcement and an effective integrated inspectorate. The same Bill makes proposals to replace NCC by three new agencies for England, Scotland and Wales. Those in Scotland and Wales are to incorporate the Countryside Commission activities whereas the NCC and the Countryside Commission are, at least initially, to remain separate entities in England. There is also the intention to set up a central GB Joint Committee to Provide advice to Government on conservation issues with a GB or International dimension and establish common standards for research, monitoring and data analysis which is particularly pertinent to marine matters[13].

Warren Spring Laboratory (WSL)

Warren Spring Laboratory is the UK Government's centre of expertise on environmental research. Its programme portfolio covers air pollution, pollution abatement, waste treatment and recycling as well as research into all aspects of response to oil and chemical spillages at sea. Much of the marine pollution work is carried out for the Department of Transport's Marine Pollution Control Unit but customers also include the Department of Energy, the European Commission and industrial companies. Warren Spring Laboratory developed many of the techniques on which response procedures in the UK are now based: aerial application of dispersant, mechanical recovery, remote sensing and beach cleaning procedures.

In 1989 extensive use was made of the laboratory's research vessel, *RV Seaspring*, to provide experimental verification of mathematical models which are being developed for both MPCU and the EEC to predict the fate of spills. Trials involved the discharge of controlled quantities of oil, samples being taken at regular intervals for analysis of both oil and the water column. Airborne remote sensing using ultraviolet and infrared line scanning systems augmented the analytical data to provide the overall picture of the oil fate. The airborne system was used in preliminary experiments to assess its ability to detect chemicals on the sea surface and also during a major collaborative trial with other North Sea states when a test spill of Middle East crude oil was monitored over several days.

The airborne remote sensing technique has been used on a regular basis for the Department of Energy to monitor pollution from offshore installations. The regular, but random nature of the flights has ensured the continued vigilance of the offshore operators in minimising their effects on the marine environment. Also for the Department of Energy, a programme of work has started which will study the factors contributing to the formation of stable oil emulsions from underwater rather than surface releases.

Work was completed during the year which examined techniques for removing oil from soft, shingle beaches and this is being incorporated into a guidance manual for widespread publication. Attention is currently being turned to the problems associated with oil clean-up from soft sediments and salt marsh terrains[14].

5.3.2 France

Note on the Activities which France undertook in 1989 in the Field of Protection of the Marine Environment

In March 1989, the *Exxon Valdez* hit a reef off the coast of Alaska, causing the most serious oil pollution which the USA ever experienced. In December 1989, *Kharg 5* exploded off the coast of Morocco; coastal pollution was avoided by a hair's breadth. On 19 March 1989, *Perintis* sank off the coast of France and lost its toxic cargo of lindane; the considerable efforts made to recover the containers were unsuccessful.

These three incidents stirred the attention of media and also reminded competent national and international authorities of the precariousness of the environment, despite laws adopted to prevent and combat marine pollution resulting from accidents. They also highlighted the necessity to intensify efforts in this field, especially by developing the resources to be put at the disposal of the Centre de Documentation Recherches et Experimentation (CEDRE) relating to accidental marine pollution.

In particular, the *Perintis* incident led France to call upon the appropriate international authority (the Internatiol Maritime Organisation) to take action with a view to improving conditions of maritime transport of hazardous substances so that the authorities concerned would have at their disposal, with minimum delay and in greatest possible detail, information relating to cargoes known to be dangerous and which are carried on board ships crossing maritime zones which have dense traffic.

Thought was also given to adapting the internal legal framework to requirements and possibilities offered by international law in maritime areas relevant to the sovereignty of France and its jurisdiction.

The colloquy 'Ecology and Power' which was held in Paris in December 1989 and was opened by President Mitterand, at the initiative of Mr Brice Lalonde, allowed one to appreciate the interministerial character of Government action relating to the environment. This audit of Government action should commence in 1990 on the application of an environmental programme.

Clean-up

Since 1984, in the field of clean-up (decontamination) the agencies opted for a contractual policy envisaging definition of study and work programmes. This applied both to collective and autonomous cleanup, with financing from both parties. This policy developed around three principal types of contract:

- contracts of agglomeration concerning large communities with a poor record of depollution;

- river contracts relating the the application of programmes and global management of a water course. Some 25 contracts have been signed to date, three are in the process of being signed whilst 19 remain at the feasibility stage;

- contracts of State/Regional plan, relating to development programmes within a region which include the aspect of decontamination.

One could quote, for example, the contracts on decontamination of regions on the western littoral zone and the contract relating to the Seine river. A specific programme for Brittany is envisaged for the year 1990.

Degradation of the quality of waters in Brittany is due to disposal of domestic waste water, pollution of agricultural origin, emission of industrial effluents and particularly the development of intensive agriculture which have led to an increase of nitrates in the underground and surface waters. This degradation has become extremely worrying in view of the water for drinking, and also for industry and for farming. This has led the State, in close collaboration with departments, the region and Agence de Bassin Loire-Bretagne to undertake a seven-year (1990–96) coordinated action and intervention plan in the selected basins which were chosen in order to accelerate decontamination of littoral local authorities and to fight pollution caused by agriculture. In addition, new types of contracts are being used (for example in the region of Somme) which include an intercommunity decontamination scheme and a scheme assessing the economic value of the sea[15].

5.3.3 Sweden

Swedish International Development Authority (SIDA)

SIDA, the Swedish International Development Authority, administrates the major part of the Swedish Government's assistance to developing countries. The assistance, through SIDA, is concentrated on 17 'recipient' countries of which eleven are situated in Africa, five in Asia and one in Latin America. SIDA also supports regional cooperation programmes and specific programmes in fields such as research and development, environmental protection, energy, population and health. In addition, SIDA provides emergency and humanitarian assistance.

The overall objective of Swedish assistance to developing countries is to improve the standard of living of poor people. The assistance should contribute to economic growth, economic and social equalisation, economic and political independence and democratic development in the recipient countries. The environmental aspects have become increasingly important in the development process over the past few years and, in future, more emphasis will be put on sound management of the natural resources and protection of the environment in the developing countries.

In the marine sector SIDA supports programmes, on a bilateral basis, for development of the artisanal fisheries in Guinea-Bissau, Angola and Mozambique as well as a Fisheries Survey Programme in Angola with the objective of strengthening the nation's ability rationally to utilise and protect her fisheries resources.

SIDA supports two FAO executed fishery programmes, one concerned with the development of small-scale fisheries in the Bay of Bengal, the other experimenting with aquaculture techniques for local community development in southern Africa.

Since 1972 SIDA has given financial support to an FAO programme on the protection of the marine environment. A number of seminars and training courses have been held, and participants from more than 60 developing countries have been trained in theoretical and practical aspects of environmental protection, analyses and surveys. As a follow-up to this programme, which has now been terminated, SIDA is preparing for further cooperation with FAO in the field of marine pollution in developing countries.

Since 1977 SIDA has been supporting the IMO programme for the protection of the marine environment. This support includes an advisory service, fellowships, seminars, training courses, expert meetings and preparation of information materials. The contents of the programme cover the role of maritime transport in regard to oil and chemical pollution as well as how legislation and technical solutions can be used to prevent or limit the extent of the pollution.

SIDA is presently funding part of the IMO Global Programme for the Protection of the Marine Environment 1990–92.[16]

5.3.4 Federal Republic of Germany

Federal Environmental Agency (Umweltbundesamt)

The Federal Environmental Agency (Umweltbundesamt) participated in the preparatory work to the Third International Conference for the Protection of the North Sea and prepared one of the national documents presented at the Conference. Furthermore, it actively collaborates in the working groups set up under the Paris Convention, the Oslo Convention, the London Dumping Convention and in the Helsinki Commission. Further activities in the field of marine pollution are concerned with chemical accidents on seagoing vessels and combating the illegal dumping of waste oil from ships (MARPOL). In addition to the work involving international bodies, several research projects are carried out in collaboration with other institutions to investigate the Wadden Sea region with respect to its flora and fauna. The knowledge gained thereby is to be used as a basis for the implementation of remedial measures.

In the FRG, the task of monitoring the North and Baltic Seas with respect to marine pollution is performed by the German Hydrographic Institute. Monitoring is directed at both sea water and sediment and includes chemical/physical parameters as well as biological parameters.

Biological monitoring is performed by the German Hydrographic Institute in collaboration with the Biological Institute Helgoland, the Federal Research Institute for Fishery, the Institute of Marine Science of Kiel University as well as with numerous other Federal and land institutes.

These are but a few examples of the activities of Federal institutions in the area of marine pollution.

The aim of the Federal Republic of Germany is to reduce pollution at source by the application of best available technology in order to reduce the pollution of the North and Baltic Seas (for example reduction of the input of pollutants via rivers by at least 50% from 1985 to 1995)[17].

5.3.5 Canada

International Centre for Ocean Development (ICOD)

The small island and coastal states that were among the chief beneficiaries of the 1982 United Nations Convention on the Law of the Sea were also among the world's least developed nations. ICOD was established in 1985 as Canada's response to their need effectively to develop and manage new ocean resources.

ICOD programmes emphasise the transfer of knowledge and skills rather than capital assistance and focus on four regions where the development of ocean resources can have a significant impact: the Caribbean Basin, the South Pacific, the South and West Indian Ocean, and West Africa. Canadian ocean experts work with their counterparts in developing nations on discrete projects which often fall below the threshold of the major donor agencies.

ICOD projects are grouped under seven sectoral themes: integrated ocean management and development; fisheries management and development; mariculture; coastal development and management; non-living resource management and development; marine transportation and ports management; and marine environmental conservation. To date, about 200 bilateral, regional, and global projects proposed by developing nations have been approved and set in motion

Programmes related to marine pollution: 1989–90

In many developing countries, marine pollution may directly affect the sustainable development of ocean resources. As a development assistance organisation with a focus on ocean resources, ICOD has identified marine environmental protection as a central concern and has adopted it as one of the seven primary sectoral themes eligible for ICOD funding. Projects directly related to this theme, such as pollution studies, the development of marine conservation strategies, and the management of marine parks form a small but significant percentage of ICOD's total support strategy.

More broadly, ICOD considers the environmental impact where applicable, as an important criterion in making project funding decisions. Depending on the specific circumstances of a developing nation, environmental issues must be balanced against a range of other priorities.

ICOD projects, such as those with the Caribbean Environmental Health Institute (CEHI), expand the environmental monitoring capability of the region. CEHI began in 1982 by establishing a subregional network to monitor bacterial pollution levels, primarily at principal sewage outflow areas. These health measures were necessary to protect swimmers, particularly around densely populated tourist areas.

With ICOD assistance, CEHI will now expand its sampling capabilities to include microbiological, physiochemical, hydrocarbon, pesticide, and herbicide pollution indicators. This information will be used to evaluate environmental degradation, particularly for influencing policy formulation and land-use regulations. Scientists may also be able to make more convincing correlations between pollution levels and fish kills, the disappearance of species, and algal blooms.

Developing nations often lack the marine management framework to control the exploitation of resources and the protection of marine ecosystems. Mauritius, for example, a small island nation in the South and West Indian Ocean, requested ICOD's assistance in providing a Canadian consultant to participate in the development of a National Environment Plan. Another project with Mauritius will also involve Canadian experts in studying land-based pollution and other threats to the lagoonal and coastal environments.

To conserve areas that are at present unspoiled, Caribbean nations have declared 27 protected areas as marine parks. To provide regional training for park managers and technical staff, the Caribbean Conservation Association, with ICOD assistance, organised a workshop attracting 30 participants from 20 countries. As a result of the workshop, project proposals will be submitted to ICOD as part of a US $475,000 programme to fund marine park management.

Central to the question of environmental decision-making is the necessity of training competent ocean management specialists who can begin to influence policy and political decision-making. ICOD scholarship programmes support the study of ocean managers from developing nations at universities in Sweden, Canada, the West Indies, and the South Pacific. ICOD assistance will allow the University of West Indies' Centre for Resource Management and Environmental Studies (CERMES) to expand beyond its diploma-level programme to grant a master's degree in marine resource and environmental management[18].

5.3.6 Australia

Marine Programs Section, Department of the Environment

Dumping at Sea

During the year there has been continued activity in dumping dredge spoil at sea, disposal of process residues, waste products of construction activities, obsolete or damaged vessels and materials for the construction of artificial reefs.

Spoil from maintenance activities in certain harbours is relatively uncontaminated and poses little threat to the marine ecosystem other than from its physical blanketing effect and a diminution of photosynthetic activity. Other ports, however, have highly contaminated areas which contain significant amounts of oils and grease, pesticides, and heavy metals such as copper, zinc and mercury. The handling and relocation of such material is cause for concern.

Monitoring and investigations have not shown any evidence for hazardous remobilisation of toxic components from the contaminated spoils which have been dumped. Careful investigation of these materials and any relevant permits is continuing in close cooperation with the relevant local and Commonwealth authorities.

The Department received numerous enquiries from individuals and agencies seeking details of the requirements of the Environment Protection (Sea Dumping) Act and advice on how any application for a permit to dump waste at sea should be made.

During 1988–89, meetings were held with NSW and Tasmanian authorities and proponents, and dumping practices, monitoring results and compliance with the Act (see 5.1.4) were kept under effective review. A number of visits and inspections were made to operations in, or from other States.

The Department was represented on the Australian delegation to the 2nd meeting of the Intergovernmental Panel of Experts on Radioactive Waste Disposal at Sea which met in London in September 1988 immediately before the 11th Consultative Meeting of Contracting parties to the London Dumping Convention.

Other Marine and Inland Water Matters

The Department participates in and provides secretariat support for the AEC Marine and Inland Waters Advisory Committee.

The Committee receives and provides advice on environmental pressures which may effect marine, estaurine and inland waters.

The Department is represented on the Maritime Services Advisory Committee (Marine Pollution) which provides scientific advice to the Commonwealth Department of Transport and Communications on pollution of the sea by oil and other ship-sourced substances, and nominates areas of study for the development of the national plan to combat pollution of the sea by oil. The Department provided environmental input to the oil spill training activities of the national plan.

Great Barrier Reef Marine Park Authority

Objective: To provide for the protection, wise use, understanding and enjoyment of the Great Barrier Reef in perpetuity through the care and development of the Great Barrier Reef Marine Park.

The Great Barrier Reef Marine Park Authority is a statutory authority established under the Great Barrier Reef Marine Park Act 1975 (see 5.1.4). It has primary responsibility for the control, care and development of the Marine Park.

1988–89 marked the implementation of the Mackay/Capricorn Section Zoning Plan, thus completing comprehensive management plans for the entire Marine Park. In cooperation with the Division of Conservation, Parks and Wildlife of the Queensland Department of Environment and Conservation, zoning plans are now applied through a variety of procedures including a permit system, education programmes and enforcement activities with reef-user groups.

In November 1988, the Authority began revision of the Cairns Section Zoning Plan. Public comments were invited and a process of consultation with local community groups occurred throughout the period of review. The draft zoning plan will be finalised during 1989 and presented to Parlia-

ment in 1990.

Continuing emphasis focused on developing education programmes and materials by the Authority. The key to successful management is to a large extent, self-regulation by an educated and supportive public. As well as the Authority's specific public participation in zoning the Marine Park, the education programme also included production of newsletters, pamphlets, books, activities, manuals and research reports; the organisation of workshops on policy issues and reef interpretation; and the production of a video newsletter.

The Great Barrier Reef Ministerial Council has endorsed in principle a system of cost recovery for assessing permits and the Authority continued investigating the application of the user-pays principle to managing the Marine Park. Discussions with the tourism industry proved both fruitful and controversial and this issue will be further developed in 1989-90.

In March 1989, the Portfolio Minister released a review into crown of thorns starfish research and management undertaken by the Authority as a significant part of its ongoing programme of research relevant to the Marine Park. The report, conducted by Professor Anderson of the University of Sydney, concluded that the Authority's policy of limiting direct intervention to areas of special interest is soundly based and that the three-year research programme has been efficiently managed and productive. The Crown of Thorns Starfish Research and Monitoring Program will be continued.

The Great Barrier Reef Marine Park Authority provides a separate annual report to the Portfolio Minister[19].

5.3.7 Saudi Arabia

Meteorology and Environmental Protection Administration (MEPA)

Within the Kingdom of Saudi Arabia, the control of pollution and protection of the environment has been the responsibility of the Meteorology and Environmental Protection Administration (MEPA). These responsibilities have been carried out by MEPA through extensive surveys of the Kingdom's natural resources and through the development of detailed environmental protection standards that set limits for environmental quality and pollution control and long range planning activities which will restructure the relationship between economic development and natural resource management.

MEPA also serves as the Kingdom's representative to the regional environmental organisations for the Red Sea (PERSGA) and the Arabian Gulf (ROPME). In this role, the agency serves as an active participant in an international coordinated process that will eventually lead to the establishment of uniform environmental standards throughout both sea areas.

MEPA (in conjunction with the IUCN) recently published the results of a detailed survey of the coastal resources on both coasts. The survey identified areas on both coasts which were environmentally sensitive. The Environmental

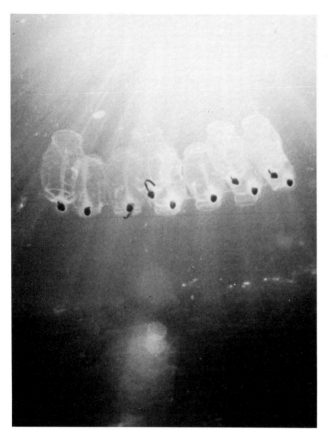

This salp chain was photographed in the entrance to the Gulf of Aqaba, Red Sea. These delicate chains of sea squirts live permanently in the plankton, and grow to vast numbers in warm, clear, relatively unpolluted waters of the Red Sea. Photo: D George

Protection Coordinating Committee (EPCCOM) designated 36 of these for special management. MEPA is currently co-ordinating with the appropriate national agencies who will be responsible for formulation of special area management plans and eventual management.

The survey forms the basis for a major effort in coastal zone management. At present management procedures are being drafted to institute a national coastal zone management programme. A series of cooperative agreements will be negotiated with all coastal municipal governments whereby local coastal zone management plans will be drafted for each municipality. MEPA will be responsible for national coordination and technical assistance in developing these plans.

As part of the coastal planning effort, MEPA is actively involved in planning for the impacts of global warming. This is particularly important in the Kingdom because of the extensive industrial development which has taken place there. The agency is actively involved in the activities of the IPCC.

MEPA is currently involved in a review of industrial compliance to environmental standards. Staff are currently being trained for monitoring. Waste water, and hazardous waste standards have been drafted and will be promulgated shortly.

During 1989, MEPA has implemented environmental impact assessment procedures. Recent development has emphasised diversification along the basic core of the petrochemical industries. As a result, there are over 300 applications under review for new industrial permits. Expansion of this load is anticipated.

The agency also maintains oil spill response capability which were called into play when the Indian tanker *Kanchenjunga* grounded and released over 7,000 tons of oil in the Jeddah area. MEPA staff were called upon in clean-up operations as well as in assessing natural resource damage.

Recent emphasis has been placed on increased coordination with other agencies within the Kingdom who have environmental responsibility. MEPA views its role as an 'honest broker' between development and conservation interests very seriously and looks towards the best interests of the environment at every turn. Realisation of that responsibility will require the best efforts of all parties if the Kingdom is to realise its responsibility to manage its resource for the good of all generations.[20]

Notes and References

1. Report by Beverley Miller, UNEP RCU, Kingston, Jamaica.
2. Report by Dr Mateo Margarinos, Associacion Uruguaya de Derecho Ambiental.
3. Report from the Ministry for the Environment, Nature Conservation and Nuclear Safety, FRG.
4. Compiled from Annual Report, Department of Environment (Australia) July 1988-June 1989.
5. Compiled by Dr Robin Churchill, Cardiff Law School, UWIST.
6. See also *NILOS Newsletters* 1 and 2.
7. *NILOS Newsletter* no. 3.
8. For more details see *NILOS Newsletters* 1 and 2.
9. See Para. 36 of the Third Declaration on the North Sea.
10. Compiled by Ton IJlstra and André Nollkaemper, Netherlands Institute for the Law of the Sea.
11. Report prepared by Dr Laura Pineschi, University of Parma.
12. Report from the ODA.
13. Report from the NCC.
14. Report from WSL.
15. Report from M J-M Massin, DEPPR.
16. Report from SIDA.
17. Report from the Ministry for the Environment, Nature Conservation and Nuclear Safety, FRG.
18. Report from ICOD.
19. Compiled from Annual Report, Department of Environment (Australia) July 1988-June 1989.
20. Report from MEPA.

6. Activities of Non-Governmental Organisations (NGOs)

Foreword

by Mostafa K. Tolba
Executive Director,
United Nations Environment Programme

As we enter the 1990s no one can escape the recognition that we are part of an historic period of global change and global awareness.

Today it is clear we have to think 'globally'. A brief examination of any of the major world issues — international security, poverty, mushrooming population, transformation of global economic activity, the technological revolution, the end of the post-World War II bipolar system — leads to the conclusion that they are not confined within the borders of any one country. Each requires multinational cooperation. We cannot escape our interdependence.

And nowhere is that interdependent global characteristic more evident than in the field of the environment. The simple fact is that human activities, numbers and technology are now great enough to seriously affect the natural environment on the planetary scale. Man rather than nature has become the primary determinant of the destruction, or the enhancement, of the earth's natural resource base and the life-sustaining environment.

Our relentless quest for more and better living conditions — for food, water, energy, shelter and other needs — to meet the requirements of an exploding population is colliding with the limitations of the natural environment. It is dramatically depleting the resource base, especially in developing countries, and it is polluting the earth's biosphere on a local, regional and global scale.

The people of the world are recognising that what used to be called environmental side effects of economic development are becoming increasingly critical and thus current forms of economic development are by no means sustainable. They know too that if the situation is critical now, it is likely to be worse in the years ahead as we move from today's population of 5.2 billion to a possible doubling by the middle of the next century and as the world economy expands more than five times.

Addressing the Environmental Components
We need deeds, not words. The world demands action. The best way to foster positive change in the interdependent world is to establish an agenda, identify the key problems, agree on goals and action programmes and provide resources and strong leadership to carry out the programme. UNEP has tried to help the environmental community to act at the local, regional and international level by promoting awareness of major problems and bringing governments and local organisations together to address them.

We have made a special effort in our programme to work with the broad communities of interest who share our concern about the environment — the NGOs, the parliamentarians, religious and spiritual leaders, industry, youth and women's groups. We are also looking to consumer organisations, film makers and artists. Responding to increased concern about environmental degradation worldwide, our Governing Council in June 1989 called for a 250 per cent increase in our voluntary operating funds to US $100 million by 1992. It called for the organisation to concentrate its efforts on eight 'areas of concentration': protection of the atmosphere (climate change, ozone layer depletion and transboundary air pollution); protection of fresh water resources; oceans and coastal areas; land resources (deforestation and desertification); biological diversity; management of biotechnology, hazardous wastes and toxic chemicals; protection of human health and quality of life.

These and other important issues will be at the centre stage when governments and non-governmental groups assemble in Brazil for the 1992 UN Conference on

Environment and Development. The Conference is expected to lay the groundwork for the next decade of action to protect the global environment. It will establish an action agenda, call for specific corrective activities, recommend strengthening new international institutions and seek to obtain or propose funding sources for future action.

The NGO Role Increases

The NGOs were one of the main driving forces behind UNEP and its programme, and those links help us today. NGOs can reach out to the poor, promote citizen participation, keep costs low, be adaptable, motivate participants and use indigenous people better. A number of our activities are conducted by or use NGO input. We back the Nairobi based Environment Liaison Centre of some 300 NGOs with close to 6,000 associate bodies and the African NGO Environmental Network of 280 member organisations. We promote networks in Asia and Latin America, and share information and expertise with both the large and small organisations worldwide.

About a quarter of the UNEP 'Global 500' awards for environmental excellence go annually to NGOs working in fields as diverse as tropical forests, climate change, soil protection, wildlife, village development, rivers, urban renewal and reforestation.

To meet the new demands of global issues NGOs will have to strengthen their management capabilities and domestic support. Because global issues bridge borders and interests, NGOs will have to broaden alliances: for example to include development and population groups, and broaden international linkages. And since governments must be the main factors in global matters, NGOs will have to develop close working partnerships with them.

Broad Action Needed

Present day threats to the global environment have opened our eyes to the need for a new agenda of cooperative action to maintain our planetary life support system. It is a broad agenda which offers expanding opportunities for worldwide cooperative action. The agenda will demand initiative at the grassroots, national regional and international levels. It is a global agenda in which NGOs must play an increasingly significant part. And it is an urgent, and interdependent agenda.

In the words of Victor Hugo: "The Challenge is great, the task is difficult, the time is now."

Dr Mostafa K Tolba
Executive Director
UNEP

NOTE: Whilst ACOPS welcomes the fact that so many NGOs have responded to our invitation by sending information which has helped the composition of this chapter, we would like to point out that expressions of opinion in this chapter may not necessarily reflect the views of ACOPS.

6.1 LOCAL AUTHORITIES

6.1.1 Council of European Municipalities (CEMR)

The Council works to promote the collective interests of local and regional authorities with the institutions of the European Communities and the Council of Europe. It supports action against marine pollution on behalf of European coastal authorities and other authorities which may be affected in any way by marine pollution.

In 1989 the headquarters of the CEMR Working Group on Marine Pollution was transferred from London to Rome, the Italian section, under the Chairmanship of Assessore Gaetano Zorzetto, Deputy Mayor of Venice and ACOPS' Vice-President for the Mediterranean. The first meeting organised by the Italian section was held in Rome on 28 April 1989. At that meeting the Chairman announced that the primary aims of the Working Group would be:

1. To observe and take part in all supranational initiatives regarding sea pollution which are of interest to Europe and to guarantee a constant presence at an intergovernmental level, for example, EEC, UNEP, and IMO.
2. To act, if necessary, as a lobbying group to:

- ensure the presence of the relevant governments at international conferences;

- seek speedy ratification of international conventions;

- guarantee effective enforcement procedures to those conventions.

3. To act as a clearing-house for the circulation of relevant documents of interest to local and regional authorities, and to support action against national and political differences.

During 1989 it was agreed that the following initiatives would be taken:

- a guide of the Basle Convention on the Control of Transboundary Movements of Hazardous Wastes and their Disposal of March 1989 with annotated comments;

- preparation of a preliminary document for the North Sea Ministerial Conference held at the Hague in March 1990;

- monitoring reports of the work of principal intergovernmental global and regional agencies relating to marine pollution;

- the Italian section undertook to translate and print 1,000 copies of the following volumes to be sent to local and regional authorities:
  — *Compensation for Pollution from Ships by Hazardous and Noxious Substances;*
  — *Guide to Dumping at Sea;*
  — *Compensation for Sea Pollution caused by Hydrocarbons;*
  — *Commentary on the Basle Convention.*

In addition to the above, the working programme for 1990 will include preparation for the United Nations Conference in Brazil in 1992 as well as preparation for the IMO Conference on Contingency Planning scheduled for autumn 1990.[1]

6.1.2 International Union of Local Authorities (IULA)

IULA, like CEMR, exists to promote the interests of local and regional authorities on a worldwide basis. IULA was the founded in Ghent (Belgium) in 1913, but now has its headquarters in The Hague (Netherlands). It also aims to raise the standards of local administration and services, to study questions concerning the activities of local authorities and the welfare of the citizens, and to promote the idea of the participation of the population in civic affairs. With regional offices and members in most continents, IULA provides a channel for the exchange of information at local government level and the monitoring of action in the marine pollution field throughout the world.

Recent activities
At the IULA world congress in Perth, Western Australia, in September 1989, the environment workshop discussed the continuing pollution of land, water and air, and called upon all Member Countries to promote by all means open to them the policies and practices which would put an end to that pollution. For the past two years IULA has been running an Environmental Dialogue Group with the European Commission to bring together local authority expertise on a range of environmental issues. Recent work has covered waste disposal processes, and soil and water pollution caused by agriculture. Developing further its concern for marine pollution caused by dirty rivers, IULA will be helping to organise a seminar in Yugoslavia in the Autumn of 1990 to highlight the concern over pollution of the Danube. Pollution and the environment also formed one of the main themes running through the IULA World Urban Development Forum held in Brussels on 4–6 April 1990.[2]

6.1.3 Local Government International Bureau

The Local Government International Bureau acts as the international arm of the British local authority associations, and also as the British Sections of the two major international local government organisations, the International Union of Local Authorities (IULA) and the Council of European Municipalities and Regions (CEMR). The Bureau seeks to facilitate UK participation in the European and international exchange of experience and ideas with the aim of providing effective local services in all parts of the world. Through the CEMR Working Group on Marine Pollution for which it previously acted as

European Secretariat, through major conferences and specialist meetings and groupings, and through its publications, the Bureau continues to support and promote the protection of the marine environment.

At the Bureau's National Conference, held in Hythe, Kent, in October 1989, at which the main theme was the significance of the Channel Tunnel project, delegates sought assurances that the work and spoil will not have deleterious effects on the Channel itself and on the coastal authorities on both sides. Its next conference in September 1990 in Newcastle is likely to be pursuing the maritime theme because of the North-East's strong associations with the sea and the effects in recent years of oil-spill disasters, dumping of mining spoil, the decline of the small fishing ports and even the devastation of the seal population.

With its monthly bulletin on European Community matters, the *European Information Service*, the Bureau has kept local authorities up-to-date on proposed and effective changes in legislation on many aspects of marine life. The many areas, ranging from bathing waters, nitrate pollution, disposal of hazardous wastes in the sea, the setting up of the Northern Seas Action Programme, the treatment of wastes on land to save the seas, are partly referred to elsewhere in this Yearbook. By constantly updating the information the Bureau keeps its readers aware of measures being taken as well as of what still needs to be achieved. Other publications available from the Bureau are the specialist publications on compensation for damage caused by oil pollution and hazardous and noxious substances, as well as on dumping, which were commissioned from ACOPS by the CEMR Working Group on Marine Pollution.[3]

6.2 ENVIRONMENTAL ORGANISATIONS

6.2.1 Advisory Committee of Pollution of the Sea (ACOPS)

During the period covered in this Yearbook, ACOPS continued to expand its international activities, as described briefly below:

International Conference on Protecting the Environment from Hazardous Substances

Six months after more than 100 countries and the EEC adopted the Basle Convention on Control of Transboundary Movements of Hazardous Wastes and their Disposal, some of the world's most senior policy makers gathered at the headquarters of IMO, 3–5 October 1989, at an International Conference on Toxic Waste organised by ACOPS. The purpose of this Conference, which marks the beginning of ACOPS' public awareness campaign on this issue, was to urge countries to give effect to the Basle Convention as soon as possible, since this umbrella agreement cannot start passing detailed technical rules until at least 20 countries ratify it. Only one ratification, from Jordan, has been received to date. (As we go to press, ratifications were also received from Saudi Arabia and Switzerland.)

ACOPS' three-day Conference was opened by its President, the former British Prime Minister, Lord Callaghan, and its Chairman, Lord Stanley Clinton-Davis. It was addressed by several European Environment Ministers, the EEC Commissioner with special responsibility for environment, as well as leading figures from Europe, Africa and South America; the Secretary-General of IMO, Mr Chandrika Srivastava; the Executive Director of UNEP, Dr Mostafa Tolba; and the Director of Environment of OECD, Mr Long. All papers presented were published in a special edition of the international journal *Marine Policy* in May 1990. Messages of support were received from the British Prime Minister, Mrs Margaret Thatcher; the Secretary General of the UN, Señor Perez de Cuellar; and the Secretary General of the Commonwealth, Sir Shridath Ramphal.

ACOPS was greatly encouraged to note that the UK Government, whose Minister addressed the Conference, signed the Basle Convention on 6 October 1989. It hopes that additional signatures and, more importantly, ratifications, will follow.

Vice Presidents

In the course of 1988-90, there were further additions to the ranks of Vice Presidents: Professor Elisabeth Mann Borgese (Canada) Professor Gennady Polikarpov (USSR) and Mr Chandrika Prasad Srivastava (India).

We were delighted to note, since the publication of our last Yearbook, that Mr Cissokho was appointed Minister for Rural Development in Senegal; that Mr Louis Le Pensec became Minister for Overseas Departments and Territories in France; whilst Professor Alexander Yankov was nominated Vice President of the Bulgarian Academy of Sciences.

Many ACOPS Vice Presidents attended their second meeting which was held in London on 4 October 1989, under the Chairmanship of ACOPS' President, Lord Callaghan. The conclusions of this meeting may be found in Annex I. Vice Presidents have continued to promote ACOPS' global policies in their respective countries and regions, through appointments and correspondence with senior government ministers and officials, and by drawing attention to their special, regional problems.

Activities with Intergovernmental Agencies

Following granting of consultative status in 1988, ACOPS started participating in the meetings of Contracting Parties to the London Dumping Convention. In particular, ACOPS' former Chairman, Baroness White, addressed the meeting of Contracting Parties on 30 October 1987 when ACOPS was invited, together with a small number of governments and intergovernmental agencies, to present its views on the implementation of Articles 1 and 2 of the London Dumping Convention.

ACOPS also participated actively in the work of IMO, including the Diplomatic Conference on Salvage (17–28 April 1989; (see 3.2.6)). It also intends to participate at the November 1990 Diplomatic Conference on International Cooperation on Oil Pollution Preparedness and Response. (see 3.2.7)

As much of ACOPS' work in 1989 focused on the problem of Toxic Waste, the Committee was represented, through its Executive Secretary, Dr Sebek, at the Euro-African Ministerial Conference on Environment and Development in January 1989.

The Chairman, Lord Clinton-Davis, and Vice President from the Mediterranean region, Mr Zorzetto, also lead ACOPS' delegation at the diplomatic conference which was held in Basle from 20–22 March 1989, and which adopted the Convention on the Control of Transboundary Movements of Hazardous Wastes and their Disposal. (see 3.3.12)

Links were also maintained throughout 1989 and in 1990 with many intergovernmental agencies, such as the European Communities, OECD, UNEP and IOPC Fund; various ministers and government officials in Europe, North and South America, Africa and Asia, especially during visits of Lord Clinton-Davis to Switzerland, Belgium, France and the United States; Rear Admiral Stacey to France and Dr Sebek to the USA, India, Egypt, Italy, France, Switzerland, USSR, Yugoslavia and Senegal.

Contacts were also maintained with research and academic institutions, as well as with industrial, wildlife and environmental agencies.

Educational Activities
As was specified in our last Yearbook (p.66), ACOPS has decided to give high priority to educational activities and exchanges of information with university and research establishments worldwide. Some examples of such activities include Dr Sebek's lectures at the University of Parma (1988), the International Ocean Institute's course to government officials from the Indian Ocean area, held in Egypt in November, 1989, and papers given to seminars organised by the European Parliament (November 1988); the Italian Supreme Court (April 1989). Dr Sebek also participated at the Pacem in Maribus XVII Session, held in Moscow in June 1989 and at the US Academy of Sciences workshop, held in Washington in February 1990.

Mr Trevor Dixon, ACOPS' Environmental Advisor, presented Papers at the Conferences in Hawaii (April 1989), San Sebastian, (June 1989) and Athens (July 1989) on beach debris and pollution. Mrs Jennie Holloway, ACOPS' Legal Officer, took part at conferences and seminars in Rome (April 1989), Moscow (June 1989),

Copenhagen (October 1989) and Vienna and Budapest (March 1990). On the home front, a seminar held in London (November 1989) for UK local authorities on problems arising from waste on UK beaches was well supported.

Working Group on Marine Pollution of CEMR
ACOPS has continued to provide technical advice to the Working Group on Marine Pollution of the Council of the European Municipalities and Regions (CEMR). ACOPS' Vice President, Mr Gaetano Zorzetto, continues as Chairman of the Working Group, whilst Dr Sebek serves as the Rapporteur. The Working Group met in Rome in April 1989 and in London in October of the same year.

6.2.2   Aktionskonferenz Nordsee e.V. (AKN)

The Aktionskonferenz Nordsee e.V. (AKN) was founded in 1985, aiming at developing and furthering environmentally and socially compatible structures for the North Sea ecosystem, its coastal areas and the rivers entering the North Sea. The basis of AKN's work is, essentially, a North Sea Memorandum which was first drawn up and published in 1984 and modified in 1987. Members of its executive committee act solely in an honorary capacity.

AKN's work mainly serves the purpose of coordinating and supporting organisations and citizens' initiatives, as well as furthering the awareness of the causes of the ecological threat and developing political pressure in order to put an end to those causes. Another focal point of AKN is to offer further education — both in general and adult education. It also issues publications of a general kind and publishes the journal *Waterkant*.

Within the framework of the International Seas at Risk Federation (see 6.2.20), AKN has been making every effort to further international cooperation of non-governmental organisations in the field of North Sea protection.

Among other things, AKN was responsible for the German participation in the activities in connection with the International Conference of Ministers on the Protection of

Litter on a beach in the UK. The Tidy Britain Group is one of the organisations in the forefront of combating beach litter (see 6.2.21). Photo: TBG

the North Sea in London in November 1987 and in Den Haag in March 1990.

One of AKN's main tasks in the future will be to press for the results of the London and Den Haag Conferences of Ministers and further measures for the protection of the North Sea to become effective.

In May 1989 AKN organised a North Sea Tribunal in Bremen where polluters of the North Sea were accused and condemned.

In June 1989 AKN, together with a Latvian partner organisation, called for an international Baltic Sea Conference to be held in Latvia. This conference will be continued in the German Democratic Republic in September 1990.

In recent years AKN's office has developed into an important information source and a centre for enquiries from both home and abroad.[2]

6.2.3 UK Centre for Economic and Environmental Development (UK CEED)

Formed five years ago on an initiative from the 1983 Conservation and Development Programme for the UK, UK CEED remains a unique organisation in Western Europe, devoted to promoting a productive partnership between environmental protection and development enterprise through the sustainable use of resources. UK CEED works with industry and commerce to ensure that development policies, investment decisions and operations incorporate the high environmental standards expected in a progressive society. UK CEED promotes economic analysis of UK environmental and development concerns through original research and dissemination of its findings and through organisation of seminars and conferences, bringing together experts from the fields of business, industry, government, academia and environmental and conservation groups.

UK CEED has a small full time staff, senior research associates and a board of directors drawn from leading figures with economic and environmental interests. The Centre provides an independent and objective advice service to many organisations and individuals. Staff represent the Centre and its aims on a wide range of committees, panels and study groups concerned with environmental policy including the Cabinet Office Advisory Council on Science and Technology. UK CEED is a registered charity funded by commercial and industrial organisations, trusts and charitable foundations and Government and public sector agencies.

Its research studies last year included a discussion paper on the definitions of sustainability, and on electricity privatisation and the environment. A distinctive element of UK CEED's research programme is a cooperative initiative with Japan aiming to compare environmental policy making within the two industrial island nations. One element of this initiative covers coastal management protection procedures.

As part of its fifth anniversary in 1989, UK CEED launched a programme entitled 'Environmental Keys to the Twenty First Century'. A series of specialist seminars covered alternative economic indicators, transport and mobility, the future economies of national parks and a major seminar on coastal management and protection in the UK and Japan.

6.2.4 Coastwatch Europe

In 1987 the Dublin Bay Environmental Group designed and tested a coastal survey around the coast of Ireland, aiming to provide an insight into coastal zone management problems and raise the level of awareness of such issues. In 1988 limited support from Directorate General XI of the EC allowed a further seven countries to adopt a limited pilot survey. The public response and results were sufficiently encouraging for DG XI to continue to fund an extended survey in the original eight pilot survey countries, plus a further two new participants.

In the UK, Coastwatch Europe was jointly organised by Farnborough College of Technology and the Marine Conservation Society, with sole sponsorship from Norwich Union Fund Managers Ltd. The UK survey targeted primarily coastline used for leisure, particularly bathing beaches. A wide spectrum of volunteers — mainly school and college groups — carried out a survey of land-use, litter and sewage pollution of their coastal units.

The coastline is coded into 5 km blocks, each comprising ten 0.5 km units. The survey is conducted via a questionnaire completed for each unit: responses are required on the association of surveyors with the site, on physical characteristics, ecology, threats, litter and pollutions status. The identical questionnaire was completed in the participating countries — UK, Ireland, Belgium, Denmark, Netherlands, West Germany, Portugal, Italy, Norway and Iceland — during the European fieldwork week from 24–30 September 1989. A simple water quality test kit was also provided exclusively in the UK, to determine nitrate levels in inflows and to note levels of faecal bacteria in the seawater.

In the UK, volunteers surveyed 5–7% of the coastline. The results indicate that questionnaires completed when cross-referenced against various quality markers in the survey provided the fairly bleak picture that virtually half of the coastline surveyed is in a poor quality state (through pollution or litter). Nearly a quarter of the coastline surveyed was affected by sewage pollution incidents.

Almost 1,200 inflows (excluding rivers and streams) were characterised, constituting an often unseen threat to coastal waters. The quality of many of the inflows was at best dubious, a quarter producing off odours, 15% with high nitrate levels, many with scums, froths, debris etc.

Foam scums were recognised in 17% of coastal units, a factor leading to some considerable disquiet. Forty-two dead birds coated in oil were recorded, suggesting that oil pollution is a constant problem and not restricted to major publicised incidents.

One in ten of the upper shore zones of the units surveyed was grossly littered, and is impossible to avoid debris in that area. Moderate littering was recorded in a further 40% plus of the units. Landfill and building debris were dumped in nearly a quarter of the upper shore sites surveyed.

General litter and debris around the shore was dominated by plastics, including fishing gear and packaging, cans and wood and paper. Other high incidences included sewage-related debris such as sanitary materials. Hazardous materials, including medical waste, were also reported in significant quantities. On a 40% response, 12% of the total units were considered to be frequently or usually polluted. This is directly related to the identification of sewage pollution to be the single most prevalent threat to coastal integrity.

Norwich Union are continuing their sponsorship into 1990. The same management structure is proposed in the UK. In the European dimension, a further five countries are proposing to join the project, reflecting the continuing and increasing support from DG XI. The long-term goal of the project is to involve all European coastal countries in an identical and simultaneous snapshot of their coastline.[5]

6.2.5 The Environmental Problems Foundation of Turkey (EPFT)

The EPFT was founded on 1 February 1978 in accordance with Law No. 903 of the Republic of Turkey. It is a private, independent volunteer organisation.

EPFT's objectives are:

- to collect and publish facts on environment and factors which have an effect on the environment;

- to conduct research on alternative solutions to Turkey's environmental problems;

- to play an advocacy role in promoting the quality of the environment;

- to assist the state and local governments in the formulation and implementation of environmental policies and programmes;

- to monitor on a continuous basis, progress made on environmental issues.

EPFT's activities are diverse and include:

- publication of a newsletter, in Turkish (circulation about 3,500) and English (circulation about 500);

- publication of studies and conference proceedings;

- organisation of conferences, discussion-meetings;

- continuous use of the media through radio and television as well as newspaper articles and press releases.

EPFT prepared an Environmental Profile of Turkey (the first since 1981) as well as the Environmental Protection Law Proposal (1981) which was accepted as a Law in 1983. It was instrumental in the inclusion of the Article of Environment in the Turkish Constitution and participated in discussions of the Five-Year Plans, successfully proposing additional articles.

During the period June 1988 to February 1990, EPFT's achievements included: publishing the *Biological Diversity Inventory of Turkey* (both in Turkish and English); organising the proceedings of a second 'Environment and Population' Conference; holding a discussion-meeting 'Biological Diversity and Development'; holding a round table discussion on the subject of 'Environment and Development Relations'; publishing *Turkish Environment Law and Some Other Related Provisions*.

The EPFT also organised the 'Mediterranean in the 1990s' Conference (MEDCON), the first time it has been organised by a Turkish NGO. This conference, held in Antalya, on the Mediterranean coast of Turkey, was attended by representatives from Mediterranean countries and international organisations. Following MEDCON, a national conference was held on 'Sustainable Development'. Proceedings of these conferences are being printed.[6]

6.2.6 European Environmental Bureau (EEB)

The EEB is a federation of environmental NGOs with 126 Member Organisations from 17 countries. In the field of sea protection the EEB is active in two major areas:

North Sea
The EEB has taken an earnest interest in North Sea environmental matters since early 1988 when it decided to be actively involved in the Seas at Risk Federation. Seas at Risk is the largest federation of North Sea and related working parties in Europe that have a certain concern for the marine environment.

In 1988 the EEB representative within Seas at Risk was elected its chairman. The EEB seeks in this way to provide input into the preparation of ministerial conferences of the North Sea riparian countries, most recently in The Hague, Netherlands, 7–8 March 1990.

The EEB has assisted Seas at Risk in the preparation of policy documents for the senior-official preparatory meetings. In particular it is the EEB's concern that a number of the decisions from the 1987 London conference are not in-effectively dealt with. The EEB therefore has pressed for:

- immediate measures for the termination of incineration at sea which for the greater part has already ceased due to the lack of profitability;

- freedom of information with regard to the contents of permits and the procedures by which they are issued. The proposal for the EEC directive in this respect should be adopted as quickly as possible;

- necessary measures to reach the 50% reduction for the pollution of the North Sea via its rivers by the scheduled date of 1995. The EEB has stressed that the adoption of the nitrate directive will be imperative.

The Mediterranean
The EEB has opened an office in Athens to establish contacts with NGOs in countries bordering the Mediterranean. It has also set up a Working Party which recently received the delegates of the Barcelona Convention at a meeting in Athens. The Working Party is also paying particular attention to a number of projects, including the MEDSPA and ENVIREG programmes which are both concerned with water treatment.[7]

The Danube Meeting, 19–21 March 1990

Following publication of the Brundtland Report, the UN General Assembly decided to hold a series of follow-up conferences to consider its implementation. The Government of Norway offered to host the Conference for the ECE region (East and West Europe and North America) which took place in Bergen, 8–16 May 1990. In view of the importance of the Bergen Conference, the EEB, at the request of the Norwegian Government, organised a meeting and agreed the principal goals and form of environmental NGO participation in Bergen and elected a Steering Committee to carry out this programme, as a result of which the Danube Meeting was convened.

The Danube Meeting was one of the largest international gatherings of NGO representatives since the Stockholm Conference in 1972 and the first opportunity for non-government groups from East and West to meet. The venue for the meeting was chosen to be on either side of the Iron Curtain to demonstrate that environmental problems do not recognise political boundaries. The meeting consisted of an opening plenary session in Vienna, a day of workshops on a boat on the Danube travelling from Vienna to Budapest, and a final plenary session in Budapest where an Agenda for Action was agreed and will be the basis for NGO input to the Bergen 1990 Conference on Environment and Development, and to the 1992 UN Conference in Brazil.

The Danube Meeting was attended by some 350 participants from the 35 countries of the ECE region as well as a small number of participants from southern countries. The Agenda for Action on sustainable development concentrated on the four main themes of the Bergen meeting: public awareness and participation, sustainable industrial activities, sustainable energy use and the economics of sustainability. The theme chosen for the NGO input to Bergen is 'Bridging the Gap': the gap between rhetoric and action, between North and South, between East and West, between Bergen and the 1992 UN Conference and between NGOs and governments.[8]

6.2.7   Field Studies Council Research Centre (FSCRC)

Established in 1967 as the Oil Pollution Research Unit, the FSCRC is a division of the Field Studies Council, an environmental education charity with ten field study centres in England and Wales. FSCRC, like the parent body FSC, operates independently of commercial or Government organisations. It offers a contract research and consultancy service in biological, chemical and physical environmental sciences using the skills of 21 scientific staff, including a diving team. FSCRC specialises in environmental assessments and monitoring and is used by industry and government agencies worldwide.

FSCRC offers a multi-disciplinary approach to the investigation of problems in the marine environments, including:

- biological monitoring;
- chemical analysis;
- coastal and offshore surveys;
- conservation evaluation;
- development appraisal;
- diving studies;
- environmental impact assessment;
- habitat mapping;
- oil spill contingency planning;
- oil spill response;
- project design and management;
- resource management and mapping;
- sediment analysis;
- sensitivity mapping;
- surveys, analysis and literature studies;
- training in field and laboratory practice.

The activities of FSCRC in the period June 1988 to February 1990 have been dominated by work on oil spills and their effects.

While the oil industry operates to high standards of environmental care, accidents do happen and oil and its products enter the environment, often as the result of human error. Providing ecological advice on the impact of such unpredictable incidents has always been an important part of the workload of FSCRC staff. In the 20 months to February 1990 FSCRC responded to oil spills in Alaska, the Mersey Estuary (a pipeline rupture), Morocco (*Kharg 5*), Southampton Water (*Worthy*) and Milford Haven (*El Omar* and *Westminster*) and to an aluminium spill into the Dart Estuary.

Timely advice from FSCRC personnel in the field can be crucial to reducing the impact of oil or chemical spills, while subsequent monitoring and research studies can generate much useful information for incorporation into beach clean-up guides and sensitivity maps.

Post-spill studies were commissioned by several industry and local authority groups in 1989. Yet persuading coastal authorities who have yet to experience a spill of the wisdom and cost-effectiveness of adequate ecological input into contingency plans is proving difficult.

Other Studies

In addition to the above, benthic fauna, hydrocarbon and sediment studies around North Sea production platforms have been undertaken for Phillips (Audrey), Britoil (Thistle), BP (Ula, Gyda, Forties, Buchan, Ravenspurn, Amethyst, West Sole and Cleeton), Hamilton Brothers (Ravenspurn gasfield) and Mobil (Beryl). These same skills have been utilised by Anglian Water (Humber), Esso Fawley (Southampton Water), SOTEAG (Sullom Voe) and South West Water (Dart) as part of routine monitoring studies.[9]

6.2.8 Finnish Association for Nature Conservation (SLL)

The Finnish Association for Nature Conservation (Suomen luonnonsuojeluliitto, SLL) is the central non-governmental organisation for the promotion of environmental conservation in Finland.

This includes:

* conservation of the indigenous nature, animals, plants and valuable natural areas for future generations;

* far-sighted use of natural resources and production methods in harmony with the environment;

* city and rural planning on a sound ecological basis.

For these purposes the SLL publishes informative material about nature and environmental conservation, works to influence decision-making politicians and officials, takes a stand on current environmental issues, organises excursions, study courses, public meetings, demonstrations, and participates in international cooperation, especially with the Scandinavian and neighbouring countries.

Organisation

The SLL has a a three-tier organisation: on the local level there are approximately 170 conservation associations belonging to 16 nature conservation districts. The latter collectively form the SLL, or central organisation. Luonto-Litto (the Finnish Youth Association for Environmental Protection) functions as a nationwide youth organisation under the SLL. The SLL has a total (1984) of 30,000 members, of whom 5,000 belong to the youth organisation.

The central office of the SLL in Helsinki employs twelve persons. In addition, there are six district secretaries working at the local level.

Information and Education

Suomen Luonto (Nature of Finland) is published by the SLL and is the only magazine devoted to environmental subjects in Finland.

The magazine contains short English summaries and can be ordered abroad.

Suoment Luonnonsuojelun Tuki Oy (Support for Nature Conservation Ltd) is a production and marketing company in which the majority of shares are owned by the SLL. The company finances, produces and distributes books, guides, nature posters, postcards and various gift articles which are necessary for the financing of activities in the SLL.[10]

6.2.9 Friends of the Earth International (FOEI)

FOEI is a federation of national non-governmental organisations, each taking positions at a domestic level on issues of environmental concern in their country. To date there are FOE groups active in 38 countries around the world and in all continents. These national groups together form an international network for information exchange and coordination of joint activities in respect of regional and global environmental issues of common concern.

The aims and purposes of FOEI are to promote understanding and appreciation of the need for the preservation, protection and restoration of the natural environment and natural beauty of areas of the world and all other aspects of man's natural environment by all lawful means.

FOEI's activities with respect to marine environmental protection are coordinated by FOEI's Ocean Environment Section in close cooperation with, amongst others, the North Sea Working Group in Amsterdam.

FOEI has observer status at several international organisations and conventions relevant to marine environmental protection: the International Maritime Organisation (IMO), the International Oil Pollution and Compensation Fund (IOPC Fund), the London Dumping Convention (LDC), the Barcelona Convention and the International Whaling Commission (IWC). Friends of the Earth International attended several meetings of these organisations during the period from June 1988 to February 1990.

International Maritime Organisation (IMO)

During the 26th Session of MEPC, from 5–9 September 1988, FOEI submitted a proposal to develop a manual for the designation of Particularly Sensitive Sea Areas. This proposal was adopted by MEPC after which the FOEI representative offered to prepare a first draft for such a manual. Other issues addressed during this meeting included air pollution from ships. At the 27th Meeting of MEPC, from 13–17 March 1989, FOEI raised the issue of the dumping of cars and wreckage of the car carrier *Reijin* which FOEI considered a questionable response to that accident.

From 17 to 28 April 1989, FOEI attended the International Diplomatic Conference on Salvage. FOEI had submitted documents to this Conference addressing issues relating to salvage and environmental protection. FOEI proposals for changes to the draft Convention were not adopted. FOEI nevertheless welcomed the new Salvage Convention adopted during this Conference as an improvement.

Other issues addressed during meetings of committees included the removal of offshore installations and environmental aspects of maritime accidents.

Currently, FOEI is preparing several documents for future submission to IMO meetings, including the draft Manual on the Designation of Particularly Sensitive Areas, and documents on air pollution from ships, enforcement of MARPOL's Annex I, and others.

London Dumping Convention (LDC)

At the 11th Consultative Meeting of the London Dumping Convention, from 3–7 October 1988, FOEI submitted an extensive document on the issue of liability for damage resulting from dumping activities and was involved in discussions relating to incineration and the removal of offshore installation.

The 12th Consultative Meeting of Contracting Parties to the London Dumping Convention met from 30 October to 3 November 1989. One of the most important issues was a discussion about the future of the Convention, an issue which also was addressed indirectly during the discussions on some other agenda items. FOEI had submitted its ideas on the future of the Convention to LDC's chairman before the meeting.

Barcelona Convention

FOEI was one of four international environmental organisations to attend the 6th Meeting of Contracting Parties to the Barcelona Convention, 2–6 October 1989. The meeting brought little, if any, progress with respect to the important environmental issues facing the Mediterranean. The fate of the monk seal, one of the most threatened species in the world, was not discussed.

International Whaling Commission (IWC)

FOEI's Ocean Environment Section was present at the 1988 and 1989 meetings of the International Whaling Commission in Auckland (New Zealand), from 30 May to 3 June 1988, and in San Diego (USA), from 12–16 June. During these periods, FOEI was one of the producers of ECO, a daily newsletter reporting on the meetings.

North Sea

FOEI's Ocean Environment Section has also joined the preparation of NGO activities surrounding the International North Sea Conference which are coordinated by the Seas at Risk Federation. FOEI's Ocean Environment Section was also co-sponsor of the Third North Sea Seminar of the North Sea Working Group, held from 31 May to 2 June 1989, which addressed the relationship between science and policy.[11]

6.2.10   Hellenic Marine Environment Protection Association (HELMEPA)

In June 1982 HELMEPA was created through the joint signing of a Declaration of Voluntary Commitment and its attached Action Plan. This document was endorsed by five international organisations: the Club of Rome, IUCN, the World Wide Fund for Nature (WWF), the International Ocean Institute (IOI) and the International Institute for Environment and Development (IIED).

The Declaration has served as HELMEPA's guideline in planning activities, as the ideas projected therein express the quintessence of man's relation to the marine environment and his direct involvement in its protection.

HELMEPA's membership presently totals about 400 Greek ocean-going vessels; associate membership numbers 105 companies; over 5,600 Greek seafarers are associate members. It is estimated that about 80% of Member seafarers are at sea at any one time.

In 1983 HELMEPA began an environmental awareness campaign with the cooperation of the State. A key part of the campaign is a permanent environmental exhibition in Piraeus and also a mobile unit. To date these have been host to over 120,000 school children. The permanent exhibition was updated and improved in 1988.

HELMEPA participates in the MEPC and MSC Sessions of the IMO as members of the Greek Delegation. HELMEPA cooperates with the US Coast Guard and UNEP supports the HELMEPA initiative. A constructive cooperation has also been established with the European Commission. HELMEPA undertook and successfully car-

ried out International (1986) and Mediterranean (1989) Workshops in Athens, Greece under the auspices of the European Commission and with the support of the IMO. These dealt with maritime legislation and Annex V (Garbage) of MARPOL 73/78, respectively.

'In 1988 HELMEPA conducted seminars on board Member vessels at Greek shipyards, as they were resuming service after a long period of lay up. Furthermore, the Association conducted a series of hands-on firefighting seminars with the kind assistance of the Greek Navy at its training installations. The 1988-89 training period was restructured to meet the special needs of each type of vessel and the speciality of the merchant marine officer. Emphasis was placed on the effects of marine pollution on oceans and shores, and general information on the marine ecosystem.

In 1989 and within the framework of HELMEPA's cooperation with the US Coast Guard, two seminars for merchant marine officers and HELMEPA Member Company executives were conducted in Piraeus by the USCG.

Voluntary clean-up of beaches by scouts on the Attica Peninsula, Greece. this was part of HELMEPA's drive to encourage public participation (see 6.2.10).

Publications include seminar hand-outs, a manual titled *MARPOL and Seafarers/Annex V — Garbage Pollution*, a special videocassette for use at the seminars and on board Member vessels.

In April and May 1989 voluntary beach clean-ups took place on the beaches of the Attica Peninsula with the participation of Boy Scouts and Girl Guides, local Authorities and Members of the Association. To support the effort for public participation, and especially children, HELMEPA produced a 3-ply leaflet in the form of a quiz.

The HELMEPA-MEDSPA Program, sponsored by the European Commission, commenced in October 1989 with a duration of 15 months. The Program provides training seminars in Piraeus and other regions of the country which address seafarers, including crew and fishermen. Three booklets *SOLAS and the Seafarers* are to be issued in 1990.

HELMEPA presented a Resolution at the Sixth Ordinary Meeting of the Contracting Parties to the Barcelona Convention, which took place in Athens under the auspices of the UNEP Mediterranean Action Plan during the first week of October 1989. The Resolution recommends that education, training and public awareness campaigns be developed through every relevant channel, as exemplified by HELMEPA, and that a regional non-governmental federation (MEDMEPA) be established, composed of national Mediterranean non-governmental associations voluntarily committed to the protection of the marine environment from pollution, and to the mobilisation of human potential. The Meeting unanimously approved HELMEPA's proposal.

Since 1987 HELMEPA has also been national operator of the EC sponsored campaign 'Blue Flags of Europe' for organised beaches and marinas. In view of the difficulties in monitoring bathing water consistent with the EC Directive, and in view of the small and medium number of bathers in large number of beaches, HELMEPA proposed to the European Commission and other national operators of the campaign to inaugurate a special distinction for remote beautiful and unspoiled beaches, in line with the European Parliament decision of 1988 and also the Ministers of Environment Council of 1989. The proposal was accepted as a 'pilot project' for 1990 and is currently being carried out in Greece and Great Britain.[12]

6.2.11 Institute for European Environmental Policy (IEEP)

IEEP is a network of four independent institutes in Bonn, Paris, London and Arnhem (the Netherlands). It analyses environmental policies in Europe and seeks to increase the awareness of the European dimension of environmental protection and to advance European policy making.

IEEP closely follows European Community policy-making and the implementation of EC directives. 1989 saw the publication of the revised second edition of Nigel Haigh's book, *EEC Environmental Policy and Britain*, which describes and analyses all EEC environmental legislation. A conference on the incineration at sea of organo-halogen compounds was held in May 1989 in Brussels, with the support of the European Commission. The Institute has continued its work in the field of Integrated Pollution Control (IPC) with a project on IPC in Europe and North America in conjunction with the Conservation Foundation (Washington). A further focus of the Institute's work has been the implications for environmental policy of '1992' and the completion of the EC's internal market. Several studies have been undertaken on this theme.[13]

6.2.12 International Institute for Environment and Development (IIED)

IIED is a global organisation established in 1971. It advocates sustainable development through the productive use of soils, water, forests and other natural resources because they are directly linked to economic growth and human needs. IIED is funded by private and corporate foundations, international organisations, governments and concerned individuals.

The Institute draws staff from around the world and operates from offices in London and Buenos Aires. Its agenda ranges from marine resources to forestry, from sustainable agriculture to human settlements. IIED's approach involves finding practical solutions to global problems by field work and grassroots collaboration, institution building and information networks.

The Renewable Resources Assessment Group — part of, and located at, Imperial College London — is closely associated with IIED. It conducts research and provides expertise on biology, economics and management methods. Begun as a joint venture between IIED and the International Union for Conservation of Nature and Natural Resources in 1984, with support from the World Wide Fund for Nature, the Group now consists of some 20 specialists under the direction of Dr John Beddington.

Amongst other things, the Group has been formulating policies for licensing and conservation of tuna, establishing a set of common data formats and methodologies to analyse fisheries resources and providing technical assistance to various governments in stock assessment and management. The Group has research programmes on marine mammal–fishery interactions, both in UK waters and with respect to the Antarctic ecosystem where models of krill, whales, squid and birds are being investigated. It also has research programmes on the parasitology and epidemology of exploited animals populations.[14]

6.2.13 International Organisation of Consumers Unions (IOCU)

In 1960, five consumer organisations — from the United States, Australia, United Kingdom, Belgium and the Netherlands — decided to pool their experience in product testing and consumer information. They set up the International Organisation of Consumers Unions (IOCU).

IOCU's independence is guaranteed by its strict membership rules. Organisations that join as Associate Members must operate exclusively in the consumer's interest. They must be non-profit organisations and non-commercial — their publications, for example, may not accept commercial advertising. They must be independent of political parties and of cash subsidies that might influence their purely consumer voice. Some organisations that cannot meet all these qualifications are eligible to join as 'Supporting members' or 'Corresponding members'.

IOCU is financed by its members' fees and by sale of its publications. For special projects, it seeks funding from governments, UN agencies or development organisations.

Under its Constitution, the ultimate source of authority is the General Assembly, where every Associate Member has a vote. Thus IOCU is governed by its members, through a Council of 20 and an Executive of six member organisations.

In the 1980s, IOCU — in parallel with consumer organisations the world over — has grown dramatically. It has a membership of over 150 organisations in more than 50 countries on all continents. Through its offices in The

Hague, Penang and Montevideo, it links the work of its members and helps to establish new consumer organisations. IOCU provides consumers with a voice at international bodies such as the United Nations. It is the consumer's watchdog over the transitional corporations and counters the power of international business lobbies.

In 1985, IOCU was instrumental in encouraging the adoption of the UN Guidelines for Consumer Protection, which were agreed to by all UN members, and which amount to a world charter on consumer rights. The Guidelines are now the basis for national consumer organisations to press their governments for better consumer protection.

One of IOCU's main functions is to speak up for the consumer on issues that cross national boundaries. Both in its own right and through its support for issue-based action networks, IOCU has been instrumental in seeing that issues such as hazardous products, baby foods, tobacco, medicines and pesticides receive the international attention they deserve.

Hazardous Consumer Products

In 1981 IOCU launched Consumer Interpol, a worldwide network of correspondents, to police the global trade in hazardous products, technologies and wastes. Besides warning consumers about newly-discovered or regulated hazardous products, it also disseminates information and generates campaigns on generic safety, problems on issues such as Bhopal, Chernobyl, food irradiation and biotechnology.[15]

6.2.14 International Union for Conservation of Nature and Natural Resources (IUCN)

IUCN is a union of sovereign states, government agencies and non-governmental organisations concerned with the initiation and promotion of scientifically-based action that will ensure the perpetuation of man's natural environment. Established in 1948, IUCN is an independent, international organisation whose membership includes 62 states, 108 governmental agencies, 406 international non-governmental organisations and 25 non-voting affiliates. IUCN also has some 700 individual and organisational supporters in 65 countries. Its Secretariat is based in Switzerland.

IUCN enjoys consultative status with ECOSOC and several UN agencies and is part of the 'Ecosystem Conservation Group'. It also maintains a close working relationship with the World Wide Fund for Nature.

Its aim is to provide international leadership for the conservation and management of living resources. In trying to fulfil its aims, IUCN undertakes eight general activities: providing advice and expertise to governments and inter-governmental bodies concerning conservation of nature and resources;

- managing networks of scientists and specialists;

- managing information from a wide variety of sources;

- assisting institutions to enhance their capacity to manage their resources effectively;

The habitat of the Triton, hunted for its magnificent shell, is tropical coral reefs, where it preys on the crown-of-thorns star-fish. Some believe that over-collecting has contributed to 'plagues' of crown-of-thorns in the Philippines, Seychelles and Pacific islands. The Marine Conservation Society has campaigned to stop imports of shells for sale as curios in the UK (see 6.2.16). Photo: D George

- promoting conservation action through working with members and collaborating in action-oriented field projects;

- providing technical support to conservation treaties and agreements;

- promoting research and new techniques relating to conservation of resources and facilitating their application at local level.

- providing an EIA support service for the governments of developing countries.

Publications

As part of its work, IUCN regularly publishes reports and books on important conservation issues, for example, *World Conservation Strategy*, *Red Data Books*, the *United Nations List of National Parks and Protected Areas* and the *IUCN Bulletin*. Recent publications include *Coral Reefs of the World* and the *Action Plan for the Conservation of Dolphins, Porpoises and Whales.*[16]

6.2.15 Lega Navale Italiana

The Italian Naval League was founded as a public organisation under the patronage of the President of the Republic. Its aim is to encourage, particularly among young people, the love for the sea, a maritime spirit and a knowledge of maritime problems, and also to stimulate the participation of citizens in the development and progress of all national activities concerning the sea.

The League encourages the protection of the marine environment, and territorial waters, and promotes cultural, nature conservation, sporting and educational initiatives in order to achieve the goals of the Organisation. It also stimulates and supports the use of pleasure crafts and other maritime activities, and cooperates with the sporting federations of the CONI and foreign Naval Maritime Leagues.

Since 1983 the League has organised an annual 'Sea and Territory' conference. The most recent were:

- 1988 — Bathing, marine parks, pleasure navigation, and archaeological submerged goods;

- 1989 — Coasts and seaports in the Mediterranean Area;

- 1990 — The protection of the marine environment and the role of the European Community Committee.[17]

6.2.16 Marine Conservation Society (MCS)

The MCS is a rapidly expanding UK-based organisation which seeks to protect the marine environment for both wildlife and future generations by promoting its sustainable and environmentally sensitive management. Despite a severe office fire in July 1989, the Society is continuing its work in all areas.

During 1989 the second edition of the best-selling book, *The Good Beach Guide*, was published and the third edition compiled, due for release in April 1990. *The Golden List of British Beaches*, formerly published by the Coastal Anti-Pollution League until their merger with the MCS in 1987, will be included in this latest edition. This publication and the associated clean beach campaign attracted mass media attention throughout the whole year.

Other pollution work focused around the concept of the precautionary approach, as adopted at the Second North Sea Ministerial Conference in 1987. MCS has lobbied for the approach to be applied to all discharges into the marine environment and highlighted the case of Nuvan (now 'Aquaguard') in particular. This chemical is widely used in the salmon farming industry to rid stock of sea lice, but its active ingredient, Dichlorvos, is listed on the Government's, 'Red List', of toxic substances. MCS is therefore highly concerned about the lack of a precautionary approach in this case and feels that Nuvan should be banned.

With two conservation officers based in Scotland, MCS has also studied the wider impacts of the salmon farming industry, including the siting of farms, predator control and benthic impacts. The Coastal Zone Management (CZM) work of the Society overlaps with this and MCS is working towards producing an effective management plan for our coastal areas and seas. This includes the lobbying of Parliament to designate further Marine Nature Reserves as the UK still only has one such statutory reserve, despite the preponderance of other unique sites which need similar protection and a Government commitment toward their designation.

Many volunteers and local group members were involved in the Basking Shark Watch campaign which seeks to attain baseline data on the largest fish which enters UK waters. MCS is highly concerned that despite the lack of any substantial data on the species, a commercial fishery still exists.

Outside UK waters, the Society has led a successful 'Let Coral Reefs Live' campaign, which seeks to halt the often illegal trade in coral and other marine curios. MCS persuaded one major UK high-street stockist to discontinue its product lines and, on a wider public level, the media interest spread the message to thousands of consumers.

The lobbying of Parliament on key policy issues has always been an important aspect of the work of MCS but several proposed new bills heightened this activity. Of most concern was the UK Government's Environment Bill, which proposes the reorganisation of the UK Government's advisor on conservation, the Nature Conservancy Council, into regional units. MCS totally opposes such moves as the marine environment does not pay homage to political boundaries and thus its protection would be weakened. On a European level too, consultation work continued on several key Bills, including a proposed EEC Habitats Directive.[18]

6.2.17 Marine Forum for Environmental Issues (MF)

The MF aims to enhance communication between the Government, NGOs and other interested parties concerned with the conservation and management of the marine resources of the North-East Atlantic region and raising the environmental awareness of all concerned, including decision makers. The Forum was set up for the Second North Sea Conference and now comprises more than 250 members from industry, pressure groups, environmental NGOs, independent scientists and Government in the UK and Northern Europe.

The MF received support from the Department of the Environment through its special grants programme, along with the WWF. The Sea Fish Industry Authority and ICI also made contributions.

In 1989 the Marine Forum changed its name to The Marine Forum for Environmental Issues and in August 1989 the MF officially separated from the Environment Council (formerly CoEnCo). Along with these changes it was necessary to give the MF a legal framework: at an MF Steering Committee meeting in October it was decided to seek the appropriate legal status and articles of association were drawn up.

Meetings
In January the first MF meeting took place at the DoE with the then Junior Environment Minister, Mrs Virginia Bottomley MP, as host. Dr P C Ried, the DoE North Sea Coordinator, gave an overview of the Government's North Sea policies with specific reference to North Sea Task Force activities. The Minister answered questions from Forum members and guests.

As a follow-up of this successful event another meeting took place between the Environment Minister and NGO representatives.

The second MF meeting took place in March at Regent's College and addressed the issue of the use of Nuvan — a red list substance which is used to combat sea lice in salmon — in the fish farming industry. Ms Alison Ross from the Marine Conservation Society and Mr Phil Dobson from Ciba-Geigy where the speakers at this meeting, which highlighted the controversy over the use of Nuvan and the problems of the fish farming industry. Dr Cato ten Hallers (Dutch MF member) from the Werkgroep Noordzee gave an update on the preparation of their Third North Sea Seminar in Rotterdam.

In June 1989 cetaceans, seals and the North Sea Task Force were the topics of the third MF meeting which was held in Regent's College. An update of the 1988 seals epidemic and a forecast for the 1989 breeding season were given, as well as a report on sighting surveys, strandings and threats to UK cetaceans and the work of the British Delegation of the North Sea Task force.

The Heriot-Watt University in Edinburgh was the venue for the fourth MF meeting. The issue of concern was fish farming in Scotland and the role of the Crown Estate Commission. Environmental problems linked to the farming of shellfish and salmon were highlighted and the licensing procedure of the CEC as well as the different approaches of industry and conservationists to fish farming were discussed.

Conferences and Seminars
At the end of May the Werkgroep Noordzee's 3rd North Sea Seminar, co-organised by the MF, took place in Rotterdam. The Administrator attended Seas at Risk meetings in Ross-on-Wye, Amsterdam, Brussels, Copenhagen and Bonn, and reported to the attending NGO representatives from different European countries on MF activities and latest UK North Sea policies.

Two working groups were set up in the UK during 1989, at which the MF is represented: a coastal zone management WG and a marine NGO WG which aims at coordinating the UK input into Seas at Risk. The MF also took part in a UK liaison working group at the DoE in July.

Marine Forum 1990 North Sea Report
In June 1989 the authors of the 1987 North Sea Forum Report were asked to update where necessary their original contributions and five new themes were chosen as additional papers to the new Marine Forum 1990 North Sea Report. The new papers relate to sea level rise, nutrients, coastal zone management, cetaceans and underwater archaeology.

Abstracts of the Marine Forum 1990 North Sea Report papers were submitted to the North Sea Secretariat in The Hague and the 6th PWG, as well as to a meeting of senior officials. The official launch of the Marine Forum 1990 North Sea Report took place at a joint conference with The Royal Geographical Society on 31 January 1990. [19]

6.2.18 US National Research Council (NRC)

The Marine Board of the NRC is an independent, non-governmental, objective advisor in the marine field. Under the Marine Board, groups of volunteer experts analyse technical and policy issues in order to:

- ensure environmental protection, safety, and competitive engineering, in marine structures and systems;

- improve the technical basis for ocean and coastal resource uses and developments;

- advance the marine engineering and technology base as a resource in policy formulation, program planning, and management.

The Board studies technical questions bearing on the nation's ability to meet its marine goals. It involves engineers, scientists and others in the NRC's marine- oriented activities by appraising the feasibility of proposals to use the oceans and their resources, reviewing government policy alternatives, and evaluating the influence of technical advances on public policy. It often reviews the state of research, engineering, and technology in relevant fields, and projects future needs, including long-range research programs. Members of the Marine Board and of its committees and staff serve as observers and participants in national and international advisory bodies and technical organisations to exchange information and technology and to improve professional communication on ocean engineering and maritime transportation issues.

1989 Activities
The widening scope of the Marine Board's studies reflects growing national concern about the ocean and its resources. Several studies completed or underway in 1989 examine marine environmental protection issues or safety of marine operations related to those issues. They share a common objective   of improving the scientific and technical

knowledge base for understanding and predicting the potential environmental effects of various uses of the ocean.

In 1989, the following studies addressed various aspects of ocean pollution:

- *Using Oil Spill Dispersants on the Sea* (published March 1989) assesses the effectiveness of chemical dispersants in removing oil from the sea surface and addresses their impact on biological resources;

- *Alternatives for Inspecting Outer Continental Shelf Operations* (published February 1990) evaluates current operating and inspection procedures for offshore oil and gas facilities and addresses concerns about environmental safety in addition to personnel safety;

- *Our Seabed Frontier: Challenges and Choices* (published November 1989) assesses existing and potential uses of the US Exclusive Economic Zone seabed and identifies policy issues and actions needed to provide sound resource management. These uses include current and potential waste disposal problems and oil and gas exploration and production;

- *Contaminated Marine Sediments: Assessment and Remediation* (published October 1989) addresses technical and engineering problems associated with cleaning up hazardous wastes and contaminated sediment at underwater sites. The study finds that solutions are variable, site-specific and costly, and urges source control of pollutants as a more effective strategy than remediation;

- *Managing Troubled Waters: The Role of Marine Environmental Monitoring* (published March 1990) examines marine environmental monitoring programs and recommends ways to improve their effectiveness and usefulness to environmental managers and decision makers;

Guillemots and other divers are most susceptible to oiling. Photo: RSPB/M.W. Richards

- *Assessment of Tank Vessel Design Alternatives*, a study in progress, is reviewing environmental, safety and economic implications of alternative designs for tank vessels, such as double hulls and segregated ballast tanks, and will update what is known about their effectiveness in preventing pollution.[20]

6.2.19 Royal Society for The Protection of Birds (RSPB)

The RSPB is a UK non-governmental organisation, with a membership of over 740,000 including its junior branch, the Young Ornithologists Club. Its principal objective is the conservation of wild birds and their habitats, and this is achieved by detecting and responding to threats; scientific research; enforcement of protection laws; and education work including the publication of literature and production of films. The Society also owns or manages 113 nature reserves throughout the United Kingdom, covering more than 72,000ha.

For many years the RSPB coordinated a survey of oiled/beached birds in the UK. However, its routine national surveys, results of which have been published in previous editions of the *ACOPS Yearbook*, were discontinued in 1986. Some regional surveys have been retained. The RSPB is maintaining much of its network of volunteer contracts, and retains the capacity to monitor and respond with advice, press comment and liaison in the event of serious oil pollution incidents. Summary details of incidents involving 50 birds or more in 1988 and 1989 are given in Table 1.

More bird casualties were recorded in 1989 than in any year since 1985, and a downward trend has reversed for the first time in six years. The number of incidents affecting over 50 birds declined steadily after the worst years in the 1970s. In 1987 there were only two such incidents, in 1988 only one. In 1989, however, the figure suddenly jumped to six incidents which killed at least 1,956 birds. Since research shows that many oiled birds are never recovered or recorded, this figure is an absolute minimum.

The impact of oil spills on birds is mainly a matter of chance, depending on time and place. The collision between two tankers which released 750 tonnes of oil into the sea off the Humber Estuary in September 1989 for example only escaped becoming a major catastrophe by luck. If the spill had occurred 20 miles to the north, seabird colonies would have been hit. The death toll from the slick which leaked from a pipeline on the Mersey in August 1989 would have been much worse in different winds, or if it had occurred in the winter when tens of thousands of wildfowl are present in the estuary. Prosecutions followed both this and the Solent incident listed in the table.

Many major seabird colonies around the UK are now slowly declining, from a peak in 1980 (see 2.5). the reasons for this are not fully understood, but avoidable mortality from oil pollution must be prevented at all costs.[21]

Table 1: Oiling incidents involving more than 50 birds in 1988-89

Date	Location	Oil type[1]	Minimum no. of birds affected[2]	Species
August 88	Clyde approaches	Medium/heavy fuel oil, little weathering, and fresh heavy fuel oil	125	Guillemot: live but heavily oiled
December 88- January 89	South coast of England	Fresh heavy fuel oil	770	Mainly guillemots: also razorbills, eider, mallard, mute swans, great crested grebe, great northern diver and red-throated diver
May- June 89	Sutherland	Probably weathered crude	216	Mainly guillemots: also razorbills, puffin, fulmar, gannet, shag and Arctic skua
August 89	Mersey Estuary	Fresh Venezuelan crude (pipeline fracture)	350	Cormorant, great crested grebe, little grebe, shag, heron, shelduck, mallard, guillemot, gulls spp.
September 89	Northumbrland	Heavy fuel oil with some weathering	150	Almost all guillemot, but also puffin, razorbill, gannet
September 89	Whitby, Yorkshire	Heavy fuel oil with some weathering	70	Almost all guillemot: also puffin, fulmar, gannet
October 89	Solent, Hampshire	Syrian crude	400	Teal, mallard, shoveler, mute swan, and some waders
			2,081	

1. Analysis courtesy of Laboratory of the Government Chemist.
2. Excludes live oiled waders, gulls and most wildfowl: includes all dead oiled birds and those live oiled species whose chances of recovery are very low (eg divers, auks, seaduck).

6.2.20 Seas at Risk Federation (SAR)

Seas at Risk is a federation of national and international European environmental organisations, working on marine issues. It was founded in 1986 with the aim of improving the exchange of information on marine pollution issues between national and international environmental organisations, so as to be able to better follow and comment upon the processes of international policy making and legislation in this area. Reports on national and international developments in the field of the marine environment are published four times a year in the North *Sea Monitor*.

Seas at Risk also coordinates international activities aimed at increasing the public awareness of marine pollution. The press-information bureau Seapress publishes press releases on developments relevant to the marine environment and provides the media with background information.

Presently 12 national and 3 international organisations participate in Seas at Risk.

Evaluation of the Second North Sea Conference
In the spring of 1989 an extensive questionnaire on the implementation of the Second North Sea Conference was sent out to the Seas at Risk participants and to civil servants of the North Sea countries and the EC. The results of the survey have been summarised in a contribution to *The North Sea; perspectives on regional environmental cooperation*, published by the *International Journal of Estuarine and Coastal Law*. The contribution is entitled *The Second North Sea Conference; national implementation*.

Preparation of the Third North Sea Conference
The following documents were submitted to meetings of Senior Officials (PMSO) and the Preparatory Working Group (PWG) of the Third North Sea Conference:

- PMSO, October 1988: general Seas at Risk Statement;
- PWG 4, July 1989: incineration of toxic wastes, policies for effluent waters, coastal zone management and wildlife;

- PWG 5, September 1989: discharges of dangerous substances, coastal zone management, shipping: enforcement of MARPOL provisions in North Sea Coastal States, Shipping (port facilities and Special Area MARPOL Annex I).

The information contained in the above documents was compiled in the recently published Seas at Risk documents *Changing Tack* and *Coastal Zone Management.*[22]

6.2.21 Tidy Britain Group (TBG)

The Tidy Britain Group is an independent, voluntary, non-profit making organisation, registered as a charity. It is recognised by the UK Government as the national agency for litter abatement, and is professionally staffed. Its purpose is to protect and enhance the amenities of town and country in the UK, particularly by promoting the prevention and control of litter and by encouraging environmental improvement schemes.

Recent Activities

Marine litter database for Canada Clean World International (CWI) is the international umbrella organisation for the various national agencies tackling the litter problem. At the present time the TBG is assisting various organisations in the development of marine litter monitoring initiatives. For example, Outdoors Unlittered (the Canadian agency) intends to develop a marine litter research programme for the Pacific coast, to assist implementation of the MARPOL Annex V regulations.

University of Malta TBG and The Central London Polytechnic are presently supervising a PhD project examining the types and distributions of marine litter in The Mediterranean Sea.

Second International Conference on Marine Debris, Honolulu, Hawaii, 2-7 April 1989 This was a major gathering of marine litter experts from across the world. TBG/ACOPS presented a paper on the various information, education and training initiatives which are required to assist implementation of the MARPOL regulations.

It was also evident from this conference that the UK will be the first country to evaluate the effectiveness of the MARPOL Annex V regulations. During August 1989, the first in a set of 'repeat' or 'after' surveys were undertaken in Scotland. The results were not very encouraging, with a tenfold increase in beach litter from Cape Wrath to Aberdeen between 1981 and 1989, much of it discarded at sea since MARPOL Annex V entered into force on 31 December 1988.

One of the more positive results of the conference is the likelihood that TBG will be contributing to the development of a marine debris survey methodology handbook in cooperation with the University of Washington.

HELMEPA workshop on the elimination of garbage from the Mediterranean Sea, under the auspices of the EEC, Athens, June 1989 The aim was to establish a Mediterranean-wide marine environmental protection-association, based upon the HELMEPA model. Most of the Mediterranean states were represented and the focus of attention was marine litter and the need to designate the

Discarded cans and a barrel on a British beach. Photo: WWF (UK)

Mediterranean as an effective Special Area for the purposes of MARPOL Annex V. TBG/ACOPS presented a paper on the types, quantities and distributions of marine litter in the Mediterranean Sea.

Since the workshop TBG/ACOPS will be contributing to a number of marine litter monitoring initiatives in Turkey and Malta (see above).

Third International Congress on Beach Preservation, San Sebastian, Spain, May 1989 TBG presented a paper concerning the need for educational initiatives to assist implementation of MARPOL Annex V.

Plastics Waste Institute, Association of Plastics Manufacturers Europe, Brussels, 14 March 1990 The Tidy Britain Group was invited to present a paper on the monitoring of plastics in the marine environment of Western Europe. This was the first meeting of the newly established Institute, and the long-term monitoring of plastics waste was high on the agenda. It is the intention of the TBG to repeat beach surveys in Europe which were initially undertaken more than 10 years ago.

The Blue Revolution This is a series of television programmes examining marine pollution issues. It is being prepared by The Australian Broadcasting Corporation in cooperation with Japanese and USA television companies. One programme will deal with pollution in the North Sea, and in this context, TBG's marine litter research programme is included. The programmes are due to be screened in the UK in 1990 on the ITV network.

QE2 During 1988 the TBG received a complaint from Australia concerning quantities of ships' garbage recovered on beaches in New South Wales. Some of this originated from the QE2. On behalf of CWI, the matter was taken up with Cunard by TBG. Cunard's response stated the vessel would be fully equipped, at great expense, to meet the MARPOL Annex V requirements by 31 December 1988. On 5 July 1989 the *Today* newspaper ran a headline story, based upon a videotape taken by two stewards showing hundreds of plastic sacks containing garbage being tossed overboard, under the cover of darkness in the Caribbean. After an investigation, the Department of Transport decided not to pursue the matter in court.

Survey of hazardous items/substances recovered on UK beaches An ACOPS/TBG survey is presently being planned on this subject with funding from DoE and DTp.

How to reduce litter discards from holidaymakers on beaches As part of its pilot project programme', the TBG has discovered the most effective means of preventing beach pollution by the holidaymakers. The amount of litter dropped on a beach in East Anglia was reduced by 90% after litter bags were issued free of charge to holidaymakers.[23]

6.2.22 World Wide Fund for Nature UK (WWF UK)

WWF is an international conservation organisation which raises funds for the conservation of wildlife and natural resources. Founded in 1961, the international headquarters are in Switzerland, and there are national organisations in 23 countries. WWF UK uses a third of the money it raises to grant aid to conservation projects in Britain and its dependent territories, with the remaining two-thirds going to WWF International for international projects.

Internationally, the seas have always featured in WWF's conservation programme, whether related to the protection of coral reefs and marine parks, conservation of monk seals and marine turtles, or by involvement in treaties and legislation such as CCAMLR and UNLOCS. WWF UK has been increasingly involved with marine conservation over the past few years, and in 1989 a marine unit was set up within the UK Conservation Department. A Marine Conservation Officer, Fisheries Conservation Officer and North Sea Officer have been appointed to address some of the major problems facing the marine environment.

WWF UK is continuing to work very closely with other non-governmental organisations concerned with various aspects of marine conservation. WWF UK has continued to work in partnership with the Marine Conservation Society and to fund various aspects of its work. It is supporting both the Society's general work and also specific areas, such as the MCS Conservation Officer based in Scotland, whose particular remit is to work on the environmental impacts of fishfarming.

An important project, funded by WWF UK, is the coastal and sea use management plan being produced by MCS. A review of approaches and techniques has already been published and detailed information on coastal zone uses, pressures and potential threats and concerns is being collated for future planning. Alongside this study, a review of marine legislation has been undertaken, which will event-

ually lead to proposals for more effective legislation. The continued funding of ACOPS by WWF UK has meant that lobbying and involvement on international legal aspects of the marine environment are also being addressed.

Investigations into the recent breeding failure of seabirds in Scotland funded by WWF is being followed up by a study to look at the interactions of seabirds and sand eels. This three year study is a joint initiative with financial or practical support from NCC, WWF UK, DoE, RSPB, DAFS and SDD and will increase knowledge about the role of the sand eel in the North Sea ecosystem. Any resulting recommendations on regulation of the sand eel stocks should help conservation of the seabirds and the fishery.

Other projects supported by WWF UK have included studies on resident dolphin populations and investigations into the interactions between fisheries and marine mammals. Voluntary marine conservation areas have also been supported by WWF UK.

Joint work with RSPB is looking at the problems of monofilament nets and associated bird deaths, the pressures for marina developments around the coast and their effects on wildlife, and an estuaries survey to create a database of information to enable the link between threats and consequences to wildlife to be made.[24]

6.3 AMENITY AND STATUTORY BODIES

6.3.1 International Transport Workers' Federation (ITF)

The ITF is a federation of over 400 transport workers' trade unions in 90 countries. Founded in 1896, it is organised in eight industrial sections: seafaring, docks, railways, road transport, civil aviation, inland navigation, fisheries and travel bureaux. It represents the interests of transport workers at a world level through its input into international organisations and by organising international solidarity.

The ITF has a deep and long-standing interest in the issue of marine pollution, particularly the direct risks to seafarers that arise from spillages and accidents at sea. It is also concerned about threats to the environment and the finite character of natural resources. The Federation and many of its affiliates have positive experience of cooperating with environmental pressure groups on issues of common interest.

The ITF maintains that profitability should not be the all-dominating factor governing maritime safety. A major part of its work consists of campaigning against the Flag of Convenience system in shipping, where out of narrow financial considerations shipowners adopt the flag of countries whose authorities do little to ensure adequate standards of ship design and maintenance, crewing levels and safety procedures.

The ITF also supports the enforcement of the best practical safety measures for all classes of shipping. For example, oil tankers should be afforded the protection of double hulls which is already a feature of other bulk carriers but which has been resisted by tanker owners on cost grounds.

In arguing for strict penalties to be levied against those who breach safety regulations and cause marine pollution, it insists that individual employees should not be made into scapegoats by the legal system where where the cost-cutting policies of big corporations are at fault.

The ITF has been vocal in condemning the use of the seas as dumping-grounds for hazardous waste from the industrialised world.

At its special inter-sectional conference on the transport of dangerous goods, held in Geneva on 30 November–1 December 1989, the Federation considered how it could increase its activities in the environmental and occupational health and safety fields. The conference adopted a resolution calling on governments to ratify and implement the Basle Convention on the Control of Transboundary Movements of Hazardous Wastes and their Disposal. It demands that the burning of Hazardous Wastes and waste at sea be prohibited as soon as possible.

The ITF seeks to highlight the issue of marine pollution through its publishing and education work, making use of its presence in every continent of the world.[25]

6.3.2   National Federation of Fishermen's Organisations (NFFO)

The NFFO is the representative body for fishermen and vessel operators in England and Wales. Comprising 60 constituent associations, the Federation undertakes a wide range of representational activities at local, national and community levels.

Pollution of the marine environment has become a matter of central concern to the Federation's members and these concerns are reflected in NFFO policy and representational efforts.

Very subtle environmental changes are crucially important in determining survival of fish larvae and juvenile fish and so dramatically affect recruitment. It is our fear that subtle environmental changes caused by small increases in dissolved nutrients combined with the synergistic effects of low levels of poisonous substances persistent in the food chain have already reduced the survival of fish larvae and juvenile fish and the fecundity of mature fish, thereby having a significant impact on recruitment.

The widespread discharge of wastes into the sea, combined with inadequate scientific data on the effect of such chemical 'cocktails', has led the Federation to advise a cautionary approach and to press for the implementation of stringent measures to control the discharge of waste that contains toxins which are persistent in the food chain and nutrients which increases the likelihood of eutrophication.

Of particular concern to the Federation is the continued discharge of sewage direct to the sea via long-sea outfalls; a practice now abandoned by other North Sea Member States. This issue is the subject of representations with the relevant water authorities, the Department of the Environment and the Ministry of Agriculture, Fisheries and Food.

The Federation has recently lobbied the British Government to support the proposed EEC Directive concerning municipal waste water treatment, stressing the potentially damaging impact of untreated sewage on the productivity

of the marine environment. With regard to nuclear power stations, the Federation maintains a monitoring brief on the level and sources of radioactive discharges through direct representation on the power station liaison committees.

In conjunction with ACOPS and other interested organisations, the Federation has campaigned to minimise the impact of the abandonment of disused offshore oil and gas installations on the marine environment and on fishing activities in particular. Close consultations with Fisheries Departments and the Department of Energy are continuing with the purpose of identifying the most sensitive areas where complete removal will be mandatory.

The dumping of colliery waste and flyash on the northeast coast of England has led to a local decline of fish and shellfish stocks. The NFFO has, therefore, supported its constituent associations in their attempts to persuade the industries responsible and the statutory bodies to seek less environmentally damaging alternatives.[26]

6.3.3   Scottish River Purification Boards' Association (SRPBA)

The Association represents the seven independent river purification boards who administer the United Kingdom legislation relating to quality on the mainland and inshore islands of Scotland. In the offshore islands water quality is the responsibility of the Islands councils in respect of their own discharges.

The purification boards are also responsible for hydrometry and are the principal source of information on rainfall, river flows and water resources in Scotland.

The Association provides a forum for discussion, and ensures coordination of approach on major issues. Each Board is, however, responsible for the quality of waters within its own area, setting appropriate quality standards and objectives for these areas. Because of this only general aims and objectives can be stated for the association namely:

- the quality of waters will be maintained at, or improved towards, standards commensurate with the ultimate use of these waters, including the conservation of the natural environment;

- Members of the Association will make full use of their powers under existing national or international legislation for the control of potentially polluting discharges, and the prevention of pollution incidents through accidents or spillages;

- Members of the Association will assist Government in their fulfilment of obligations under relevant international conventions;

The Association will initiate or participate in the promotion of action or legislation necessary for the elimination of pollution of the seas or other natural waters.

Current Activities
1989 saw the implementation of the Water Act, Schedule 23 of which amended the Control of Pollution Act 1974 as it applies to Scotland. Under the new Act the Boards

received unequivocal confirmation of their powers to control discharges from cage fish farms.

All of the Boards have been developing their expertise for the analysis of Red List substances, under the EC Dangerous Substances Directive, although some are more advanced than others. Where monitoring of Red List substances has been carried out, contaminant levels were generally found to be very low, often at or below detection limits. Significant sources of PCP from three identified sources and cadmium from one major source were all eliminated from the Forth catchment. The Clyde RPB identified cadmium, mercury, HCH (Lindane), DDT, PCP and carbon tetrachloride in discharges to saline waters, but in no case did these exceed the appropriate EQS.

Results from the 1989 Bathing Waters surveys (under the Bathing Waters Directive) showed some improvement over the previous year. In 1988 12 (52%) of the designated bathing beaches passed the mandatory coliform standards, while in 1989 16 (70%) beaches passed. This may have been due at least in part to the warm, dry weather enjoyed in 1989. Two of the Clyde RPB's seven bathing waters passed the coliform standards, whereas all had failed the previous year. Clyde RPB also tested twice for enterovirus and salmonella; enterovirus was detected at five, and salmonella at six, of the seven beaches.

Under the Shellfish Directive compliance with bacterial standards again proved more difficult than compliance with physico-chemical standards. The Clyde RPB reported that all their sites failed the coliform standard and some failed for chromium. SDD has allowed a reduction in those determinants which complied.

River flows were generally low in 1989. Forth RPB reported its driest ever period with river flows over 50% below the long-term average. Nevertheless, estuarine water quality was maintained. Reductions in organic loading from Stirling sewage treatment works and DCL Menstrie helped in this. The capture in the estuary, for the first time in several decades, of the sparling, or cucumber smelt (*Osmerus eperlanus*), was regarded as a welcome sign, as this species is relatively pollution-sensitive. In summer 1989 in the uppermost reaches of the Clyde Estuary dissolved oxygen levels fell to below 10% saturation, although the rest of the estuary was in fair condition.

All Boards carried out regular monitoring work, as well as special surveys, both hydrographical and biological. Both the Clyde and Highland RPBs have significant numbers of open water cage units and land-based fish farms in their areas, and each carried out a number of surveys on both water and sediment quality at cage fish farm sites. In general the only detectable effects on the seabed were limited to a zone no greater than 30m around each cage. Monitoring programmes will be continued. The North East RPB carried out a number of investigations into the effectiveness of existing sewer outfalls. In Irvine Bay the Clyde RPB have made extensive investigations into the effects of discharges from the Beechams Pharmaceutical and Irvine Valley Sewer long sea outfalls. Results suggested that settling sewage solids had the greatest effects on the benthic communities up to 300m from the outfalls. Clyde RPB tested a random walk computer model to predict the likely distribution of faecal coliform bacteria discharged from long sea outfalls. Using Irvine Valley Sewer as a test case they found that the model provided useful information.

The effect of TBT on the dog whelk reproductive system has been used for monitoring purposes at a range of sites by the Solway RPB in Loch Ryan and by the Highland RPB. Highland RPB found various levels of contamination in areas used by shipping and around some fish farms.

A number of improvements in sewage disposal systems helped to maintain or improve water quality. The Kirkcaldy sewage disposal scheme became operational and resulted in significant improvements in the bacteriological quality of beaches. The Aberdeen long sea outfall came fully into commission, and the Abercrombie sewer outfall at the mouth of the R. Dee is programmed to be connected into the long sea outfall system in 1991. Regional Councils were encouraged by the Boards, partly as a result of bathing beach surveys, to improve sewage disposal schemes. An environmental impact assessment is currently being undertaken on the new Inverness sewage disposal system.[27]

6.3.4 Water Research Centre (WRC)

WRC is an independent research-based company whose scientific and engineering skills are applied to help governments and other regulatory agencies develop environmental policies and help utilities meet the required standards in the most cost-effective way.

WRC's principal UK customers are Central Government, the National Rivers Authority and the water utilities. It is active worldwide with a client list that includes the European Commission, overseas WHO, development banks, municipal authorities and industrial corporations.

The company operates from two bases; at Medmenham, Buckinghamshire (environmental management) and Swindon, Wiltshire (water and sewage management) with a current turnover of about £26 million. In total, WRC employs over 650 staff and maintains liaison offices in Stirling, Warrington, Hong Kong and Bologna, Italy. Its United States subsidiary is based in Huntingdon Valley, Pennsylvania. The Water Byelaws Advisory Service provides advice and testing services from its base in Slough.

Recent Activities
In 1989 WRC completed a major programme of research, sponsored by the Department of the Environment, to evaluate the use of physiological and biochemical indices of pollution induced stress in the marine environment. The common mussel, *Mytilus edulis*, was selected as the test organism and was deployed in estuaries, at sewage sludge disposal grounds, and in the vicinity of coastal discharges.

WRC has, on behalf of the UK water industry, undertaken intensive four-year research programmes into the environmental impact of discharges from long sea outfalls and the disposal of sewage sludge at sea. The work on outfalls was centred on two outfalls, at Tenby and Weymouth, and used both traditional methods of pollution detection and novel bioassay techniques.

The outer Thames Estuary was used as the case study site for research on the environmental impact of sewage sludge disposal at sea. Methods have included assessment of sediment quality and macrobenthic community structure, sludge dispersion studies, bioassays and studies into the prevalence of fish diseases. In 1989 additional work on fish diseases was carried out in Liverpool Bay.

The water quality of estuaries has been of major concern. Research has been undertaken into the factors influencing the oxygen balances of estuaries; and mathematical water quality models have been developed for a number of estuaries in the UK. This expertise has been utilised commercially in Hong Kong, Portugal and New Zealand with the use of predictive models to determine the impact on water quality of long sea outfalls and land reclamation.

Another related aspect is the setting of discharge consents using the Environmental Quality Objective/ Environmental Quality Standard approach which requires the establishment of a 'mixing zone' around the discharge. This often entails the use of mathematical models to predict the dilution and dispersion of the discharge. In 1988 this approach was tested at an outfall discharging industrial effluent into the Severn Estuary.

A continuing source of debate is the potential health risk associated with bathing in sea water. WRC managed a study, carried out on behalf of the DoE in September 1989, which for the first time measured the incidence of medically-significant symptoms in bathers. Following reporting of the results, this pilot study may be expanded to other sites in 1990.

WRC's marine research has culminated in the compilation of a *Design Guide for Marine Treatment Schemes*, which is due to be published in 1990. It presents in four volumes a planning framework to allow schemes to be effectively designed in both environmental and engineering terms.[28]

6.3.5   Water Services Association (WSA)

The WSA was established on 1 September 1989 and is an Association of water and sewerage undertakers appointed under sections 11 or 12 of the Water Act 1989. Currently its Members are the ten water service companies (WSCs) which are the successor companies to the water authorities.

The main functions of the WSA are:

* to promote and protect the common interests of the Members of the Association, ensuring that UK and EC institutions are made aware of Members' common needs and interests;

* to distribute accurate and relevant information about the water industry and provide public relations and public information services;

* to provide a forum for Members to discuss among themselves and with representatives of government and outside organisations matters relevant to the water industry.

Thre are 401 bathing waters identified in England and Wales under the EC Bathing Water Directive. Of these, 76% to currently comply with the Directive compared with 55% in 1985. Under current investment plans most of the designated waters should comply by the mid-1990s. The

full programme of compliance will be completed within ten years.

The WSCs play an important role in ensuring that the waste-water discharges from sewage treatment works and sea outfalls which affect the quality of water used by bathers meet the required standards. These standards are set and monitored by the UK Government.

In the UK 83% of sewage/waste-water is treated at land based sewage treatment works. About 13% of waste-water is pumped out to sea through sea outfalls. More than half of these are modern properly-designed long sea outfalls. Modern sea outfalls are now being built which take advantage of favourable marine conditions such as water depth and current. Headworks provide preliminary treatment and screening to remove grit, gross solids and plastics. The location of the actual discharge points from the outfall are fixed only after completing comprehensive investigations of the seabed, tides and currents and detailed site-specific mathematical models are made of both the effluent and receiving water quality.

The construction on long sea outfalls have been endorsed by the Royal Commission on Environmental Pollution, which stated in 1984: "With well-designed sewage outfalls we believe that discharge of sewage to the sea is not only acceptable but, in many cases, environmentally preferable to alternative means of disposal" (Cmnd 9149, 10th Report, February 1984). The alternatives are usually a land-based sewage treatment works, which can be unsightly and which require large areas of land.

Installing long sea outfalls is not just a matter of extending a pipe out to sea. The landward sewerage system has to be remodelled, pumping stations and screening/degritting plants have to be built. The water service companies have committed £1.1 billion over the next ten years to this costly engineering exercise.[29]

6.3.6   National Rivers Authority (NRA)

The NRA was formed as a result of the Water Act 1989; responsibility for the control of pollution in controlled waters of England and Wales under the Act was vested in the Authority as from 1 September 1989. The NRA has been created from what were, briefly, the ten 'Rivers' units of the Regional Water Authorities of England and Wales. It has therefore inherited the staff and resources of those units, but has quite new statutory powers under the Water Act. These include responsibilities for a range of matters concerning the water environment, only one of which includes that of water quality; the others relate to water resources, flood defence, salmon and freshwater fisheries, plus some navigation, conservancy and harbour authority functions. The Water Act also places general duties on the NRA to promote conservation and enhancement of the natural beauty and amenity of inland and coastal waters, and of the land associated with them, and to promote their use for recreational purposes. Furthermore, it places a duty on the NRA to make arrangements for the carrying out of research activities in support of all of its functions.

The NRA has retained the regional structure of the previous Water Authorities; there are therefore 10 regional

NRA units, plus a central Head Office. A national network of 11 laboratories is being established to undertake all of the pollution control work.

Under the Water Act (1989), the NRA has statutory duties and responsibilities relating to the aquatic environment which are both general and specific. It is responsible for the quality of freshwaters, estuarine waters, and territorial waters to a distance of 3 nautical miles from the shore. The NRA thus consents discharges from land-based sources, monitors both effluents and receiving waters, and has widespread powers to prosecute for pollution offences . It is also the competent authority for monitoring for compliance with EC Directives, relating to matters affecting water quality, on behalf of the Department of the Environment.

In the first few months of its existence, the NRA has made a wide-ranging review of the basis for consenting discharges and of demonstrating compliance. It has also undertaken reviews of the basis upon which statutory water quality objectives may be introduced in the future, paying particular regard to the use to which waters may be put — including estuarine and coastal waters.

Criteria for the development of a classification scheme for all waters have similarly been reviewed. This work has been carried out in order to assist the UK Government in deriving water quality objectives; the NRA has a statutory duty to ensure that these will be met, both by its consenting of discharges and enforcement powers. Whilst the latter are not the only means of combating pollution, the rate of prosecutions has greatly increased. A record fine of £1M for Shell UK resulted from a prosecution brought about by the NRA in connection with the release of 150 tonnes of crude oil into the tidal reaches of the River Mersey in August 1989.

Monitoring

The NRA has taken steps to harmonise the monitoring of bathing waters in England and Wales in connection with Directive 76/160/EEC, particularly with regard to testing for salmonella. The results of the 1989 survey indicated that 304 of the 401 designated beaches (76%) met the mandatory coliform bacteria standards, compared with only 66% in 1988. Although this improvement was, in part, the result of capital improvement schemes, the principal factor was considered to be the good summer weather, which reduced the operation of storm water overflows and increased the rate of bacterial die-off.

The NRA has a large R&D programme in support of all of its functions. Projects currently funded in relation to pollution of estuaries include those on: sediment/water contaminants, estuarine modelling, aerial inputs to the Severn Estuary, and biological monitoring to evaluate the clean-up of the Mersey Estuary. Coastal water research includes studies on the environmental impact of long-sea outfalls for sewage discharges, and on the monitoring of nutrient levels in the Irish Sea. These research activities will be increased during 1990/1991, priorities being assessed as the NRA settles down into its new and important role in controlling pollution of marine, estuarine and inland waters.[30]

6.4 SHIPPING AND INDUSTRY

6.4.1 Contract Regarding Supplement to Tanker Liability for Oil Pollution (CRISTAL Ltd)

CRISTAL Ltd was established in 1971 to administer a contract now entitled Contract Regarding a Supplement to Tanker Liability for Oil Pollution. This Contract, often referred to as 'the CRISTAL Contract', provides for compensation on a voluntary basis in respect of oil pollution damage, caused by persistent oils carried on tankers, over and above that paid by tanker owners up to limits as stated in the Supplement to the TOVALOP Agreement (see 6.4.6). This additional compensation is funded by the parties the CRISTAL Contract, such parties in the main comprising oil companies, but including some other users of oil, such as power companies. Currently, well over 700 companies are parties to the Contract under which compensation totalling some US $70 million has been paid since 1971.[31]

The current version of the Contract will not apply to incidents occurring after 1992, unless the parties to it decide to extend its operation for a further period. The Contract applies to incidents wherever they may occur throughout the world and the only prerequisites for its application are: (1) that the cargo on the tanker involved must be owned by a party to the Contract; and (2) that the tanker owner must meet his financial obligation up to the limit as stated in the Supplement to the TOVALOP Agreement. The amount of compensation which can be made available under the CRISTAL Contract is a function of the size of the tanker involved in an incident; aggregated with the compensation paid by the tanker owner, the maximum can range between US $36 million and US $135 million.

During the period under review, compensation claimed and paid under the Contract has been low in amount, but CRISTAL Ltd now faces a number of outstanding claims the amounts of which in total could equal, or even exceed, the total of all claims paid since 1971.

6.4.2 Institute of Petroleum (IP)

The IP has long established roots in the marine environmental area and has sponsored associated programmes of research for over 20 years under the general coordination of its Marine and Freshwater Environment Committee. However, until 1989 there was no focus in the Institute for activities in the atmospheric environmental area. In view of the current importance of atmospheric matters to the industry (e.g. vehicle exhaust emissions, the greenhouse effect, climatic change etc.) an Environment Committee was set up by the Institute on 1 December 1989. The new committee will appoint and control Task Groups which will address environmental issues including, atmospheric, open sea, inshore, estuaries, inland water and ground and groundwater contamination by oil.

Dispersants Working Group

During 1989 a research project on the accelerated storage stability of oil spill dispersants sponsored by the Institute was completed at Warren Spring Laboratory (WSL). Although some deterioration had occurred in the commercial dispersants tested they were all still fit for purpose at the end of the research programme.

The main recommendations arising from the project were that dispersants should, wherever possible, be kept in the manufacturers' containers, and according to manufacturers' instructions and, as far as possible, the bulk storage of dispersants in mild steel containers should be avoided.

An executive summary of the findings of the project was published in the January 1990 issue of *Petroleum Review* whilst copies of the final report are available from WSL. An ambient temperature storage programme of work on the same eight oil spill dispersants will continue until 1992 when a separate short report will be issued.

With the advent of the single Europe in 1992, and the attendant requirement for the Member States of the Community to accept each others tests and standards, subject to safety considerations, Warren Spring Laboratory, with Institute support, is putting forward its efficiency test for oil spill dispersants to the British Standards Institution (BSI) for approval as a British Standard which can then be submitted for consideration as a CEN (European Committee for Standardisation) standard. In a similar way, the Ministry of Agriculture, Fisheries and Food (MAFF) is seeking BSI approval for its toxicity test for oil spill dispersants.

Environment Research Advisory Group

In order to reflect the wider remit of the Institute's new Environment Committee, the Marine Environment Research Advisory Group (MERAG) was renamed the Environment Research Advisory Group (ERAG).

In 1989, MERAG/ERAG monitored the following Institute-sponsored environment projects within a budget of £48,000:

- importance of hydrocarbon input as an energy source for benthic production near natural oil and gas seeps;

- supralittoral lichens and atmospheric pollution;

- growth and population dynamics of *Echinus esculentus* around oil installations in Sullom Voe;

- use of an amphipod (sand hopper) as a heavy metal monitor species.

Plans for 1990

The Environment Committee will undertake the following activities in 1990:

- agree procedures for formulating an Institute response to environmental issues in the future;

- sponsor a review of ground and groundwater contamination by oil seepage and spillage in the UK;

- establish a Safety and Environmental Discussion Group to

- organise four early evening meetings per year at which specialist speakers will give presentations on new safety and environmental issues;

- organise an autumn conference on Climatic Change.

Only a few new research projects will be taken on in 1990 because during the year the new Committee will be establishing its remit and assigning work to Task Groups. However, through ERAG, the Committee will continue to monitor sponsored projects which started in 1989 and have a lifespan of more than one year.[32]

6.4.3   International Association of Independent Tanker Owners (INTERTANKO)

INTERTANKO's uniqueness is that it only works for independent tanker owners from 33 maritime countries and represents 85% of the world's independent tanker fleet. The Association cooperates with other organisations on issues of mutual interest.

INTERTANKO's role as a service association, apart from its task as a spokesman for the industry, is well known and widely appreciated. The information service is the platform upon which many other activities rest. Its established Freight and Demurrage Informations Pool has secured considerable savings for the members by speeding up settlement of long overdue claims. INTERTANKO will continue to direct its attention to the problems being discussed by IMO on safety at sea, as this work also has an important consequence in the prevention of marine pollution.

Activities during 1989

INTERTANKO has been at the forefront of promoting US ratification of the internationally agreed treaties regarding compensation for oil pollution damage, the so-called 1984 Protocols. In 1989, the Government and the public of the United States turned their attention to oil pollution liability and compensation problems and examined an array of possible solutions. INTERTANKO was an active participant in the process.

It is essential that the US ratify these Protocols because they are structured in such a way that, without the participation of the United States, which receives far more oil transported by ship than any other country, the criteria for entry into force for these treaties would be difficult to meet.

The Bush Administration has pressed for Senate advice and consent to ratification of the Protocols in the new Congress. This was in line with pronouncements by President Bush, who made it clear that environmental issues were to be given priority by his Administration.

The *Exxon Valdez* accident on 24 March 1989 gave a new impetus to the interest for oil pollution compensation in the States. The cumulative result was legislative uproar. A large number of Bills were prepared. The Bush Administration prepared its own version of comprehensive legislation to Congress and stated that it believed ratification of the Protocols should receive priority attention by the Senate. In the months that followed, a number of committees of both House and Senate held hearings and INTERTANKO participated by submitting statements in support of the Protocols to four of these committees.

At the time of going to press, the situation is that the Senate and the House of Representatives have adopted Bills which differ on the issue of US ratification of the Protocols; the House version envisages US ratification whereas the Senate version does not. A Conference of members of both chambers will meet to produce a joint version. The two Bills also contain language regarding mandatory double bottoms and hulls. Following an INTER-TANKO initiative, a message setting out the industry's main concerns has been prepared by ITOPF, INTERTAN-KO, ICS and OCIMF. The thrust of this message is to urge Congressmen to adopt legislation which does not commit the US to a future course of action based on decisions made in haste, and without the benefit of results of detailed investigations and international discussion.[33]

6.4.4 International Chamber of Shipping (ICS)

Some few months prior to the period covered by this Yearbook, a new Constitution was adopted by ICS introducing changes in the qualifications for membership. While the full members continue to be national associations of shipowners and ship operators, a new category of associate membership includes representative regional associations and, in certain circumstances, individual companies. Membership presently comprises 38 full members in 33 countries with associate members. A further important develpment is currently underway with the merger of the Secretariats of ICS and the International Shipping Federation (ISF). The organisational structure of ICS has also been streamlined to ensure that the technical committees through which consultation with members is maintained are effective in meeting the needs of the industry.

As with other international organisations, the activities of ICS in the area of pollution prevention have to some extent been influenced by the decision of the IMO Council in November 1988 to curtail the 1989 programme of meetings in the face of the reduction of funding which has built up in recent years.

Activities during 1989
The black spot of 1989 was undoubtedly the grounding of the *Exxon Valdez* in Prince William Sound, Alaska on 24 March. Inevitably there followed renewed calls for more stringent safety standards involving the design, construction and operation of oil tankers, together with coordinated oil spill response measures worldwide. During the following months the July Summit meeting of the heads of state of the world's seven wealthiest nations and the special meeting of IMO in September to agree the foundations of a programme of action were two visible signs of the moves set in train by the events in Alaska.

A major preoccupation of ICS has been the monitoring of US domestic legislation on oil spill liability and compensation. Intense efforts have been made by many parties to encourage the United States to ratify the 1984 Protocols to the 1969 Civil Liability and 1971 Fund Conventions and to reject calls for unlimited liability on the part of the carrier.

In close consultation with other representative organisations in the oil and tanker operating industries, ICS has also been anxious to see that the calls for measures to minimise oil spills, especially those aimed at modificaitons of tanker hull structures, will be discussed in IMO — the correct forum for effective implementaion of agreed action — and not become the subject of unilateral or regional legislation.

ICS is presently involved in studies on these subjects.

During the preparation of the draft Basle Convention on the Transboundary Movements of Hazardous Wastes, ICS was grateful for the opportunities extended by UNEP to present the views of the ship operator on relevant Articles in the new instrument.

Through the offices of ACOPS, ICS assisted the World Bank in carrying out an environmental study programme in the Mediterranean Sea area. During the first half of 1988 a survey was carried out into the availability of reception facilities in ports in the region for oily residues, chemical tank washings and garbage. The results of that survey, which were presented to the 26th Session of the IMO Marine Environment Protection Committee in September 1988, suggested that the position regarding oily residues and garbage appeared to be improving although the picture remained far from complete.

ICS began a worldwide survey into the availability of reception facilities at the beginning of 1990. From the results of past surveys it is expected that this exercise will reveal a continuing ambivalence in some administrations in their attitudes towards stricter oil pollution controls on ships without apparent willingness to provide reception facilities to meet the needs of those ships. The results of the survey will be presented to the IMO at the 30th Session of the Marine Environment Protection Committee in November 1990.

The publication of manuals for the guidance of shipping companies and seafarers on safe ship operation and pollution prevention remains among the foremost of ICS activities. During 1988–89 the third edition of *Peril at Sea and Salvage* was published, jointly with the Oil Companies International Marine Forum, as was a revised third edition of the *Clean Seas Guide for Oil Tankers*, taking account of changes in MARPOL.[34]

6.4.5 International Petroleum Industry Environmental Conservation Association (IPIECA)

IPIECA is an association of international oil companies and petroleum industry associations that provides a focal point of contact with the United Nations Environment Programme (UNEP) and other bodies on the impact of petroleum industry operations and on environmental protection. IPIECA has observer status with the UN Economic and Social Council (ECOSOC).

During 1989, the occurrence of several major marine pollution incidents has catalysed an industry-wide activity concerned with reviewing its capacity to respond. IPIECA considered its contribution through its Oil Spill Working Group membership. It concluded that it would seek further guidance through organising an International Workshop. This Workshop took place in Edinburgh during the week of 8 October. Seventy-five delegates attended drawn from oil spill cooperatives and from those within the international oil industry responsible for oil spill response.

Through a series of syndicated sessions the Workshop identified needs for guidelines on assessment of oil spill response facilities, contingency planning and an improved basis for credible communication to the concerned public relating to realistic expectations of industry in responding to major oil pollution incidents.

Through 1990 the IPIECA Working Group will be developing these proposals and will consider whether any final output is reported back to a future Workshop during 1991. The Working Group will also provide the IPIECA membership with an ongoing appreciation of international developments and the means to participate as appropriate. The Group will monitor the developments of the UN/IMO Convention on Oil Spill Response.[35]

6.4.6 International Tanker Owners Pollution Federation Limited (ITOPF)

The International Tanker Owners Pollution Federation (ITOPF) was established as a non-profit making service organisation in 1968 for the principal purpose of administering the Tanker Owners Voluntary Agreement concerning Liability for Oil Pollution (TOVALOP), a voluntary compensation regime offered by the world's tanker owners to meet the costs of cleaning up oil spills and compensating any damage caused. Whilst this remains an important function, the Federation also gives great emphasis to the provision of technical services.

Administration of TOVALOP

TOVALOP came into effect in October 1969 when it was applied to 50% of the world tanker tonnage. Today, just over twenty years later, some 6,200 tankers with a total tonnage of over 149 million gross tons are entered in TOVALOP, representing in excess of 95% of the present world tanker tonnage. In combination with its companion voluntary industry agreement, CRISTAL, (see 6.4.1) and the two international conventions developed under the auspices of IMO, TOVALOP provides an international system of compensation that is unique in the field environmental pollution.

TOVALOP is regularly reviewed to ensure that it remains responsive to current demands. One such review was completed in 1989, and at the AGM members of the Federation voted in favour of Special Resolution to amend TOVALOP with effect from 20 February 1990. Many of the amendments simplify or otherwise clarify existing provisions, whereas others are a consequence of the decision of the House of Lords in October 1988 on the *Esso Bernicia* case [Esso Petroleum Co. Ltd v. Hall Russell & Co. Ltd (Shetland Islands Council, third party) — The *Esso Bernicia* — [1989] 1 ALL ER 37]. At the same time the definition of Pollution Damage in the TOVALOP Supplement (which applies only when the tanker involved in an incident is carrying a cargo owned by a party to CRISTAL) is extended to exclude costs actually incurred in taking reasonable and necessary measures to restore or replace natural resources damaged as a direct result of the incident.

Response to Spillage

The Federation's small team of technical experts is ready to respond to marine oil spills anywhere in the world to advise on the most effective clean-up techniques and the mitigation of damage. This service, whilst not limited to cases where compensation may be sought under TOVALOP, is normally provided at the request of tanker owners and their P & I Clubs. However, both Cristal Ltd and the IOPC Fund also usually rely on the technical services of the Federation when they are faced with a major spill.

Calls on the Federation staff to attend on site at pollution incidents have increased steadily in recent years, despite a significant reduction in the incidence of major tanker spills in the 1980s as compared with the previous decade. The total number of spills attended on-site by Federation staff since 1970 now exceeds 200.

1989 was a particularly busy year for the Federation's technical staff with on-site attendance at 24 incidents, including the *Exxon Valdez* in Prince William Sound, Alaska, USA in March, which occupied three members of staff for much of the spring and early summer. A list of all the oil spill incidents attended on site by ITOPF during 1988, 1989 and the first two months of 1990 appears in Table 2.

Advisory Work

Because of the experience gained through attending major oil spills around the world, the Federation is often asked by governments, international agencies, tanker owners, oil companies and various other organisations to assist with the preparation of contingency plans and to undertake advisory assignments.

At the beginning of 1989 the third and final phase of a worldwide survey of oil spill response arrangements for the US Navy was completed. During the course of the whole study, Federation staff visited 121 different ports and produced reports describing the oil spill response arrangements in 128 different countries.

A number of contingency planning studies were completed during the period covered by this Yearbook, and a further major assignment for the CEC commenced. This involves reviewing and improving the format and content of the Community Information System for combating spills of hydrocarbons and other harmful substances.

Training and Information

The Federation has recently devoted considerable effort to providing and disseminating practical information on oil spill response techniques. Examples include a series of 12 Technical Information Papers and the production of a series of five 20-minute training videos in cooperation with IMO, EDC and Videotel Marine International. Books, produced originally by the Federation as supporting material for the videos, have now also been published in English, French and Spanish by Witherby & Co. Ltd of London, entitled, as are the videos, *Response to Marine Oil Spills*.

Regular training is vital if personnel are to respond effectively to an oil spill incident. The Federation has over the years run, and participated in, numerous training courses for government and industry personnel around the World. During 1988 and 1989, training courses involving Federation staff were held in Brazil, Colombia, Egypt, France, Denmark, Italy, Norway, the People's Republic of China, Singapore, South Korea, Spain, Venezuela and at various locations in the UK.[36]

6.4.7 Oil Companies' European Organisation for Environment and Health Protection (CONCAWE)

CONCAWE, founded in 1963, is the oil companies' European organisation for environmental and health protection. There are 36 member companies, representing 90% of the oil refining capacity in the European OECD countries. These companies coordinate the work programme and provide both funding and sources of expertise to carry out CONCAWE's scientific and technical functions. The original remit of air and water pollution has been expanded to include consumer and employee health protection and, most recently, certain safety aspects. In January 1990 the CONCAWE Secretariat relocated from The Hague to Brussels.

The Water Quality Management Group is one of the eight different groups with which CONCAWE spans its field. Technical advice is given to maintain readiness for oil spills, contained in three field guides supported by ten other publications. These continue to be widely used among organisations concerned in spillage incidents, for reference and training. The CONCAWE Oil Spill Workshop held in focusing the increased activity in the light of

Table 2
Oil spills attended on site by ITOPF from January 1988 to February 1990

Date	Vessel	Source/Location
1988		
14.01.88	Cobiere*	Portsmouth, UK
04.01.88	Borcea*	Schelde Estuary, Netherlands
30.01.88	Amazzone	Off Brest, France
24.02.88	Kyung Shin	Pohang, South Korea
02.04.88	Mataram*	Elbe River, FRG
08.05.88	Czantoria	Quebec, Canada
19.06.88	Fulgur	Puerto la Cruz, Venezuela
06.07.88	Piper Alpha*	North Sea, UK
13.07.88	Nord Pacific	Corpus Christi, Texas, USA
10.10.88	Century Dawn	Off Singapore
21.10.88	Jupiter *	Piraeus, Greece
05.12.88	El Omar	Milford Haven, UK
10.12.88	Kasuga Marn	Maizuru, Japan
23.12.88	Nestucca	Grays Harbour, USA
1989		
16.02.89	Unicorn Derek	Virgin Gorda, Br. Virgin Is.
25.02.89	Baba Gurgur	Bakar, Yugoslavia
28.02.89	Swallow*	Dutch Harbour, Alaska, USA
24.03.89	Exxon Valdez	Prince William Sound, Alaska, USA
30.03.89	Esso Picardie	Sidi Kerir, Egypt
13.04.89	Gas Enterprise	Mina Ahmadi, Kuwait
25.04.89	Tropical Lion	Arzenah Island off Abu Dhabi, UAE
26.04.89	Kanchenjunga	Jeddah, Saudi Arabia
11.05.89	Bukom	Yosu, South Korea
28.05.89	Vigour Pioneer*/ Qing Chuan	Mokpo, South Korea
07.06.89	Ciudad de Berisso*	Port Alfred, Quebec, Canada
23.06.89	World Prodigy	Newport, Rhode Is., USA
24.06.89	Presidente Rivera	Delaware River, Philadelphia, USA
10.07.89	Makhachkala	Talara, Peru
14.07.89	Marao	Sines, Portugal
25.07.89	Nancy Orr Gaucher	Hamilton, Ontario, Canada
17.09.89	Phillips Oklahoma	Humberside, UK
01.10.89	Worthy	Southampton, UK
21.10.89	Merchantile Maricah*	Bergen, Norway
11.11.89	Norrona*	Milford Haven, UK
14.11.89	Milos Reefer*	St. Matthew Island, Alaska, USA
03.12.89	Eyal	Tel Aviv, Israel
29.12.89	Aragon	Madeira
18.12.89	Khark 5	Off Morocco
1990		
05.02.90	Tribulus*	Bantry Bay, Ireland
07.02.90	American Trader	Long Beach, Califonia, USA

* Non-tanker/barge spills.

the oil spill events in 1989. Particularly, contingency planning and preparedness on dispersant utilisation have been targeted by CONCAWE for action in conjunction with the initiatives of other organisations.

Another emphasis of CONCAWE is on the quality of aqueous effluents from oil refineries. The 1989 published survey of the European refineries performance, covering the year 1987, confirms that the long-standing trend of improvement has continued since the previous survey in 1984. The ongoing dialogue of CONCAWE with the Paris Commission, and others, has aimed to ensure informed consideration of environmental responses.

CONCAWE reports are produced to advise on good scientific, engineering and operating practice, and efficiencies and cost-effectiveness of emission control alternatives. Catalogues of reports and annual reports are provided by the CONCAWE Secretariat.[37]

6.4.8 Oil Companies International Marine Forum (OCIMF)

The OCIMF is a voluntary association of 35 oil companies and groups worldwide having an interest in the transportation by sea and marine terminalling of crude oil and its products, including gas and petrochemicals. Essentially concerned with the safe conduct of these operations and the prevention of pollution, OCIMF provides a medium whereby its members can formulate their views and present these to intergovernmental bodies, national governments and international industry organisations.

OCIMF, which is incorporated in Bermuda, was formed in 1970 as the oil industry's response to increasing public awareness of pollution. National legislation and international conventions were being developed and the oil industry sought to play its part by making its views known and its professional expertise available to governments and intergovernmental bodies.

The primary objectives of OCIMF are the promotion of safety and prevention of pollution from tankers and at terminals. To this end OCIMF, having been granted consultative status by IMO, actively participates in and contributes to the work of IMO through the Marine Safety Committee and Marine Environment Protection Committee, together with their associated sub-committees.

During the review period OCIMF participated in the ongoing work at IMO, representing the interests of its members. Among the subjects addressed were:

• the International Convention on Salvage;

• the Guidelines on Management for Safe Ship Operation and Pollution Prevention;

• the Guide to International Assistance on Marine Pollution Emergencies;

• arrangements for emergency cargo transfer;

• cargo vapour emission control systems;

• fuel oil quality and air emissions from ships.

OCIMF has sponsored industry research projects addressing safety and pollution prevention resulting in the publication of guidelines for both ships and terminals, many of which have become accepted as the normal operating practices in their fields. The subjects covered include cargo hoses, moorings and related equipment for use at offshore loading terminals, tanker and gas carrier loading arms and manifolds, the effect of wind and currents on ships, the safe mooring of ships at terminals, and the handling of disabled ships.

In conjunction With ICS and IAPH, OCIMF produces the *International Safety Guide for Oil Tankers and Terminals* (ISGOTT) which has become the accepted safety manual for both tankers and terminals. The third edition of ISGOTT was published in November 1988.

Several new or revised publications have been prepared including *Recommendations for Equipment Employed in Mooring Ships at Single Point Moorings, Inspection Guidelines for Bulk Oil Carriers, Information paper on Marine Vapour Recovery Systems, Guidelines for the Control of Drugs and Alcohol On Board Ship* and *Effective Mooring.* OCIMF has also produced *Peril at Sea* — a Guide for Masters in association with ICS and Planning and *Crew Response Guide for Gas Carrying Liquefied Gasses in Bulk* and revision of the *Malacca and Singapore Straits Transit Guide.* A work group is preparing guidelines for mooring equipment which will establish recommendations for the safe mooring of tankers. Another work group is updating the series of booklets concerning offshore hose systems and equipment.[38]

6.4.9 Society of International Gas Tanker and Terminal Operators Ltd (SIGTTO)

SIGTTO is a non-profit making organisation dedicated to the protection and promotion of the mutual interests of its Members in the safe operation of liquefied gas tankers and liquefied gas loading and receiving terminals. In pursuit of this aim, the Society serves as a forum for the exchange of technical information and experience on safety and reliability and conducts studies and research relating to safety or protection of the environment in the ocean transportation or bulk storage of liquefied gases.

SIGTTO celebrated its 10th Birthday in October 1989 at the fitting location of the LNG 9 Conference. From an initial membership of 10 founding companies, SIGTTO has grown in terms of membership (currently 67 companies representing some 80% of world LNG interests and 40% of world LPG interests) and stature and its contribution to safety in gas tanker and terminal operations is well regarded.

Research coordinated by SIGTTO and financed by Members' fees has resulted in a number of authoritative publications being produced for open sale, the first of which is a recent revision:

• a Contingency Planning and Crew Response Guide for Gas Carrier damage at Sea and in Port Approaches;

• a Guide to Contingency Planning for the Gas Carriers Alongside and within Port Limits;

- a Guide to Contingency Planning for Marine Terminals Handling Liquefied Gases in Bulk;
- Hydrates in LPG Cargoes — a Technological Review;
- Prediction of Wind Loads on Large Liquefied Gas Carriers
- a Review of LPG Cargo Quantity Calculations;
- Cargo Firefighting on Liquefied Gas Carriers (Film/Video and Study Notes);
- Liquefied Gas Handling Principles on Ships and in Terminals;
- Recommendations and Guidelines for Linked Ship/Shore Emergency Shut-Down of Liquefied Gas Cargo Transfer;
- Guidelines for the Alleviation of Excessive Surge Pressures on ESD.

SIGTTO represents its Members and enjoys consultative status at the IMO and acts on behalf of its Members in liaising with such bodies as the International Association of Classification Societies, ICS, OCIMF and other industry associations in so far as their own activities bear upon the interest of members of SIGTTO.

The Society is very conscious of its international membership and holds two meetings each year which, wherever possible, are held in Eastern and Western Hemisphere locations. Recent meetings of the Society have been held in Abu Dhabi, Osaka, Amsterdam, Stavanger, Perth (Western Australia) and Edinburgh. Future meetings are scheduled for Veracruz, Mexico in March 1990 and a Far East location in October 1990.

The format of these meetings generally includes a technical visit to a marine terminal or gas-related research organisation followed by a Panel Meeting which is open to all Members for the free exchange of views, concerns and problems of safety and reliability in relation to terminal and gas carrier operations. The General Purposes Committee also meets at the same time as the Panel Meetings and this represents the central technical body of the Society and manages its work. The Society is directed by a Board, elected annually by the Members and, as with the GPC, comprises a balance of representation of LNG, LPG, gas carrier and gas terminal operator interests to ensure that no particular bias enters the Society's activities.[39]

6.4.10 United Kingdom Offshore Operators Association (UKOOA)

The UK Offshore Operators Association (UKOOA) is an organisation whose Members are the oil and gas companies which are designated operators of licences on the UK Continental Shelf. At present, UKOOA has 36 Members. Its main objectives are to provide an industry forum for discussion of matters of mutual interest to its members and, as appropriate, to provide its members with a means of communication with Government and other relevant bodies, the press and other news media. UKOOA's Council consists of 25 Members, 20 of which are producing companies. Its Representatives are the

Exploration and Production Chief Executives of the Member Companies and they meet regularly each month to discuss UKOOA policy. The Executive Officers of UKOOA are appointed annually, and recommend UKOOA policy to the Council, at its monthly meetings.

The UKOOA Clean Seas and Environment Committee, in conjunction with the Drilling Practices Committee, have continued to study the environmental effects of oil-based mud cuttings.

To provide a fisheries biological input to abandonment discussions, UKOOA has carried out a pilot fish tagging and monitoring exercise to judge the effectiveness of the techniques of acoustic tracking.

Studies have also been conducted on fish flavour and tissue hydrocarbon of fish caught near offshore installations. The UKOOA dispersant equipment has been containerised following experience of its use during an oil spill. This will speed up mobilisation, reduce the possibility of handling damage and add protection when in use at sea.[40]

6.4.11 Petroleum Industry Association (UKPIA)

UKPIA is a trade association comprising private oil companies involved in the downstream oil industry activities of supply, refining and distribution of petroleum products in the UK. It was formed in 1979 to represent its members in communications with government, other industrial and commercial associations, the media and the general public. It monitors and advises on regulatory and other changes and developments affecting the industry. A major section of its work concerns environment, health and safety activities which are directed by Steering and Management Committees comprising senior employees from Member Companies.

UKPIA has nine regional oil spill coordinators who provide contact points for liaison between the oil industry and the local and national authorities in the event of an oil spill. Each coordinator has responsibility for a particular area so that the whole of the UK coastline is covered by the group. In the event of an oil spill, the relevant coordinator will provide advice to the local and national authorities on request and assist any UKPIA Member Company involved in the clean-up activity by making available his expertise and knowledge of the area.

UKPIA activities over the recent period have included promoting contacts with local authority pollution control officers, assisting at various oil spill emulation exercises and training schemes and reviewing local oil spill contingency plans and available resources for response.[41]

6.5 ACADEMIC INSTITUTIONS

6.5.1 The Fridtjof Nansen Institute

The Fridtjof Nansen Institute was established in Lysaker, Norway as an independent research foundation in 1958.

Today it is engaged in selected areas of international politics on the international market, ocean management, ocean mining, polar matters and Soviet affairs. The staff numbers approximately 30 Members, mostly researchers for the fields of political science, economics and international law.

For many years the Institute's funds were raised through donations, but as it has grown the situation has changed and today most of its financial basis is derived from research assignments. Approximately 35% of its budget is an annual grant from the Norwegian Government.

The Energy Programme

This focuses on the interaction between political and economic factors in the international oil and gas markets. Special attention is devoted to the organisation of the European gas market. The goal is to be able to predict market developments under different economic, institutional and political assumptions. The 'petropolitical' cycles of the world oil market is studied by analysing internal OPEC relations, the relationship between OPEC and non-OPEC exporters, the organisational restructuring of the oil industry, the role of geopolitical factors and energy prospects in the Third World. Special attention is paid to Norway as an oil and gas exporter.

The Commons Programme

This Programme studies the possibilities and constraints of effective management of specific geographical areas and natural resources known as 'commons', in particular, certain parts of the oceans, the atmosphere, living resources, Antarctica and the deep seabed. The political potential for the management of commons is determined partly by the interests involved, partly by characteristic features of the commons in question and also by organisational structure.

The Polar Programme

The Institute has a longstanding commitment to research on the polar regions. It is circumpolar as well as bipolar in scope and concentrates on both political and military-strategic developments in the polar region as well as international and national political, economic, legal and environmental aspects of the exploitation of living and mineral resources in the polar regions. Current projects include an analysis of the military–strategic, economic and political significance of the northern areas in the international system; an examination of the changing conditions under which long-distance fleets in northern and southern areas are operating; a survey of current territorial boundary disputes in the Arctic; and an analysis of the negotiation process of the Antarctic minerals treaty.

The Soviet Programme

This examines the role of the Soviet Union as an energy producer and exporter, the factors which influence Soviet behaviour in the oil and gas markets, and the importance of energy exports to the Soviet economy. Emphasis is placed on the use of original Soviet sources as well as Western research. Soviet policy with regard to the Barents Sea is also studied from the perspectives of both energy and security.

The Ocean Mining Programme

This Programme aims to inform the business community, research institutions and government authorities on developments and possibilities in the field of ocean min-ing. Information is disseminated by publishing reports, arranging seminars and initiating and coordinating efforts to establish technological–industrial collaboration with companies and countries with development projects in this field.

The Programme analyses perspectives for the exploitation of polymetallic nodule deposits in the pacific and Indian Oceans, polymetallic sulphide deposits along the east Pacific Rise, polymetallic crust deposits in Pacific, metalliferous mud in the Red Sea, phosphorite nodules and placer deposits. The main reports published are given the form of market analyses, and cover economic, technological, political, legal and environmental factors. The Programme has established an extensive network of contacts in India, China, South Korea, the US, Canada, France, Germany, Italy, Finland, Sweden, Great Britain and the Soviet Union.

Research from this Programme is reported through the Information and Research Secretariat on Ocean Mining which was founded in 1981 by the Institute and the SINTEF-group to provide the basis for cooperation between the business community, research institutions and government authorities.[42]

6.5.2 Indonesian Center for the Law of the Sea (ICLOS)

ICLOS was established in 1985 at the Faculty of Law, University of Padjadjaran (UNPAD), in Bandung, Indonesia, by the Consortium of Legal Sciences of the Indonesian Ministry of Education and Culture. The Director of ICLOS is Professor Komar Kantaatmadja, while its Executive Secretary is Dr Etty R Agoes. Professor Mochtar Kusumaatmadja, the architect of an archipelagic (mid-ocean) state regime and former Indodesian Minister of Justice and Foreign Affairs, is Advisor to ICLOS. He is also the Director of the new Center of Law, Development and Archipelagic State Regime in Bandung. In addition, several Indonesian authorities in ocean affairs and the law of the sea serve as Special Consultants to ICLOS.

The establishment of ICLOS was a result of the importance of Indonesia in ocean affairs and the law of the sea. While the Indonesian struggle to gain international recognition for the archipelagic state regime has been successfully completed with the incorporation of this regime into the 1982 UN Convention on Law of the Sea, Indonesia now faces the complex task of implementing this Convention which it ratified in 1985. Consequently, Indonesia attaches vital importance to the development of ocean affairs and the law of the sea, as evidenced by the Action Plan for Sustainable Development of Marine and Coastal Resources, which was prepared by the Indonesian National Development Planning Board and which includes medium term policies and programmes to complement the Indonesian Development Plan for the period 1989-94.

Since 1986, ICLOS has carried out a wide range of co-operative research and training activities with the Netherlands Institute for the Law of the Sea (NILOS of the University of Utrecht, within the framework of the Netherlands-Indonesian Cooperation on Legal Matters (Phase I 1986-1990).

Since its establishment, ICLOS staff members have co-operated with various other institutions dealing with ocean affairs, such as the International Ocean Institute of Canada and the South-East Asian Programme on Ocean Policy, Law and Management (SEAPOL) in Bangkok, Thailand. ICLOS staff members also carry out many consulting activities. A significant recent development was a Workshop on Managing Potential Conflicts in the South China Sea held in Bali, Indonesia 22–25 January 1990, and co-sponsored by the IOI.[43]

6.5.3 International Juridical Organisation for Environment and Development (IJO)

The IJO is a non-governmental and non-profit making international organisation, founded in 1964, with its seat in Rome, Italy. IJO conducts legal research concerning economic and social issues in developing countries, human rights, East-West cooperation and environmental and development law. Since 1981, IJO has enjoyed consultative status with the United Nations Economic and Social Council (ECOSOC). IJO is a consultant for the United Nations Environment Programme (UNEP) and has observer status with that Programme.

For many years the President of IJO, Dr Mario Guttieres, together with his staff and other legal experts, has undertaken legal research and held seminars and conferences in the field of marine protection.

The Draft Fifth Protocol for the Protection of the Mediterranean Sea against Pollution Resulting from Exploitation and Exploitation of the Continental Shelf and the Sea-bed and its sub-soil.

By virtue of the contract between the UNEP Mediterranean Action Plan (MAP) and IJO, the IOJ is in charge of preparing the above mentioned Draft Protocol to implement the provisions of Article 7 of the Convention for the Protection of the Mediterranean Sea against Pollution (Barcelona Convention, 1976).

The text is scheduled to be reviewed and finalised by the Contracting Parties' experts in May 1990 before its submission to a plenipotentiaries conference to be held in 1991.

Objectives — The main consideration, throughout the draft Protocol, is to address the concern of the Coastal States to protect and improve the Mediterranean environment through preventing, abating, combating and controlling pollution resulting from exploration and exploitation of the continental shelf and the sea-bed and its subsoil.

Seminar on Exploration and Exploitation of Oil and Gas in the Mediterranean Sea

This Seminar was held in Capri from 15-16 April 1988. It was promoted and organised by the Committee of Operators for the Exploration and Exploitation of Oil and Gas in the Territorial Sea and in the Italian Continental Shelf. The object of the Seminar was to present the draft of the Fifth Protocol to the Barcelona Convention to the Italian administrators and operators, in order to verify the impact of the draft on the national legislation and its Operational effectiveness.

Other reports concerned the following themes: 'Geophysics, Drilling and Production' — operations, environmental aspects and the contingency plan against pollution oil and gas-provided by Elf Italiana S.p.A; the national contingency plan to protect the sea and coastal areas against pollution caused by accident; environmental impact of exploration and exploitation activities; operational offshore problems in relation to existing legislation and future regulations; an overview of the Protocol: Inventory of the law, jurisdiction and administrative procedures in Italy for the protection of the marine environment against offshore activities.

A strong consensus was expressed on the Draft Protocol and the text was considered to be in line with the expectations of the Italian administrators and the operators and with recent Italian legal guarantees for the protection of the Mediterranean Sea.[44]

6.5.4 International Maritime Law Institute (IMLI)

IMLI was established under the auspices of the IMO as an international centre for the training of specialists in maritime law, including the international legal regime of merchant shipping and the law of the sea. There is special emphasis on protecting the marine environment by implementing the numerous international conventions, regulations and procedures for furthering vessel safety and prevention of pollution concluded by the IMO. All students are law graduates.

The Institute is situated on the island of Malta, which a has a rich maritime heritage and interest in the development of the maritime legal regime. Under an Agreement concluded between IMO and the Government of Malta, the Government has provided attractive premises and facilities for the establishment and operation of the Institute. Accommodation for up to 20 students is available on campus.

The Institute opened on 2 October 1989. The official inauguration ceremony of the Institute by Dr CP Srivastava, then Secretary-General of the IMO, took place on 4 November 1989. At present 17 students are in residence, from Sri Lanka, Somalia, Western Samoa, Malta, Seychelles, Panama, Philippines, Malawi, Ethiopia, Malaysia, Nigeria, Hong Kong, Bangladesh, Commonwealth of Dominica, Trinidad and Tobago, Nigeria, Hong Kong, Pakistan and Bahrain. The students range in age from 25 to 47 years and five of them are women. About half of the students have a maritime related background, the rest are involved in their government legal departments.

The academic programme consists of two main areas of study, namely, law of the sea, including environmental law aspects, and shipping law. The topics covered include public international law, international institutions, marine environmental law, international law of the sea, and commercial and regulatory shipping law. In addition, training is provided in the development and drafting of maritime legislation; in particular, the incorporation of IMO and other maritime conventions into domestic legislation. The final evaluation of students will be based on two written examinations, the submission of a major essay and a drafting project. Successful candidates will be awarded a Master Degree in Maritime Law. The field of study is highly specialised and the programme as designed is intensive and demanding. It is expected that upon returning to their respective countries, IMLI graduates will be able to facilitate the incorporation of the many IMO and related international maritime and marine environment protection conventions and codes into their domestic legislation, and thus establish sound maritime regimes comparable to those in developed countries.

The first Director of the Institute is Professor Patricia Birnie, formerly of the London School of Economics, a specialist in international law, especially law of the sea and environmental law. The Senior Deputy Director is Professor PK Mukherjee of Canada, a Master Mariner and maritime lawyer, who is also experienced in the drafting of maritime legislation. Professors Birnie and Mukherjee constitute the permanent faculty. Professor David A Attard of the University of Malta is Special Adviser to the Director and a part-time member of the Faculty teaching certain topics in the law of the sea. In addition, there are three part-time lecturers who are Maltese maritime law practitioners with postgraduate qualifications in maritime law. They teach selected topics in shipping law and assist with tutorials.

Needless to say, marine pollution is an important and major topic in the programme. Several students are expected to write their essays or base their drafting projects on this topic.[45]

6.5.5 International Ocean Institute (IOI)

The IOI was formally established in 1972, in cooperation with the Government and the University of Malta, assisted by the United Nations Development Programme (UNDP). This was the culmination of five-year effort, initiated in 1967 by the Centre for the Study of Democratic Institutions in Santa Barbara, California, focused on the need for a new ocean regime, based on the concept of the Common Heritage of Mankind. The potential of ocean development, both in economic and in political/institutional terms, was awesome; but it was clear from the outset that ocean development was possible only if there was an end to the mindless degradation of the marine environment. It is simply impossible to develop marine resources while destroying the environment within which such development is to take place. In the oceans, a synthesis must be found between economics and ecology;

development must be sustainable or there cannot be any development at all.

The IOI is an independent international non-governmental organisation. Its Statutes are inscribed as a Chapter in the Statutes of the University of Malta, but it is governed by its own governing body, a prestigious Planning Council of 24 members from North, South, East and West, and a Board of Directors of 12, including such personalities as the Prime Minister of Tanzania, the Minister for Development Cooperation of the Government of the Netherlands, and a leading member of the new Government of Romania.

The activities of the IOI consist of four major components:

Policy Research — A great number of projects have been completed. Presently IOI research is concentrating on feasibility studies for the establishment of regional centres for Research and Development in Marine Industrial Technology. As in all IOI projects, these studies emphasise the need for development of 'clean' technologies and for the development of pollution monitoring and combating technologies. These projects are sponsored by UNIDO and UNEP and endorsed by the Government of Malta.

Other projects deal with alternative modes of funding the UNEP Trust Fund for the effective implementation of the Mediterranean Action Plan (this study is financed by the Ford Foundation) and with the next phase of implementation and development of the Law of the Sea (in cooperation with the Asian African Legal Consultative Committee.

Publications — The IOI publishes an Ocean Yearbook, about 600 pages per issue through the University of Chicago Press. Seven volumes have appeared so far; Volume 8 will be published later this year. Conference proceedings, occasional papers, and newsletters make up the remaining publications programme.

Conferences — The IOI organises a major annual conference, called 'Pacem in Maribus', Peace in the Oceans. Each one deals with a different aspect of ocean management and law of the sea. Seventeen conferences have been held. Pacem in Maribus XVIII will be held in Rotterdam on 27-31 August 1990 on 'Ports and Harbours' as nodal points in global communication. The conference is co-sponsored by the Government of the Netherlands, and the City and Port Authority of Rotterdam. Obviously, the environmental problems of ports and harbours from the impacts of dredging to those of coastal congestion and waste disposal, etc will be given due attention.

Besides Pacem in Maribus conferences, the IOI also organises regional seminars.

Training — Most of the IOI's financial resources go into the training of civil servants in developing countries, to assist those countries in the formation of cadres of persons expert in the new science of 'ocean management'. To date 29 programmes have been completed for participants form about 100 developing countries. Each programme lasts 10 weeks.

The activities of the IOI are funded through voluntary contributions, mostly from governments and intergovernmental institutions. A campaign is now under way to establish an Endowment Fund for the IOI which should ensure the continuation of these activities in perpetuity.[46]

6.5.6 The Netherlands Institute for the Law of the Sea (NILOS)

NILOS was established in September 1984 at the Faculty of Law, University of Utrecht, and operates under the Directorship of Professor Alfred H.A. Soons (Head of the Institute of Public International Law at the Faculty). It has two principal objectives:

- to conduct research on all issues related to the development of ocean affairs and the law of the sea generated by the 1982 UN Convention on Law of the Sea, with special emphasis on questions of resource management and institutional cooperation of states, and including those of marine environmental protection;

- to assist states, in particular the developing countries, in dealing with the law of the sea issues which confront them.

Apart from a wide range of issues covered by individual research projects of staff members, NILOS edits an annual publication, *The International Organisations and the Law of the Sea Documentary Yearbook*, which systematically reproduces the most important documents related to the law of the sea issued each year by international organisations. The Yearbook aims at improving access by scholars and practitioners to such documents. It covers organisations of the United Nations system, including the UN (General Assembly, ECOSOC, PrepCom), FAO, IAEA, ICAO, ILO, IMO, UNCTAD, UNEP, UNESCO-IOC and WMO, and also some regional organisations of the developing states from outside the UN system, such as ASEAN, AALCC, IOMAC, OECD and South Pacific Forum.

Its specialisation in institutional marine affairs cooperation led NILOS to organise and co-sponsor the 23rd Annual Conference of the Law of the Sea Institute on Implementation of the Law of the Sea Convention Through International Institutions, held in Noordwijk aan Zee, the Netherlands, on 12–15 June 1989. Other NILOS joint research projects included those on sea use planning for the North Sea (1988), research for fisheries organisations within the framework of the EEC's Common Fisheries Policy (1982 and 1989) and preparation of the First Supplement (concerned with Law of the Sea) to the series, *International Organisation and Integration*. In addition, the NILOS staff members published several individual research projects and books.

The cooperative research and training projects undertaken by NILOS in the period June 1988-February 1990 covered:

- cooperation with the Indonesian Center for the Law of the Sea (ICLOS);

- cooperation with the Indian Ocean Marine Affairs Cooperation Conference (IOMAC);

- cooperation with the South-east Asian Programme on Ocean Policy, Law and Management (SEAPOL);

- co-organising 4th (1988) and 5th (1989) Annual Course on the Law of the Sea in the Inter-University Centre of Dubrovnik, Yugoslavia;

- participating in the Erasmus Program of the European Communities providing scholarships to students from EC Member States;

Scientists from the Bermuda Biological Research Station tend to a coral respirometer. (See 6.6.1.) Photo: M Doty

- co-organising an International Policy Seminar on the Management of the Exclusive Economic Zone for Development to be held in the Institute of Social Studies in the Hague, in May-June 1991.

In 1989 NILOS also undertook activities related to drafting of fisheries legislation for the Netherlands Antilles (and Aruba), and to organising joint ESCAP/NILOS seminars in South-East Asia. Moreover within their consulting activities, NILOS staff members continued to assist various institutions on various problems pertaining to their activities. Professor Soons prepared a working paper for the UN Office of Ocean Affairs and the Law of the Sea (OALOS) on the implementation of the provisions on the consent regime for marine scientific research in the 1982 Convention on Law of the Sea which served a basis for discussion at the UN OALOS Meeting Of a Group of Experts on Marine Scientific Research held in New York, in September 1989.

In addition, each year NILOS staff members jointly run a specialised course on the international law of the sea at the University of Utrecht.[47]

6.5.7 The World Maritime University (WMU)

The World Maritime University provides advanced training for senior personnel from developing countries who are involved in various maritime activities. Training of this type is not available in the developing countries and there is no comparable institution anywhere in the world. Based in Malmo, Sweden, the WMU was officially opened on 4 July 1983. Its creation was largely due to the vision of the then Secretary-General of the International Maritime Organisation, Mr C.P. Srivastava, and present Chancellor of the University until 1991.

The WMU is one of the most ambitious projects ever undertaken by the IMO. It fills an important gap in maritime training and will have a marked and beneficial effect on the two areas of greatest concern to the IMO: the improvement of maritime safety and the prevention of marine pollution from ships.

Since 1987 the annual intake has been around 100 students, sent from 105 different countries. The students come from administrations, shipping companies, port authorities and teaching positions at maritime academies. All have high professional or academic qualifications before going to Malmo, and many have spent years at sea, gaining a master's or chief engineer's certificate. In almost every case the students have been selected by their governments because of qualities which have already been proven. The average age is around 35–40 years.

The courses offered by the WMU are all two years in length and graduates receive a Master of Science degree.

The courses are as follows: General Maritime Administration, Ports and Shipping Administration, Maritime Safety Administration (Nautical), Maritime Safety Administration (Engineering), Maritime Education and Training (Nautical), Maritime Education and Training (Marine Engineering), Technical Management of Shipping Companies.

Much of the teaching at the WMU is conducted on traditional lines, with lectures, tutorials and regular examinations, culminating in the production of a final written paper on a chosen subject. A special feature of the University is the use made of field trips and on-the-job-training in other European countries, North America, Japan or Egypt. This additional experience and exposure to practical problems is of great value and graduates often make contacts with organisations or people which may be useful when they return to their home countries.

The academic and administrative work of the WMU is directed by the Rector, assisted by the Vice-Rector, seven full-time academic professors each specialising in a different field, and a number of lecturers. The University also has three programme officers who are involved in organising field trips and the practical aspects of the on-the-job training. In addition to the full-time academic staff, the University benefits from short-term visiting professors and lecturers; during 1988 more than 80 visiting professors and 10 senior IMO officials lectured there.

Funding for the WMU is provided by contributions from the United Nations Development Programme (UNDP), from the host country Sweden as well as from Norway, France and many other countries. About 40% of the budget has come in the form of fellowships. Most of the fellowship financing has been allocated by countries of their UNDP resources; the next largest source has been governments, followed by companies and national organisations of the countries sending students to Malmo. The Commonwealth Secretariat and the European Community have also provided fellowships.[48]

6.6 OTHER REPORTS

6.6.1 Bermuda Biological Station for Research (BBSR)

The BBSR was established in Bermuda in 1903 through the joint efforts of individuals with the Bermuda Natural History Society, New York University, Harvard, and Princeton. As a US non-profit organisation incorporated in the State of New York in 1926, funding comes from private gifts and grants with a large portion of its research funded by the US and Bermudian governments.

BBSR projects include research on oil spills, atmospheric pollution and transport, coral bleaching, coral reef biology, geology, and biological chemical, and physical oceanography.

Its involvement in marine pollution began in the early 70s with BBSR's Vice President Dr James Butler's work on oil pollution. From 1981 to 1984 the COROIL project under Dr Anthony Knap (presently BBSR's Director) examined the effect of oil and oil dispersants on corals and spawned many pollution-related projects

In 1980, Drs Knap and Butler combined their expertise to create one of BBSR's most successful courses entitled Analysis of Marine Pollution. Since its inception, it has trained over 100 students from 29 different countries in biological and chemical techniques for assessing hazard levels of organic and other marine pollutants.

Most recently, BBSR scientist have been involved with the two largest oil spills in the Americas: one in Panama (*Galeta* 1986), and the one in Alaska (*Valdez* 1989). At the Galeta site, BBSR scientists have recently discovered drastic differences in the oil degradation rates between tropic and temperate zones. In August 1989, Dr Kathryn Burns (BBSR) was asked by officials from the state of Alaska to grade their sampling program and assess quality assurance, quality control for lab procedures and data gathering.

Dr Susan Cook, BBSR Assistant Director for Education, also investigates marine pollutants but from an ecotoxicological perspective using limpet larvae. She uses the larvae as biomonitors to determine environmentally safe levels of metals, oil, and dispersants. Dr Cook also oversees BBSR's education program which offered seven summer courses in 1989.

The BBSR's investigation of trace metals in the marine environment has helped the Bermuda government establish standards for hazardous levels of metals. It has also examined effects of pollutants on marine organisms. This research will add to an environmental impact assessment being funded by the Bermuda government in their consideration of a mass burn incinerator for 1990.

Last year its education program handled 37 academic groups, taught 29 scientific seminars, and offered 15 lectures by guest speakers. It also provided 166 visiting scientists with equipment and facilities while accommodating some 1,600 users ranging from high school students to private industry researchers.[49]

6.6.2   International Maritime Bureau

The IMB is a non-profit membership organisation founded in 1981 by the International Chamber of Commerce. It is a focal point for the global shipping and transportation industry in the fight against fraud, violence and malpractice.

On 1 September 1988 the IMB created the Toxic Waste Hotline. Its launch was a response to a concern that increasing amounts of toxic wastes were dumped at sea due to difficulties in finding legitimate land disposal sites. The intention behind the Hotline is that it should provide a confidential service for the receipt and monitoring of information about possible or actual irregular dumping of waste.

The Hotline has attracted a large amount of media interest with enquiries from around the world:

- information received — over 200 independent calls;
- information on 160 different vessels;
- information on 54 individuals;
- information on 19 different companies.

The IMB also intends to offer a Waste Inspection Service in the near future. The global framework for waste shipments has and will change considerably, due to stricter international regulations mainly provided by the Basle Convention. The present difficulties and the need for a measure of independent control have created the concept of an inspection service within the framework of the International Chamber of Commerce.

The service will make the market for transboundary movements of waste more regulated and accountable and will bring it in accordance with the hazardous waste treaty.

The benefits to the industry will be that its function will not only become more organised and less speculative but that the existence of an independent external agency will ensure the removal of present and future problems like costly re-import obligations and bad publicity in a critical public arena.

The Waste Inspection Service will be structured in three independent phases which would normally operate in sequence but which could be taken as individual part services:

The initial phase will be the supervision of the packaging and labelling of the hazardous cargo in order to comply with global safety standards (IMDG-Code, MARPOL Annex V, etc.) as will be required by the Basle Convention. At this point it will also be possible to analyse the cargo from samples;

The second phase will consist of a pre-shipment check and certification to ensure that the proposed destination country has adequate facilities for the disposal of the waste in an environmentally sound manner;

As a third phase the IMB offers monitoring of the voyage in order to keep control of the shippers and to ensure all requirements of the charter party are adhered to.[50]

6.6.3   Seatrade Annual Awards for Achievement

The Seatrade Annual Awards scheme was launched in January 1988 and first presentations were made to representatives of successful companies at an inaugural Awards Ceremony Gala Dinner in the City of London's Guildhall in April 1989. All proceeds from this Dinner were donated to the Royal National Lifeboat Institution.

The Seatrade Organisation, publishers of the international shipping publications *Seatrade Business Review* and *Seatrade Week*, and organisers of maritime exhibitions and conferences throughout the world, had been looking for ways to help stimulate and encourage technical innovation in the shipping industry by giving recognition to those companies and individuals who make substantial contributions. It was recognised that the shipping industry, unlike most others of comparable size, had no international awards structure. Seatrade decided to institute a series of three international awards, which would be given for outstanding contributions in the areas of countering marine pollution, safety of life at sea, and the search for greater efficiency in vessel operations.

In presenting the Seatrade Awards 1989, C P Srivastava, at that time Secretary-General of the IMO and Chairman of the Judges, said "I am convinced that in the the the years to come this will become on the most prestigious and eagerly awaited events in the maritime calendar."

The winner of the Award in the Countering Marine Pollution category was Unitor Marine Chemicals, UK (now Unitor Rochem) for 'Enviroclean'. This product was the result of two and a half years' research work into the cleaning of engine rooms and problems regarding pollution, the cleaning of oil residues from tankers, and the problems associated with discharging slops ashore to slop receivers and oil refineries. In developing 'Enviroclean', Unitor considered that the pollution and environmental aspects were the most important criteria in producing a cleaner that was to be safe to the user and the environment as well as cleaning efficiently too. 'Enviroclean' is non-toxic to the environment and user, and a non pollutant.

The winner of the Award in the Safety At Sea category was Norpol of Denmark, for their safety release hook for rescue boats and life rafts. One feature on its coupling system is an automatic release mechanism when the craft is waterborne.

The 1990 Seatrade Annual Awards for Achievement
The 1990 Awards which include an additional category — Seatrade personality of the Year Award, will be presented at an Awards Ceremony Dinner, with proceeds in aid of King George's Fund for Sailors at the Guildhall, on 30 April 1990.

The Chairman of the judges for the 1990 Awards is Mr William O'Neil, Secretary-General of the IMO.

The 1991 Seatrade Annual Awards for Achievement
Details of the 1991 Seatrade Annual Awards for Achievements, together with entry forms, are available and the deadline for all entries/nominations will be 31 December 1990.[51]

6.6.4 Tanker Advisory Center, Inc.

This New York based organisation publishes figures of accidental oil spills, total losses and deaths for tankers and combination carriers. The current table is shown below (Table 3).[52]

Table 3 Trends in Oil Spills, Total Losses and Deaths (Excludes Hostilities)

Year	Accidental Oil Spills		Total Losses			Dead	Tankers	
	No.	Tons	No.	Dwts (000,000's)	No.	No.	Dwts* (000,000's)	Av. Dwt (000's)
1990						3,305		
1989	29	162,969	8	0.800	74	3,206	271.4	84.7
1988	13	178,265	3	0.257	63	3,147	267.2	
1987	12	8,700	5	0.392	12	3,132	267.8	
1986	8	5,035	8	0.723	23	3,139	273.1	
1985	9	79,830	12	0.762	53	3,285	299.9	
1984	15	24,184	14	0.857	68	3,424	319.0	
1983	17	387,773	11	0.885	14	3,582	341.7	95.4
1982	9	1,716	21	1.107	72	3,950	364.7	
1981	33	45,285	21	1.166	73	3,937	372.5	
1980	32	135,635	15	1.703	132	3,898	375.7	
1979	65	723,533	26	2.501	306	3,945	376.0	
1978	35	260,488	27	0.913	148	4,137	380.4	92.0
1977	49	213,080	20	1.000	113	4,229	369.0	
1976	29	204,235	20	1.172	226	4,237	336.7	
1975	45	188,042	22	0.815	90	4,140	296.2	
1974	48	67,155	14	0.536	94	3,928	253.6	
1973	36	84,458	12	0.679	70	3,750	218.9	58.4

Revised 30 January 1990.
*10,000 dwts and over except, 6,000 dwts and over prior to 1983.

Source: Lloyd's List, Tanker Register and Institute of London underwriters.

Sources of information which have been included in chapter 6 are shown below and acknowledged with gratitude.

1. ACOPS delegation.
2. Local Government International Bureau.
3. Local Government International Bureau.
4. AKN.
5. Coastwatch UK.
6. EPFT.
7. EEB.
8. ACOPS delegation.
9. FSCRC.
10. From SLL information.
11. FOEI.
12. HELMEPA.
13. IEEP.
14. IIED.
15. From IOCU information.
16. IUCN.
17. Lega Navale Italiana.
18. MCS.
19. MF.
20. NRC.
21. RSPB.
22. SAR.
23. TBG.
24. WWF UK.
25. ITF.
26. NFFO.
27. RSPBA.
28. WRC.
29. WSA.
30. NRA.
31. CRISTAL Ltd.
32. IP.
33. INTERTANKO.
34. ICS.
35. IPIECA.
36. ITOPF.
37. CONCAWE.
38. OCIMF.
39. SIGTTO.
40. UKOOA.
41. UKPIA.
42. Fridtjof Nansen Institute.
43. ICLOS.
44. IJO.
45. IMLI.
46. IOI.
47. NILOS.
48. from WMU News.
49. BBSR.
50. IMB.
51. Seatrade Organisation.
52. TAC.

7. Legal

Foreword

by Professor Patricia Birnie
Director
International Maritime Law Institute, Malta

The past year has evidenced a remarkable heightening of interest in marine environmental protection stimulated by such events as the *Exxon Valdez* oil spill in Alaska in March 1989, which followed immediately upon the oil spillage in the Antarctic from the grounded *Bahia Paraiso*, and more recently, in December 1989, the oil slick flowing from the Iranian supertanker, the *Kharg 5* after an explosion 100 miles off the Moroccan coast (see 2.1). Concern for the possible adverse effects on the environment of rising sea levels and cooling of coastal waters should global warming result in accelerating climatic change have also focused attention on marine environment problems. These developments illustrate both gaps in the application and enforcement of the existing regime and, in the latter case, the need to think ahead and consider what new measures, if any, might be required.

The past year has, however, also witnessed a number of encouraging advances in strengthening the legal regime. First, a Convention on the Control of Transboundary Movements of Hazardous Waste and their Disposal was concluded in Basle on 22 March 1989 under UNEP's auspices and was signed on the spot by 35 states, including both developing and developed countries, although it has not yet obtained enough ratifications to enter into force (see 7.2.1). When it does so, it will fill an important gap in international law, since state parties will no longer be able to ship hazardous wastes to non-state parties; the consent of importing states must be obtained before shipments can be exported and the exporter must provide sufficiently full information for proper risk assessment to be made. Nor can parties ship such waste to others that lack the requisite disposal facilities for environmentally sound disposal.

Second, in April 1989, a new IMO Salvage Convention was at last adopted, though it too has not yet entered into force (see 7.9.1). When it does, it will make a major contribution to marine environment protection since, unlike the previous convention, it recognises that speedy and effective salvage prevents pollution. Both owners and masters of wrecked vessels must take the action appropriate to ensure and facilitate salvage and coastal states' rights to protect their coasts are recognised, they too being allowed to instruct the salvor. A vital encouragement to salvors to act is the provision of special compensation to salvors who, though unsuccessful in saving the ship, nonetheless do succeed in minimising environmental damage.

Thirdly, Annex V of the 1973/83 MARPOL Convention is now in force (see 7.3.2). For the first time, this past year the ships of states' parties have been under an obligation not to dispose of their garbage at sea. Moreover, the North Sea has now been added to the list of Special Areas under that Convention, which are subject to more stringent discharge standards. In March 1990, the further protection of its marine environment was the object of the Third North Sea Ministerial Conference to be held at the Hague, at which we saw a move to a concerted declaration of Exclusive Economic Zones by North Sea States, a move may lead to expanded scope for enforcement.

In the last decade enormous progress has been made in augmenting the treaty law governing prevention of marine pollution but there remains a huge gap between the law now available and the number of states willing and technically and economically able to apply and, especially, to enforce it. Some states are becoming frustrated by this situation and are even going so far as to take enforcement into their own hands, contrary to the generally accepted rules of international law — not entirely a desirable solution for the strengthening rule of law, since international law, given the equality of states must be based on some form of consent, whether expressed or implied. Haiti, for example, has purported to ban transport of hazardous materials through its territorial waters and some coastal states, and during the Basle Convention negotiations argued that their permission was required for this. Proposals for a protocol to the Caribbean Convention on Special Areas and Wildlife Protection seem to contemplate regulation of vessels in such areas by Coastal States. An improvement in international standards of regulation and enforcement by Flag States is required if further inroads into the doctrine of innocent passage negotiated at the UNCLOS are not to be made and the balance of environment and development achieved in the UNCLOS is to be preserved.

IMO Conventions and codes in particular are often very detailed and technical; developing states, under whose flags vessels are being registered in increasing numbers, frequently lack the resources not only of skilled officers and port officials but of lawyers with both knowledge of the conventions' requirements and interpretive problems and of the techniques of legislative drafting. To remedy the first part of the problem, the IMO, following its success in establishing the World Maritime University in Malmö, Sweden, in 1983, has now established in Malta an International Maritime Law Institute (IMLI). The Institute was built in conjunction with the Government of Malta and here graduate lawyers, each from a different developing country and two Maltese lawyers, all with previous experience in legal or maritime departments, are being trained intensively on a course leading to a Masters degree in Maritime Law to help remedy the second part of the problem. IMLI opened

on 2 October 1989 and its first graduates will begin to facilitate the implementation of IMO conventions in their respective countries later this year, since uniquely, they will not only be versed in the necessary details of every aspect of maritime and environmental law but will also be familiar with the problems of legislative drafting. The IMO conventions and codes are internationally negotiated and enable the sustainable development urged by the Brundtland Commission.

I was therefore proud to be asked to become the first Director of this new enterprise and contribute in this way to the more effective global implementation of the legal regime that now exists to protect the marine environment.

Professor Patricia Birnie
Director
International Maritime Law Institute Malta

THE LEGAL REGIME

The aim of this chapter is to present an introductory overview of the main international and regional legal instruments concerning marine pollution. It has been revised and updated from the previous *1987-88 Yearbook*. The legislation of a few countries other than the UK is again incorporated in this section of the *Yearbook* and information on other states will be added in future editions as it becomes available.

Within each topical sub-section, the order of presentation maintains the same outline: international conventions (those of global application), followed by regional conventions, then EEC Directives and, finally, national legislation. Each group is presented in reverse chronological order.

7.1 POLLUTION FROM SHIPS — OIL

International

7.1.1 United Nations Convention on the Law of the Sea (UNCLOS) 1982

This Convention, which is still not in force, is a comprehensive umbrella text covering all aspects of maritime law. Here we give a brief overview of the main aspects of the Convention which relate to marine pollution, of any origin.

The Convention comprises 320 articles and nine annexes, governing all aspects of ocean space from delimitations to environmental control, scientific research, economic and commercial activities, technology and the settlement of disputes relating to ocean matters. The Convention represents not only the codification of customary norms, but also and more significantly the progressive development of international law. Part XII of the Convention is entitled "Protection and Preservation of the Marine Environment" (Articles 192-237) and Section 5 of this part refers to the international rules and national legislation to prevent, reduce and control pollution of the marine environment (Articles 207-212). At the same time there are a number of articles dealing with marine pollution in other sections of the Convention. Requirements of these articles include:

- adoption of measures to prevent accidents and deal with emergencies, preventing intentional and unintentional discharges and regulating the design, construction, equipment, operation and manning of vessels (Art. 194(3)(b));

- action though the competent international organisation or general diplomatic conference to establish international rules and standards to prevent, reduce and control pollution of the marine environment from vessels and national laws and regulations must have the same effect as that of generally accepted international rules and standards (Art. 211(1)(2));

- foreign vessels could be refused entry to ports if they do not comply with particular requirements for the control, prevention and reduction of marine pollution as established by the port state (Art. 211(3));

- coastal states may adopt laws and regulations for the prevention, reduction and control of marine pollution from foreign vessels in their territorial waters, including vessels in innocent passage (Art, 211(4));

- laws and regulations may also be adopted in respect of a state's Exclusive Economic Zone but only in conformity the generally accepted international rules and standards (Art. 211(5));

- flag states must ensure compliance with applicable international rules and standards, seaworthiness of their vessels, periodical inspections, investigations of any violations of international standards and severe penalties for violations (Art. 217);

- port states may undertake investigations and prosecutions of discharges outside their jurisdiction and prevent vessels from sailing if they are not seaworthy (Arts. 218, 219);

- strait states may also adopt laws and regulations giving effect to applicable international regulations regarding the discharge of oil, oily wastes and other noxious substances in the strait (Art. 42(1)(b));

- laws and regulations to prevent, reduce and control pollution from sea-bed mining in the Area beyond the limits of national jurisdiction, undertaken by vessels, installations, structures and other devices, not less effective than the international rules and regulations (Art. 209 and A/3/17/(2) (f)).

The Convention is due to enter into force 12 months after the deposition of the 60th instrument of ratification or accession. At the end of 1989, 43 States had ratified the Convention. For current developments see 3.1.

7.1.2 International Convention for the Prevention of Pollution from Ships 1973 (MARPOL 1973/78)

The Convention, which entered into force in October 1983, is the first to regulate all forms of marine pollution from ships except dumping. Only the provisions concerning oil are given in this section. It differs from the 1954 Convention (see 7.1.4) in a number of important respects:

- restrictions on discharge apply to all petroleum oil except petro-chemicals;

- while the 1954 discharge criteria are retained for existing tankers, for new tankers the total quantity permissible is reduced by half;

- no discharges are permissible from either tanker or non-tanker vessels in 'special areas' designated under the Convention, ie. the Baltic, Mediterranean, Black and Red Seas and the Gulf area;

- otherwise discharges from non-tanker vessels may be made only at a distance of more than 12 miles from the nearest land;

- all tankers of over 150 gross tonnage are required to

have monitors that will provide details of the rate and amount of discharge (existing tankers are exempt until three years after entry into force of the Convention) and in new tankers include an automatic shut-off device;

- a more detailed Oil Record Book covering a wider range of operations is specified;

- all new tankers of 70,000 tons DWT or above must be provided with segregated ballast tanks of sufficient capacity to allow the ship to operate safely on ballast voyages without using oil tanks for water ballast;

- the need to provide adequate reception facilities is extended to oil loading terminals, repair ports and other ports in which ships have oily residues to discharge;

- port states may inspect ships to verify whether the ship has discharged any harmful substance in violation of the Regulations and detain sub-standard vessels until faults are corrected, though the flag state's competence extends to all offences except for those committed within waters under the coastal state's jurisdiction.

States become parties to the Convention on ratifying both Annex I on oil and Annex II on noxious liquid substances ie. chemicals, in bulk.

The 1978 Protocol extends the structural requirements for new tankers to new vessels of 20,000 tons or over and introduces requirements relating to the protective location of segregated ballast tanks and for crude oil washing (COW).

7.1.3 International Convention relating to Intervention on the High Seas in Cases of Oil Pollution Casualties 1969 (Cmnd 6056)

The Convention, which entered into force in 1975, gives a coastal state the right to intervene on the high seas only after a casualty has occurred and there is grave and imminent danger of pollution of its coastline or territorial waters, or related interests.

The coastal state may take such action as is necessary to avoid the pollution danger after having first consulted the flag state of the vessel concerned.

7.1.4 International Convention for the Prevention of Pollution of the Sea by Oil 1954 (OILPOL as amended in 1962 and 1969 (Cmnd 595)

The Convention, which entered into force in 1958, marked the first major step towards international control of marine pollution. The amendments made to it in 1962 and 1969 entered into force in 1967 and 1978 respectively.

The Convention applies only to persistent oils, ie. crude, fuel, heavy, diesel and lubricating oils.

Tankers over 150 gross tons may discharge oil only if the vessel is more than 50 miles from the nearest land, is under way, the rate of discharge is limited to 60 litres per mile, and the total quantity discharged is restricted to 1/15,000 of total cargo carrying capacity.

Other seagoing ships over 500 gross tons are subject to

similar restrictions, except that the discharge must be made as far as practicable from land and the oil content of the discharge must not exceed 100ppm (parts per million).

Other requirements are

- vessels must maintain an oil record book and record certain operations;

- contracting parties must provide at their main ports adequate reception facilities for the residues from oily ballast water and tank washings from non-tanker vessels.

These requirements (together with the amendments) enabled the operation of the Load on Top (LOT) system for oily water discharges, and improved port state checking.
N.B. These provisions have been largely superseded by the more stringent regulations of MARPOL 73/78 (see 7.1.2).

Regional

7.1.5 Convention for the Protection and Development of the Natural Resources and Environment of the South Pacific Region (Noumea Convention) 1986

This treaty requires the contracting parties to prevent, reduce and control pollution in the Convention area from any sources, including *inter alia*, that resulting from the storage of toxic and hazardous wastes and disposal into the seabed and subsoil of the Convention area of radioactive waste or other radioactive matter.

7.1.6 Convention for the Protection, Management and Development of the Marine and Coastal Environment of the Eastern African Region (Nairobi Convention) and Protocol concerning Cooperation in Combating Marine Pollution in Cases of Emergency etc. 1985

The area covered by the Convention is defined as the marine and coastal environment of that part of the Indian Ocean situated within the Eastern African region and falling in the Jurisdiction of the contracting parties. The terms of the Convention and Protocol, which are still not in force, are broadly similar to those adopted for the Mediterranean (see 7.1.13).

7.1.7 Bonn Agreement for Cooperation in Dealing with Pollution of the North Sea by Oil and Other Harmful Substances 1983

This agreement, between eight North Sea states and the EEC, superseded the Bonn Agreement for Cooperation in Dealing with Pollution of the North Sea by Oil 1969 on its entry into force in September 1989. In addition to requiring exchanges of information and cooperation among competent national authorities, the Agreement provides for active cooperation to protect against pollution presenting a grave and imminent danger to the coast or related interests of the parties. For this purpose the North Sea area is divided into six national zones and three zones of joint responsibility. Zonal parties are responsible for assessing threats and reporting their assessment and action taken to other parties. In responding to requests for assistance to deal with pollution, parties have to use their best endeavours within the technological means available to them.

7.1.8 Convention for the Protection of the Marine Environment of the Wider Caribbean 1983 (Cartagena Convention) and Protocol concerning Cooperation in Combating Oil Spills in the Wider Caribbean Region 1983

The Convention and Protocol, which cover pollution from ships in the Gulf of Mexico, the Caribbean Sea and areas of the Atlantic south of 30°N latitude and within 200 nautical miles or the coasts of the states invited to take part in the Cartagena Conference that led to the signing of the Convention, follow broadly the lines of those for other UNEP Conventions.

7.1.9 Convention for the Conservation of the Red Sea and the Gulf of Aden Environment 1982 (Jeddah Convention) and Protocol concerning Regional Cooperation in Combating Pollution by Oil, etc. . . . 1982

The Convention and Protocol, which cover vessel-source pollution, cooperation during pollution emergencies and scientific and technical cooperation, was adopted in Jeddah in February 1982 and entered into force in 1985.

The area covered by the Convention consists of the Red Sea, Gulf of Aqaba, Gulf of Suez, the Suez Canal and the Gulf of Aden. The terms are broadly similar to those for the Mediterranean (see 7.1.13).

7.1.10 Convention on the Protection of the Marine Environment of the South-East Pacific 1981 (Lima Convention) and Complementary Protocol to the Agreement on Regional Cooperation to Combat Pollution of the South-East Pacific by Hydrocarbons, in Cases of Emergency 1983

The Convention, which was adopted at Lima in 1981, entered into force in May 1986. The Protocol affecting oil pollution which was adopted at Quito, Ecuador, in 1983, entered into force 20 May 1987. These are broadly similar to those adopted for the Mediterranean (see 7.1.13)

7.1.11 Convention for Cooperation in the Protection and Development of the Marine and Coastal Environment of the West and Central African Region 1981 (Abidjan Convention) and Protocol concerning Cooperation in Combating Pollution in Cases of Emergency 1981 (Lomé Protocol)

The area covered by the Convention is defined as the marine environment, coastal zones and related inland waters falling within the jurisdiction of the states of the West and Central African Region from Mauritania to Namibia inclusive. The terms of the Convention and Protocol, which entered into force in August 1984 after ratification by six countries, are broadly similar to those adopted for the Mediterranean (see 7.1.13).

7.1.12 Kuwait Regional Convention for Cooperation on the Protection of the Marine Environment from Pollution 1978 and Protocol concerning Regional Cooperation in Combating Pollution by Oil and Other Harmful Substances in Cases of Emergency 1978

The Convention and Protocol entered into force in June 1979. The terms are generally similar to those of the Barcelona Convention (see 7.1.13). Subsequent implementation has been mainly through the Kuwait Action Plan (KAP) under the auspices of the Council of the Regional Organisation for the Protection of the Marine Environment (ROPME).

7.1.13 Convention for the Protection of the Mediterranean against Pollution 1976 (Barcelona Convention)

The Convention was signed by all Mediterranean coastal states (except Albania) in 1976, was joined by the EEC in 1977 and entered into force in February 1978. The area covered is the Mediterranean from the Straits of Gibraltar to the Dardanelles except for the internal waters of the states concerned.

Under the 1973 MARPOL Convention the Mediterranean Sea is designated a 'special area' in which all discharges of oil are forbidden (7.1.2). The Convention is a comprehensive document covering general provisions by which the states concerned agree to act both to reduce pollution and to 'enhance the environment'. Detailed rules for applying these general principles are left to additional Protocols.

Protocol 2 concerning **Cooperating in Combating Pollution of the Mediterranean Sea by Oil and Other Harmful Substances in Cases of Emergency** was signed and entered into force on the same date as the Convention. The contracting parties agree to cooperate in dealing with emergencies in the area in order to reduce or eliminate any damage caused by an incident. They must also notify UNEP and any other state likely to be affected. Under the

Protocol UNEP established the Regional Marine Pollution Emergency Response Centre for the Mediterranean Sea in Malta (see 3.9) to receive reports and coordinate action.

7.1.14 Convention of the Protection of the Marine Environment of the Baltic Sea Area 1974 (Helsinki Convention)

Under the 1973 MARPOL Convention (see 7.1.2) the Baltic Sea was designated a special area. The 1974 Helsinki Convention, which entered into force in May 1980 after ratification by seven signatory states, provides the legal framework for ensuring that the contracting parties take, individually and jointly, all appropriate legislative, administrative or other relevant measures to prevent and abate pollution and enhance the marine environment of the Baltic Sea area. States accept particular obligations to:

* counteract the introduction into the area of hazardous substances;
* take measures to prevent pollution from ships and pleasure craft.

The Convention set up the Helsinki Commission (HELCOM) to administer its operation and all states are requested to cooperate with others within the framework provided by its activities.

The six-pack ring: handy for man, lethal to wildlife. It is estimated that this form of plastic has a life-span of 450 years. Photo: CEE

UK Legislation

7.1.15 Merchant Shipping Act 1988

Under section 35 of this Act, regulations may be made to regulate the transfer of cargo, bunker fuel or ballast between ships in the UK territorial sea to prevent pollution.

7.1.16 Territorial Sea Act 1987

This Act, which came into force on 1 October 1987, extends the breadth of the UK territorial sea from three to twelve nautical miles. The extension of the territorial sea

expands the regulatory powers of the UK. Vessels that discharge oil into the extended territorial sea are liable to prosecution in UK ports.

7.1.17 The Merchant Shipping (Reporting of Pollution Incidents) Regulations 1987 (SI 586/1987)

These Regulations, which came into force in April 1987, give effect to Protocol I of MARPOL 73/78 as amended by amendments adopted by the MEPC in 1985. The principal purpose of these amendments is to specify in detail the incidents required to be reported. These Regulations replace Regulation 31 of the Merchant Shipping (Prevention of Oil Pollution) Regulations 1983.

7.1.18 Prevention of Oil Pollution Act (POPA) 1986

This UK Act came into force in May 1986. It serves to prohibit discharges from vessels of oil or mixtures containing oil in territorial waters. In the case of such an offence the owner or master of the vessel is responsible, unless he can prove the discharge took place during the course of a transfer or oil to another vessel or to a place on land and that the other vessel or place was responsible for the discharge.

This Act amends Section 2 of the Prevention of Oil Pollution Act 1971 (7.1.25) and Regulations 12 and 13 of the Merchant Shipping (prevention of oil pollution) Regulations 1983 (control of discharge of oil by ships) SI 1398/1983 are repealed (7.1.20).

7.1.19 Prevention of Pollution (Reception Facilities) Order 1984 (SI 862/1984)

This Order implements the MARPOL requirements on the provision of reception facilities in harbours and terminals for the discharge of oily mixtures and residues. Harbour authorities or terminal operators are empowered to provide reception facilities that meet with MARPOL requirements by 2 October 1984 unless exempted by the Secretary of State and inspectors are given the power to report on the adequacy or otherwise of the facilities provided, as are masters of ships.

7.1.20 Merchant Shipping (Prevention of Oil Pollution) Regulations 1983

These Regulations came into effect in October 1983 and apply to all UK ships, other ships in UK waters and to Government ships registered or held for Government purposes in the UK, but not to warships and similar non-commercial vessels of any state.

The Secretary of State may exempt from the Regulations, either wholly or in part, both new and existing ships whose construction makes it difficult to apply them, provided equivalent protection is provided against pollution by oil.

The Regulations lay down detailed provisions for:

* surveying of oil tankers of certain sizes before they are issued with an International Oil Pollution Prevention Certificate (IOPP Certificate);
* reporting of accidents at the earliest opportunity;

- carrying of an Oil Record Book for recording specified operations;
- discharging oil, which must be regulated according to MARPOL standards;
- the fitting of segregated ballast tanks or operation of a crude oil washing system (COW);
- inspection of ships on entering UK ports;
- penalties for infringement of the Regulations.

7.1.21 Merchant Shipping (Prevention of Oil Pollution) Order 1983 (SI 1106/1983)

This Order, which came into effect in October 1983, is specifically to give effect to MARPOL 73/78 and enumerates those matters which the Secretary of State is expected to cover in subsequent Regulations.

7.1.22 Merchant Shipping Act 1979

The Act deals with a number of measures but, from the point of oil pollution from ships, is important in giving the Secretary of State powers to make Regulations to enable the implementation of MARPOL 73/78 (7.1.2). The Act also empowers an inspector to board any UK ship or any other ship at that time in UK waters for the purpose of ensuring that any Regulations in force at the time are complied with.

7.1.23 Petroleum and Submarine Pipelines Act 1975

The Act amends POPA 1971 (7.1.25) by giving the Secretary of State powers to exempt from its provisions any ship or class of ships or any discharge of oil, or mixture containing oil, either absolutely or under such conditions as he thinks fit.

7.1.24 Oil in Navigable Waters (Records) Regulations 1972

These Regulations, which originally laid down the occasions on which entries had to be made in Oil Record Books, have been largely superseded by the **Merchant Shipping (Prevention of Oil Pollution) Regulations 1983** (7.1.21) but still apply to oil tankers of less than 150 GRT and to other ships of less than 400 GRT. Entries must be made when oil is pumped into the sea for safety purposes or has escaped owing to damage or a leak.

7.1.25 Prevention of Oil Pollution Act 1971 (POPA 1971)

This Act consolidated (and repealed) the Oil in Navigable Waters Acts of 1955, 1963 and 1971 and enabled the UK to apply the provisions of OILPOL 1954 (7.1.4) to the extent of prohibiting discharges of oil from ships of any nationality in UK waters. As OILPOL 1954 has now been largely superseded by MARPOL 73/78 (7.1.2), most of the provisions of the Act concerning discharges from ships have been superseded by the **Merchant Shipping (Prevention of Oil Pollution) Regulations 1983** (see 7.1.20) and the new POPA 1986 (7.1.18).

Sections of the Act still in force prohibit transfers of oil in a UK harbour without giving appropriate notice to the harbourmaster and make it necessary to report any discharges or escapes of oil into UK waters to the harbourmaster.

The Act also enables the UK to apply the provisions of the International Convention relating to Intervention on the High Seas in Cases of Oil Pollution Casualties 1969 (7.1.3) by empowering the Secretary of State to give such instructions as are necessary to prevent or reduce pollution and, if these are inadequate, to take any necessary action, including the sinking or destruction of the ship.

The Act provides for the carrying of Oil Record Books for recording specified operations and for the appointment of inspectors to determine whether the restrictions and obligations of the Act are carried out.

7.1.26 Merchant Shipping Act 1970

As amended by the Merchant Shipping Act 1979, this Act gives the Secretary of State for Transport the power to order either a preliminary or a formal inquiry into a shipping casualty, if the ship was either registered in the UK or the incident happened in UK territorial waters. **The Merchant Shipping (Prevention of Oil Pollution) Regulations 1983** (see 7.1.20 extend this power to discharges of oil.

Netherlands Legislation

7.1.27 Decree of the Minister of Merchant Shipping 8 March 1988

This Decree makes provision for the implementation of some technical assessments required by Regulation 8 of Annex II of MARPOL 73/78 (7.1.2).

7.1.28 Decree of the Minister of Merchant Shipping 3 December 1988

This Decree contains provisions on IOPP Certificate required by Annex I of MARPOL 73/78 (7.1.2).

7.1.29 Ministerial Regulation Concerning Reception Facilities

This Regulation, which entered into force in April 1986, appoints 26 harbours in which reception facilities have to be maintained, in compliance with MARPOL 73/78.

7.1.30 Prevention of Oil Pollution by Ships Decree 1986

This Decree implements Annex I of MARPOL 73/78 (7.1.2) and provides in general for reception facilities.

7.1.31 Prevention of Pollution by Ships Act 1983

This Act, which entered into force in April 1986, implements MARPOL 73/78 (7.1.2). The Act applies to all Dutch and foreign ships in the Netherlands territorial sea and provides for a general prohibition of the discharge of harmful substances from ships into the sea. The power to designate substances exempt from this prohibition and to designate reception facilities is delegated by the government. The Act replaces the 1958 Pollution of the Sea by Oil Act.

7.1.32 Royal Decree Concerning the Returns of Received Oil 1982
This Decree contains specific provisions on the return of received oil.

Italian Legislation

7.1.33 Decree of the Minister of Merchant Shipping 25 August 1987
This Decree provides that oil tankers and other ships, which are not bound to be provided with an International Oil Pollution Prevention (IOPP) Certificate should comply with the requirements set forth in Annex I of MARPOL 73/78 (7.1.2).

7.1.34 Decree of the Minister of Merchant Shipping 11 June 1986
This Decree sets out a programme of training courses on crude oil washing of tanks according to rule 13B of MARPOL 73/78 (7.1.2).

7.1.35 Regulation of Maritime Traffic through the Strait of Messina, Decree of the Minister for Merchant Shipping 8 May 1985
Following the collision of two tankers in the Strait of Messina in March 1985, this Decree contains interim measures for the prevention of maritime accidents. Ships of 50,000 GRT or above and carrying petroleum are prevented from passing through the Strait.

7.1.36 Decree of the Minister of Merchant Shipping 25 January 1985
This Decree approves the form of the IOPP Certificate provided for by Annex I to MARPOL 73/78 (7.1.2).

7.1.37 Decree of the Minister of Merchant Shipping 11 January 1984
This Decree approves the form of oil record book according to Annex 1 to MARPOL 73/78 (7.1.2).

7.1.38 Provisions for the Protection of the Sea, Law No. 979, 1982
The purpose of these provisions is to promote a rational system for the protection of the marine environment to be drawn up by the Ministry for Merchant Shipping. Other provisions relate to emergency measures in cases of accidental pollution.

7.1.39 Laws No. 662, 438, 1980
These laws implement MARPOL 73/78 which came into force for Italy in October 1983.

7.2 POLLUTION FROM SHIPS — HAZARDOUS AND NOXIOUS SUBSTANCES OTHER THAN OIL
(including the Carriage of Dangerous Goods)

International

7.2.1 Basle Convention on the Control of Transboundary Movements of Hazardous Wastes and their Disposal 1989
This Convention, which is not yet in force, imposes obligations upon exporting and importing states to tighten up waste management; exports can take place only if the written consent of the importing state and of any transit state has been obtained. Consent is not, however, required for movement of hazardous waste through the territorial sea or EEZ of a foreign state in exercise of rights of innocent passage or freedom of navigation.

The exporting State is required to re-import the waste if the importing State is unable to dispose of it as agreed. The Convention, if widely implemented, will help to end the shuttling of ships from port to port in search of a dump for hazardous cargoes.

7.2.2 United Nations Convention on the Law of the Sea UNCLOS
- states must take measures to minimise the fullest possible effect the release of toxic, harmful or noxious substances, especially persistent ones, to the marine environment (Art. 194(3)(a));

- coastal states may with hold their consent to the conduct of a marine scientific research if it involves the introduction of harmful substances into the marine environment (Art. 246(5)(b)).

See also 7.1.1 for vessel-source pollution in general.

7.2.3 International Convention for the Safety of Life at Sea 1974 (SOLAS)
Although primarily concerned with the safety and welfare of those on board a ship, this Convention, which entered into force in May 1980, prohibits the carriage of dangerous goods on ships unless this is done according to its provisions. These apply not only to states that have accepted the Convention but also to flag states that have not, in order to ensure similar treatment. These provisions do not apply to a ship's own stores or equipment or to purpose-built chemical tankers.

7.2.4 International Convention for the Prevention of Pollution from Ships 1973 and Protocol I: Provisions concerning Reports on Incidents involving Harmful Substances 1979 (MARPOL 73/78; see also 7.1.2)

Annex II of the MARPOL Convention which entered into force in April 1987 lays down regulations for the 'Control of Pollution by Noxious Liquid Substances in Bulk', ie. the carriage of liquid noxious substances by chemical or oil tankers, and the conditions under which chemicals can be discharged into the sea from tank cleaning or deballasting operations.

Appendix II to the Annex lists noxious substances, which are divided into four categories, A, B, C and D, according to their toxicity. Category A substances, discharge of which into the sea would present a major hazard to marine resources or human health, are prohibited. Similar, but less stringent provisions, apply to the other categories according to the extent of the potential danger.

Ships are exempted from the Regulations if the discharge is made to save life or for the safety of the ship.

Chemical tankers must carry a Cargo Record Book. Port states are to provide adequate reception facilities for the discharge of tank washings and may appoint inspectors to supervise tank washing operations. Flag states are to survey ships and issue an International Pollution Prevention Certificate for the Carriage of Noxious Liquid Substances in Bulk on similar conditions to the issue of an IOPP Certificate.

While MARPOL states that a report of an incident involving harmful substances must be made without delay, the Protocol places the responsibility for this on the master of the ship and lays down the method of reporting.

7.2.5 Protocol relating to Intervention on the High Seas in Cases of Marine Pollution by Substances Other than Oil 1973

This Protocol to the 1969 Convention (see 7.1.3), which entered into force on 30 March 1983, extends the provisions of that Convention to the threat of pollution by substances other than oil.

7.2.6 International Atomic Energy Agency Regulations for the Safe Transport of Radioactive Materials 1973 (as amended in 1979)

These Regulations provide the basis for Class 7 substances of the International Maritime Dangerous Goods (IMDG) Code dealing with the carriage of radioactive substances at sea.

7.2.7 International Convention on Safe Containers 1972

The Convention, which entered into force in 1977, lays down the requirements to which containers must conform, if they are to be used for carrying dangerous goods at sea.

Regional

7.2.8 Protocol to the Noumea Convention concerning Cooperation in Combating Marine Pollution Emergencies in the South Pacific Region 1986 (see 7.1.5)

The Protocol was adopted in November 1986 in Noumea.

7.2.9 Protocol to the Nairobi Convention concerning Cooperation in Combating Marine Pollution in Cases of Emergency in the Eastern African Region 1985 (see 7.1.6)

The Protocol also follows the broad principles of the Barcelona Convention and marine pollution is defined to include harmful substances other than oil.

7.2.10 Protocol to the Cartagena Convention concerning Cooperation in Combating Oil Spills in the Wider Caribbean Region 1983 (see 7.1.8)

The Protocol follows the broad principles of the Barcelona Convention but has been extended by an Annex to cover hazardous substances other than oil.

7.2.11 Bonn Agreement for Cooperation in Dealing with Pollution of the North Sea by Oil and Other Harmful Substances 1983 (see 7.1.7)

Unlike the Bonn Agreement of 1969 which this Agreement superseded on its entry into force in 1989, this treaty covers not only oil but other harmful substances also.

7.2.12 Agreement on Regional Cooperation in Combating Pollution of the South-East Pacific by Hydrocarbons and Other Harmful Substances in Cases of Emergency 1981 (see 7.1.10) and Supplementary Protocol to the Agreement on Regional Cooperation in Combating Pollution of the South-East Pacific by Hydrocarbons or Other Harmful Substances 1983

The Agreement covers contingency plans, monitoring, etc. for harmful substances in containers, portable tanks or tank vehicles. The Supplementary Protocol lays down methods of cooperation, the contents of national contingency plans and training programmes.

7.2.13 Protocol to the Jeddah Convention concerning Regional Cooperation in Combating Pollution by Oil and Other Harmful Substances in Cases of Emergency 1982 (see 7.1.9)

The Protocol is broadly similar to others of its kind and lays down plans for cooperation and the establishment of a Marine Emergency Mutual Aid Centre.

7.2.14 Protocol 1 (Lomé Protocol) to the Abidjan Convention (see 7.1.11) concerning Cooperation in Combating Pollution in Cases of Emergency 1981

The Protocol, which entered into force in August 1984, covers both oil and other harmful substances and calls for contingency planning, monitoring, dissemination of information and reporting of accidents along the lines of the Mediterranean Protocol.

7.2.15 Protocol to the Kuwait Convention (see 7.1.12) concerning Regional Cooperation in Combating Pollution by Oil and Other Harmful Substances in Cases of Emergency 1978

The Protocol is broadly similar to that for the Mediterranean, but does not contain the detailed provisions relating to packages, etc. It includes provision for setting up a Marine Emergency Mutual Aid Centre and lays down its functions.

7.2.16 Protocol 2 to the Barcelona Convention (see 7.1.13) concerning Cooperating in Combating Pollution of the Mediterranean Sea by Oil and Other Harmful Substances in Cases of Emergency 1976

The Protocol, which is concerned with situations of a grave or imminent danger to the marine environment, provides for contingency plans, monitoring activities, the dissemination of information and the reporting by the most rapid means of accidents involving harmful substances. A unique feature is the requirement concerning the release or loss overboard or harmful substances in packages, freight containers, portable tanks or road or rail tank wagons, when those concerned are required to cooperate in salvage and recovery to reduce the danger of pollution.

7.2.17 Convention on the Protection of the Marine Environment of the Baltic Sea Area 1974 (Helsinki Convention) (see 7.1.14)

Under Article 5, States undertake to counteract the introduction of certain hazardous substances listed in Annex I into the Baltic Sea, whether from land, sea or air. In practice the article has been mainly applied to pollution from the land.

Article 7 deals specifically with pollution from ships and Annex IV gives provisions for the discharge from ships of oil and noxious liquid substances carried in bulk as well as making it obligatory for states to provide adequate reception facilities in ports receiving residues containing such harmful substances.

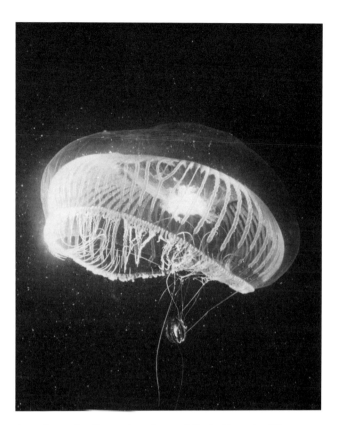

A medusa, the dispersive phase of the bottom-dwelling Aequorea aequorea, off Sherkin Island, SW Ireland. Delicate planktonic organisms such as this require unpolluted water. Photo: D George

UK Legislation

7.2.18 Merchant Shipping (Reporting of Pollution Incidents) Regulations 1987 (SI 586/1987)

These Regulations came into force in April 1987 and give effect to Protocol 1 of MARPOL 73/78 as amended by amendments adopted by MEPC in 1985. The principal purpose or these amendments is to specify in detail the incidents required to be reported. These Regulations replace Regulation 31 of the Merchant Shipping (Prevention of Oil Pollution) Regulations 1983.

7.2.19 The Merchant Shipping (IBC Code) Regulations 1987 (SI 549/1987)

These Regulations came into force in April 1987 and require chemical tankers built on or after 1 July 1986 and carrying polluting or dangerous liquid substances in bulk and converted ships after that date to comply with the International Code for the Construction and Equipment of Ships

Carrying Dangerous Chemicals in Bulk which was adopted by the MEPC in December 1985. This Code is an amended version of the Code adopted by the Maritime Safety Committee in 1983 covering dangerous substances only.

7.2.20 The Merchant Shipping (BCH) Code Regulations 1987 (SI 550/1987)

These Regulations, which came into force in April 1987, give effect to Regulation 3(3) of Annex II to MARPOL 73/78 as set out in the amendments to that Annex which were adopted by the MEPC on the same day as the Code for the Construction and Equipment of Ships Carrying Dangerous Chemicals in Bulk (see above).

7.2.21 The Merchant Shipping (Control of Pollution by Noxious Liquid Substances in Bulk) Regulations 1987 (SI 551/1987)

These Regulations, which came into force in April 1987, are in accordance with SI 1987 Nos 549 and 551 (see above) and apply to ships carrying noxious liquid substances in bulk. Noxious liquid substances are divided into four categories, A, B, C and D, in accordance with the severity of the hazard which they present to human health and the marine environment (see 7.2.4).

7.2.22 The Classification, Packaging and Labelling of Dangerous Substances (Amendment) Regulations 1986 (SI 1922/1986)

These Regulations, which came into operation in December 1986, amended the Classification Packaging and Labelling of Dangerous Substances Regulations 1984 ('the principal Regulations') to give effect to United Kingdom of the Commission Directive 83/467/EEC (OJ No L 257, 16.9.83) and 84/291/EEC (OJ No L 144, 30.5.84). This revision provides additional entries relating to substances which are dangerous for supply within the meaning of the conventional toxicity values for use in classifying pesticides in accordance with Schedule 3 to the principal Regulations.

7.2.23 The Dangerous Substances in Harbour Areas Regulations 1987 (SI 37/1987)

These Regulations, which came into force in June 1987, provide for the control of carriage, loading, unloading and storage of dangerous substances in harbours and harbour areas. In particular they give a harbourmaster authority to prohibit or regulate the entry of a dangerous substance; take precautions avoid fire or explosion; instigate duties relating to dangerous substances in bulk; enforce packaging and labelling; apply emergency plans for dealing with accidents; designate storage areas and facilities.

7.2.24 Merchant Shipping (Fire Protection) Regulations 1984

These specify the requirements with which passenger ships and ships of 500 tons or more must comply if they are to carry dangerous goods on international voyages unless the quantities are limited in accordance with the IMDG Code.

7.2.25 Prevention of Pollution (Reception Facilities) Order 1984 (SI 862/1984)

This Order implements the Regulation of Annex II of MARPOL requiring the provision of reception facilities in harbours or terminals for the discharge of residues or mixtures containing noxious liquid substances. The Secretary of State is given the power to demand information from those providing the facilities, to appoint inspectors to report on their adequacy, and to order improvements if he considers them inadequate.

7.2.26 Merchant Shipping (Dangerous Goods) Regulations 1981 (SI 1747/1981)

These Regulations, made in December 1981 and effective from January 1982, do not relate to pollution as such but lay down general duties so far as the health and safety of those on board are concerned. They apply to all UK ships and to other ships in UK waters.

The rules over documentation, packaging, marking, stowage and packaging, carriage of explosives (other than safety explosives) and carriage of dangerous goods in passenger ships for packaged goods, and documentation and carriage for chemicals in bulk. In practice these are a general application of regulations given in either the 'Blue Book', ie. the 1978 Report of the DTp's Standing Advisory Committee on the Carriage of Dangerous Goods in Ships, or the IMDG Code, ie. the 1977 International Maritime Dangerous Goods Code.

7.2.27 Merchant Shipping (Prevention of Pollution) (Intervention) Order 1980 (SI 1093/1980)

This Order, made under the Merchant Shipping Act 1979 (7.2.28) in July 1980 and taking effect from August 1980, implements the Protocol to the 'Intervention' Convention (7.2.5) and enables the Secretary of State to apply powers conferred on him by POPA 1971 (7.1.25) with respect to the threat of oil pollution to similar threats from substances other than oil. The definition of these is broader than the lists of substances given as it includes other substances "liable to create hazards to human health, to harm living resources and marine life, to damage amenities or to interfere with other legitimate uses of the sea" without specification.

7.2.28 Merchant Shipping Act 1979

The Act empowers inspectors to enter ships to determine whether the Dangerous Goods Regulations (7.2.26) are being complied with. It also provides the legal basis for implementation in the UK of a number of international conventions:

- the 1973 Protocol to the 'Intervention' Convention (see 7.2.5)

- the appropriate sections of the SOLAS Convention (see 7.2.3)

  Annex II of MARPOL 73/78 (see 7.2.4)

A sample of litter from a 5-m wide transect of UK beach. With the entry into force of Annex V of MARPOL on 31 December 1988 (see 7.3.2) this type of litter should begin to decrease. Photo: TBG

7.2.29

Health and Safety at Work Act 1974
This Act gives an inspector the right to enter a ship for carrying out his duties and the power to seize and make harmless any article or substance he reasonably believes to be a cause of imminent danger or personal injury.

7.2.30 Merchant Shipping (Safety Convention) Act 1949
This Act provides the legal basis for issuing the Dangerous Goods Regulations (7.2.26).

7.2.31 Merchant Shipping Act 1894

Among its numerous provisions the Act lays down restrictions on the carriage of dangerous goods which, so far as notification, marking and supplying a false description are concerned, are similar to the Dangerous Goods Regulations (7.2.26).

Netherlands Legislation

7.2.32 Prevention of Pollution by Ships Act 1983
This Act provides for a general prohibition of discharges of harmful substances from all ships in the territorial sea (see 7.1.31).

Italian Legislation

7.2.33 Ordinances and Decrees on Control and Elimination of Noxious Wastes
During 1988 and 1989 a large number of Ordinances and Decrees were issued to deal with the emergencies created by the toxic industrial wastes of Italian origin carried from Nigeria and Lebanon by the ships *Karin B*, *Deep Sea Carrier* and others. Most of this legislation was concerned with arrangements for mooring the vessels and cataloguing, storing and eventually eliminating the wastes. Particular mention may be made of the Law of 9 November 1988 making urgent provision for the elimination of industrial wastes and regulating the transfrontier movement of wastes produced in Italy; the Decree of 22 October 1988 on rules for the export and import of wastes; the Law of 10 February 1989 on urgent measures for disposal of industrial waste; and the Decree of 26 April 1989 establishing a national register of special wastes.

7.2.34 Decrees of the Minister for Shipping 1987
A number of Decrees were issued by the Minister for Shipping during 1987 which implement Annex II of MARPOL 73/78 (7.2.4):

* 3 April 1987. This Decree contains the approval of the form of loading record books for ships carrying noxious liquid substances in bulk;

* 14 and 16 March 1987. These Decrees contain a revised list of the substances to which the regulation for the construction and equipment of ships carrying dan-

gerous chemicals in bulk and the carriage of these products as approved by earlier Decrees is not applicable;

- 13 March 1987. This Decree contains technical rules
for the construction and equipment of ships carrying
dangerous chemicals in bulk as well as provisions for
their loading, transport and unloading;

- 8 January 1987. This Decree provides for approval of
the certificate of fitness for the carriage of dangerous
chemicals in bulk according to the IMO Code for the
Construction and Equipment of Ships Carrying Dangerous Chemicals in Bulk;

- 8 January 1987. This Decree provides for the approval
of the International Pollution Prevention Certificate for
the carriage of noxious liquid substances in bulk.

7.2.35 Regulation of Maritime Traffic through the Strait of Messina 1985 (see 7.1.35)

This regulation also applies to ships of 50,000 GRT or
more carrying harmful substances.

7.3 POLLUTION FROM SHIPS: MISCELLANEOUS — SEWAGE, GARBAGE AND DISCHARGES DURING NORMAL OPERATIONS

International

7.3.1 United Nations Convention on the Law of the Sea (UNCLOS) 1982

- coastal states may characterise a clearly defined area as
a special area with the consent of the competent international organisation and adopt additional laws and
regulations relating to discharges in this area. This
should include the requirement of prompt notification of
incidents, including maritime casualties, which involve
discharges or probability of discharges (Art. 211
(6)(a)(c) (7));

- a coastal state may undertake physical inspection of a
vessel navigating in its EEZ if it has committed a violation resulting in a substantial discharge causing or
threatening significant pollution of the marine environment. If there is clear objective evidence of such an
offence the state may institute proceedings, including
detention of the vessel in accordance with its laws (Art.
220(5)(6)).

See also 7.1.1.

7.3.2 International Convention for the Prevention of Pollution from Ships 1973 and 1978 Protocol (MARPOL 73/78)

Annex IV, which is not yet in operation, lays down conditions to prevent pollution by sewage from ships which are

similar to those for oil and chemicals. The regulations will
apply to new ships of 200 GRT or more; existing ships will
have ten years from the date on which they become effective.

- no sewage may be discharged into the sea less than four
nautical miles from land (except for disinfected or
treated sewage) and no untreated sewage may be discharged less than 12 nautical miles from the nearest
land;

- sewage stored in a holding tank can be discharged only
when the ship is proceeding at at least four knots and at
a specified rate of discharge;

- ships will be surveyed to ensure that they have equipment to treat sewage before discharge, or a holding
tank, and, if satisfactory, issued with an International
Sewage Prevention Certificate;

- the regulations will not apply when a ship is in the territorial waters of a state with less stringent restrictions on
operation;

- governments party to the agreement must provide adequate reception facilities for sewage at their ports.

Annex V, which entered into force on 31 December
1988, lays down conditions to prevent pollution by garbage
from ships.

- in the special areas, ie. the Mediterranean, etc., disposal
of plastics and other durable wastes is forbidden. Food
wastes may be discharged when the ship is 12 nautical
miles from land. If garbage is mixed with other wastes,
the more stringent rules will apply;

- the same rules apply to other areas, though dunnage,
lining and packing materials that will float must not be
discharged less than 25 nautical miles from land;

- governments must provide reception facilities for garbage at ports or terminals that will not cause undue delay
to ships using them.

7.3.3 International Health Regulations 1969

These Regulations require every port to be provided with
an effective system for removing and disposing of excre-
ment, refuse, waste water, condemned food and other
matters dangerous to health. They also empower a health
authority to take all practicable measures to control the discharge from a ship of sewage and refuse that may
contaminate the waters of a port, river or canal.

Regional

7.3.4 Helsinki Convention 1974 (see 7.1.14)

The Convention contains provisions to control discharges
of both sewage and garbage and for the provision of reception facilities at ports. HELCOM has made a number of
recommendations to Member States in their efforts to implement these aspects of the Convention.

UK Legislation

7.3.5 Control of Pollution (Landed Ships' Waste) (Amendment) Regulations (SI 65/1989)

These regulations amend SI 402/1987 to apply to ships' garbage landed in Great Britain the same requirements that apply to tank washings landed there.

7.3.6 The Merchant Shipping (Prevention of Pollution by Garbage) Order 1988 (SI 2252/1988), the Merchant Shipping (Prevention of Pollution by Garbage) Regulations 1988 (SI 2292/1988) and the Merchant Shipping (Reception Facilities for Garbage) Regulations 1988 (SI 2293/1988)

These three statutory instruments implement Annex V of MARPOL 73/78. In particular, the third instrument imposes a duty on harbour authorities and terminal operators to provide the necessary reception facilities for garbage from ships.

7.3.7 The Merchant Shipping (Prevention and Control of Pollution) Order 1987 (SI 470/1987)

This Order, which came into force in April 1987, gives effect to Annex II and Protocol I MARPOL 73/78 and amendments adopted by the MEPC in December 1985. The Order empowers the Secretary of State to make regulations in relation to this Annex and Protocol (see 7.2.4).

7.3.8 Public Health (Ships) Regulations 1979

These Regulations implement the 1969 International Health Regulations (7.3.3) but deal mainly with disease control. They do not provide for the treatment of sewage in a port.

7.3.9 Control of Pollution Act 1974 (COPA 74)

The Act prohibits the discharge of sewage from ships using inland waters but discharges into territorial waters are specifically exempted. While the discharge of trade effluent, ie. any liquid other than surface water and domestic sewage from a ship in port is not an offence under the Act, it is an offence to discharge poisonous, noxious or polluting solid matter into inland, controlled, or restricted waters, ie. areas in which ships lie at moorings close to one another.

7.3.10 Dark Smoke (Permitted Periods) (Vessels) Regulations 1958

The Regulations exempt certain emissions of dark smoke from ships (see 7.3.11) as long as they do not exceed specified time limits which vary according to the source.

7.3.11 Clean Air Act 1956

The Act makes it an offence for dark smoke (as defined by comparison with a Ringelmann Chart) to be emitted from the funnel of a ship in the tidal waters of ports, harbours, rivers and estuaries under the control of an authority (but see 7.3.10).

7.3.12 Public Health Act 1936

Port authorities and district councils are given the power to take action against a ship lying at anchor within their jurisdiction if its *condition* is that of a 'statutory nuisance', ie. prejudicial to health, or a nuisance, or if it emits 'dust or effluvia' that is likely to be harmful to health.

Netherlands Legislation

7.3.13 Prevention of Pollution by Ships Act 1983

See 7.1.30

Italian Legislation

7.3.14 Provisions for the Protection of the Sea 1982

See 7.1.38.

7.4 POLLUTION FROM SOURCES ON LAND (excluding radioactive discharges — see 7.5)

International

7.4.1 United Nations Convention of the Law of the Sea (UNCLOS) 1982

States may adopt laws and regulations to prevent, reduce and control pollution of the marine environment from land-based sources, including rivers, estuaries, pipelines and outfall structures taking into account internationally agreed rules, standards and recommended practices and procedures. In particular these must include laws and regulations designed to minimise, to the fullest extent possible, the release of toxic, harmful or noxious substances (Art. 207).

Regional

7.4.2 Convention for the Protection, Management and Development of the Marine and Coastal Environment of the Eastern African Region 1985 (Nairobi Convention)

The Convention covers pollution from land-based sources and includes provisions similar to those of the Barcelona Convention (7.4.8).

7.4.3   1983 Cartagena Convention (see 7.1.8)

The Convention covers pollution from land-based sources and includes provisions similar to those of the Barcelona Convention (7.4.8).

7.4.4   Jeddah Convention (see 7.1.9)

The Convention covers pollution from land-based sources and includes provisions similar to those of the Barcelona Convention (7.4.8).

7.4.5   Lima Convention (see 7.1.10) and Protocol for the Protection of the South-East Pacific against Pollution from Land-Based Sources 1983 (Quito Protocol)

The provisions of the Convention for dealing with pollution from land-based sources are amplified in the Protocol in terms similar to those of the Athens Protocol to the Barcelona Convention (7.4.8), including the area covered, atmospheric pollution, lists of substances and provisions for cooperation in monitoring and research.

7.4.6   Abidjan Convention (see 7.1.11)

The Convention contains provisions for dealing with pollution from land-based sources which are broadly similar to those of the Barcelona Convention (7.4.8).

7.4.7   1978 Kuwait Regional Convention (see 7.1.12)

The Convention includes provisions on pollution for dealing with land-based sources similar to those of the Barcelona Convention (7.4.8).

7.4.8   1976 Barcelona Convention (see 7.1.13) and (Athens) Protocol 3 concerning the Protection of the Mediterranean Sea against Pollution from Land-Based Sources 1980

The Convention places contracting parties under an obligation to take all appropriate measures to prevent and combat pollution from sources on land. The Protocol, which applies both to the sea and to inland waters up to the freshwater limit, gives two lists of substances, the first of which are to be eliminated and the second strictly limited through suitable programmes.

7.4.9   1974 Helsinki Convention (see 7.1.14)

The Convention covers pollution from land by direct discharges through sewers, by solid substances carried in rivers and by agricultural operations, and sets out the obligations for measures to control and minimise the effects of 16 substance groups, including heavy metals, persistent plastics, etc., significant quantities of which cannot be discharged without special permits from national authorities, who collaborate in finding common criteria for potential inputs.

7.4.10   Convention for the Prevention of Marine Pollution from Land-Based Sources 1974 (Paris Convention)

The Convention, which entered into force in May 1978, applies to all land-based discharges into the North Sea and North-East Atlantic. Contracting states are required to exercise controls of varying degrees over substances listed in annexes to the Convention on the principle of 'black' and 'grey' lists.

The Convention is administered by the Paris Commission (PARCOM), which consists of government representatives from contracting states together with observers from specialist intergovernmental organisations. The Commission's technical working groups have concentrated on measures to eliminate pollution by mercury, cadmium, PCBs and other substances, and on the question of setting an environmental standard for water.

European Economic Community (EEC) Directives

7.4.11   Directive 88/347/EEC

This Directive amends Annex II to Directive 86/280/EEC on limit values and quality objectives for discharges of certain dangerous substances included in List I of the Annex to Directive 76/464/EEC. It adds to the list of products for which emission values and quality objectives for the aquatic environment have to be observed, aldrin, dieldrin, endrin, isodrin, hexachlorobenzene, hexachlorobutadiene and chloroform. In September 1988 the Commission proposed a number of further additions to the list: dichloroethane, trichloroethylene, perchloroethylene and trichlorobenzine (OJ 1988 C253/4).

7.4.12   85/467/EEC Directive on Dangerous Substances in Water

This Directive includes amendments to the 76/769/EEC Directive on Dangerous Substances, especially the residues of polychlorinated biphenyls (PCBs) and polychlorinated terphenyls (PCTs), and restricts the marketing and use of these dangerous substances. Under the new Directive the sale and use of new and secondhand appliances containing one of these substances will be prohibited. Existing appliances are exempt from this provision until the end of their service life, though the maximum limit of either PCBs or PCTs in preparations and mixtures is reduced from 1,000 ppm to 100 ppm.

7.4.13   85/337/EEC Directive on Environmental Impact Assessment

This Directive of March 1985 ensures that authorisation for any development likely to affect the environment is subject to prior assessment of its direct effects on man, flora, fauna, soil, water, climate, landscape, material goods and cultural heritage. Projects which will have to be submitted to the appropriate authority for assessment include: crude oil refineries; power stations; nuclear storage or disposal installations; integrated chemical installations; sea ports; waste incineration plants; and plants for the chemical treatment or storage of dangerous and toxic wastes.

7.4.14 84/156/EEC Directive on Limit Values and Quality Objectives for Mercury Discharges by Sectors other than the Chloro-Alkali Industry

This 1984 Directive is a particular application stemming from the 'framework' Directive 76/464 (7.4.24) and has similar provisions to Directive 82/176 (7.4.18) but applies to all industries (excluding the chloro-alkali industry) which discharge mercury as part of their production process.

7.4.15 83/513/EEC Directive on Limit Values and Quality Objectives for Cadmium Discharges

This Directive of September 1983 has the aim of applying Directive 76/464/EEC (7.4.24) to cadmium discharges. It sets out quality objectives and describes the monitoring procedure to be used to achieve them. New plants may be authorised only if they use the best technical means available for the elimination of pollution.

7.4.16 82/883/EEC Directive on Procedures for the Surveillance and Monitoring of Environments Concerned by Waste from the Titanium Dioxide Industry

This Directive of December 1982 lays down the methods to be used in controlling the application of Directive 78/176/EEC (7.4.23) and specifies the way in which the effects of discharges into the sea are to be measured and the results reported to the Commission every three years.

7.4.17 82/838/EEC Directive on Limit Values and Quality Objectives for Mercury Discharges by Sectors other than the Chlor-Alkali Electrolysis Industry

This Directive of March 1982 applies Directive 76/464/EEC (7.4.24) to the industrial sector specified. It sets limit values for certain industrial processes, while quality objectives must be the same as those specified in Directive 82/176/EEC (7.4.18).

7.4.18 82/176/EEC Directive on Limit Values and Quality Objectives for Mercury from the Chlor-Alkali Industry

This Directive of March 1982 is a particular application stemming from the 'framework' Directive 76/464 (see 7.4.24). It applies only to discharges of mercury from plants using alkali chlorides in the production of chlorine. Members may authorise discharges of mercury and its compounds in conformity with either limit values or quality objectives as specified in the Directive. Authorisations for new plant must contain a reference to use of the best technical means available for preventing discharges of mercury.

Member States are responsible for monitoring the waters affected.

7.4.19 79/923/EEC Directive on the Quality Required of Shellfish Waters

This 1979 Directive applies to coastal and brackish waters designated by Member States as needing protection or improvement in order to support shellfish life and growth. The Directive recommends values for a number of parameters used to measure water quality and lays down methods to be used in sampling operations.

7.4.20 79/869/EEC Directive concerning the Methods of Measurement and Frequencies of Sampling and Analysis of Surface Water Intended for the Abstraction of Drinking Water

This 1979 Directive, which operates within the framework of Directive 76/464 (7.4.24), fixes common reference methods of measurement to determine the values of parameters used to determine the quality of surface water. Minimum annual frequencies of sampling and analysis are also recommended (see also 7.4.28).

7.4.21 79/831/EEC 'Sixth Amendment' on the Classification Packaging and Labelling of Dangerous Substances

This 'Sixth Amendment' to the parent Directive 67/548 replaces and extends both the original Directive and five subsequent amendments. While the aim of the original Directive and amendments was to protect man, particularly in the workplace, this Directive adds both a new classification of 'dangerous to the environment' and a scheme of prior notification involving tests for potential hazards before a substance is marketed. Notification thus becomes a prerequisite for classification; packaging and labelling follow from notification and classification. Medicinal products, narcotics, radioactive substances, foodstuffs and feedingstuffs, and waste covered by Directives 75/442 and 78/319 (see 7.4.27; 7.4.22) are excluded.

7.4.22 78/319/EEC Directive on Toxic and Dangerous Waste

This Directive of March 1978 operates within the framework of Directive 75/442 (see 7.4.27) and lays down more stringent controls for toxic and dangerous wastes. Such wastes may be stored, treated and/or deposited only by authorised undertakings and anyone producing or holding such waste without an appropriate permit must hand it over to such an undertaking. Member States have the general duty of ensuring that toxic and dangerous waste is disposed of without harming human health or the environment, and in particular without risk to water, air, soil, plants or animals. Certain wastes, ie. radioactive waste, some agricultural wastes, explosives and hospital waste are excluded from the scope of this Directive.

Pollution caused by detergent at Portpatrick, Scotland. Stricter legislation on the dumping of wastes at sea must be followed up by effective enforcement. Photo: WWF (UK)

7.4.23 78/176/EEC Directive on Waste from the Titanium Dioxide Industry

The aim of this Directive of February 1978 is the prevention and progressive reduction of pollution caused by waste from the titanium dioxide industry, leading eventually to its elimination. Member States have the duty of ensuring that waste is disposed of without endangering human health or harming the environment, and to encourage recycling. All discharges, dumping, storage, etc. of wastes are made subject to prior authorisation and Member States are required to produce programmes for progressive reduction and eventual elimination of pollution.

7.4.24 76/464/EEC Directive on Pollution caused by Certain Dangerous Substances in Water

This Directive of May 1976 sets the framework for the elimination or reduction of pollution of inland, coastal and territorial waters by particularly dangerous substances. It seeks to establish the extent of such pollution by requiring an inventory of discharges, and to ensure consistency in implementing certain international conventions through the operation of 'black' and 'grey' lists. The aim is to eliminate pollution from substances on the first list and to reduce it from those on the second.

7.4.25 76/403/EEC Directive on the Disposal of Polychlorinated Biphenyls and Polychlorinated Terphenyls (PCBs and PCTs)

This Directive on April 1976, which took effect from April 1978, aims at making the harmless disposal of PCBs compulsory, at promoting their reclamation, providing a system of authorisation for firms responsible for their disposal and prohibiting their discharge, discarding and unsupervised dumping.

7.4.26 76/160/EEC Directive concerning the Quality of Bathing Water

This Directive of 1975 aims at ensuring that the quality of bathing water is raised over time, or maintained, not just to protect public health but also for reasons of amenity. This is to be done largely by ensuring that sewage is not present or has been adequately diluted or destroyed.

7.4.27 75/442/EEC Directive on Waste

This Directive of July 1975 places a general duty on Member States to ensure that waste is disposed of without endangering human health and without harming the environment. Under the Directive competent authorities with responsibility for waste are to be appointed, and waste disposal plans prepared by these authorities, who will issue permits to installations or undertakings handling waste. The Directive is sometimes referred to as a 'framework' Directive, more detailed measures being provided by other Directives, eg. toxic waste.

7.4.28 75/440/EEC Directive concerning the Quality Required of Surface Water Intended for the Abstraction of Drinking Water

This 1975 Directive fixes a minimum quality requirement for surface water for drinking and sets standard methods of treatment for transforming surface water to drinkable water (see also 7.4.20).

7.4.29 75/439/EEC Directive on the Disposal of Waste Oils

This Directive, which was adopted in June 1975 and took effect from June 1977, aims at determining arrangements to be made for the collection and harmless disposal of waste oils, recommends that waste oils be re-used, and provides for a system of authorisations for firms responsible for collection and disposal, and also for a register of waste oils.

UK Legislation

7.4.30 Environmental Protection Bill (Bill 14/1989)

Part I of this Bill, published in December 1989, establishes a new pollution control regime, Integrated Pollution Control, which will apply to prescribed industrial, commercial and other processes discharging into any environmental medium (including internal waters and the territorial sea). The Secretary of State will be empowered to issue regulations setting limits on the concentrations or amounts of released substances and other requirements relating to the operation of prescribed processes; establishing environmental quality objectives and standards; and making national plans for controlling releases. Every prescribed process will require an authorisation before it allowed to operate. Such authorisations will include conditions requiring the use of best available technology not entailing excessive cost to prevent and minimise releases of substances, and to render harmless any substances which are released; and compliance with limits and plans, and achievement of quality standards and objectives, set by the Secretary of State.

7.4.31 Statutory Instruments implementing Water Act 1989

The Statutory Instruments so far issued to give further effect to the pollution control provisions of this Act (see 7.4.32) include the following, all of which entered into force on 1 September 1989:

* The Control of Pollution (Registers) Regulations 1989 (SI 1160/1989)

* The Control of Pollution (Radioactive Waste) Regulations 1989 (SI 1158/1989)

* The Control of Pollution (Discharges by the National Rivers Authority) Regulations 1989 (SI 1157/1989)

* The Control of Pollution (Consents for Discharges etc.) (Secretary of State Functions) Regulations 1989 (SI 1151/1989)

* The Controlled Waters (Lakes and Ponds) Order 1989 (SI 1149/1989)

* The Surface Waters (Classification) Regulations 1989 (SI 1148/1989)

7.4.32 The Water Act 1989

The main purpose of this Act, which received the Royal Assent on 6 July 1989, is to privatise the water industry in England and Wales. Nevertheless, the Act also contains, in Chapter I of Part III, some important provisions on pollution. These provisions replace Part II of the Control of Pollution Act 1974 (although that Act continues to apply in Scotland, subject to amendments contained in Schedule 23 of the Water Act). The pollution controls of the new Act are largely based on the consent system for discharges contained in the 1974 Act, though with important modifications. First, implementation and enforcement of the system is by an independent body, the National Rivers Authority, which is established by the Act, and not by the water industry itself, as was formerly the case. Second, water quality objectives and targets for achieving them are for the first time given statutory form. Third, the Act puts more emphasis on precautionary measures and on controlling diffuse as well as point sources of pollution: thus, for example, a new power is created of preventing and regulating the entry of nitrates into the waters to which the Act applies from agricultural activities by the establishment of nitrate sensitive areas. The Act's pollution controls apply to rivers, coastal and underground waters, and the inner three miles of the territorial sea (though orders may be made extending the controls to other parts of the territorial sea).

7.4.33 Harbour Works (Assessment of Environmental Effects) (No. 2) Regulations 1989 (SI 424/1989)

7.4.34 Harbour Works (Assessment of Environmental Effects) Regulations 1988 (SI 1336/1988)

7.4.35 Environmental Assessment (Salmon Farming in Marine Waters) Regulations 1988 (SI 1218/1988)

These three instruments implement EEC Directive 85/337 and relate to harbour works and the assessment of the environmental effects of certain salmon farming projects in marine waters.

7.4.36 The Control of Pollution (Exemption of Certain Discharges from Control) (Variation) Order 1986 (SI 1623/1986)

This order, which came into operation in October 1986 (Article 2(a)) and the Schedule in October 1987, varies the Control of Pollution (Exemption of Certain Discharges from Control) Order 1983. Section 32(1) of the Control of Pollution Act 1974 makes certain discharges of trade or sewage effluent unlawful if made without the consent of a water authority. This order withdraws many of the exemptions from 15 October 1987 by substituting a new and more limited Schedule. In particular, discharges direct to rivers or to the sea will now require a consent under the 1974 Act (see also 7.4.30-32).

7.4.37 Statutory Instruments on implementing COPA II

Implementation of COPA II (7.4.41) is proceeding by means of a series of Regulations concerning different aspects. The Control of Pollution (Consents for Discharges) (Notices) Regulations 1984 prescribe the form in which notices concerning applications for consents are to be published in the press. The Control of Pollution (Consents for Discharges) (Secretary of State Functions) Regulations 1984 lay down the procedure to be followed when there is controversy over the granting of consents; and the Control of Pollution (Discharges by Authorities) Regulations 1984 give the Secretary of State the power to give the necessary consent to a water authority wishing to make discharges within its own area. (See also 7.4.30-32.)

7.4.38 SI 1182/1983 on Implementation of COPA II

In order to ensure a smooth transition to full implementation of COPA 74 (7.4.41), the Secretary of State made this order exempting from the provisions of COPA II all existing discharges of trade and sewage effluent into controlled waters or beyond the three-mile limit begun on or before 30 April 1974. (See also 7.4.30-32.)

7.4.39 The Water Act 1983

The Act introduces an administrative reorganisation by which the authorities established under the 1973 Act (7.4.42) are replaced by a Water Authorities Association and area authorities for England and Wales. These will be responsible *inter alia* for implementation of COPA 74 (7.4.41). Members of water authorities are appointed solely by Ministers as opposed to the previous boards which contained a number of local authority nominees. Each authority is required to make arrangements for the representation of consumer interests according to guidelines issued by the Secretary of State for the Environment.

The guidelines issued in October 1983 (W12/1983) contain no explicit references to pollution and are concerned mainly with charging principles and investment planning (See also 7.4.30-32.)

7.4.40 SI 1980 No. 1709 on Polychlorinated Biphenyls

The Statutory Instrument gives effect to the EEC Directive on polychlorinated biphenyls so far as the UK is concerned (see 7.4.25).

7.4.41 Control of Pollution Act 1974 (COPA 74)

The Act is a comprehensive measure regulating not only pollution of water but also atmospheric pollution, noise and waste on land. Part II (COPA II) is concerned with water. Under the Act is is an offence to allow any poisonous, noxious or polluting matter to enter a stream or controlled waters, ie. the territorial sea. It is also an offence to discharge trade or sewage effluent into streams and coastal waters or through a pipe into the sea outside controlled waters unless this is done with the consent of the water authority or river purification board and in accordance with conditions laid down by them. Applications for consents to discharge effluent must give details of the proposed discharge and be sent to the appropriate authority who must publicise the application in a manner laid down in the Act. The Secretary of State may grant exemption from publication to protect trade secrecy or if he considers that publicity is not in the public interest. (See also 7.4.30-32.)

7.4.42 The Water Act 1973

The Act established the nine English water authorities and the Welsh National Water Development Authority to take over responsibilities for *inter alia* water supply and water conservation, sewage disposal and the issue of consents subject to specified conditions for discharges into water courses. In formulating or considering any proposals relating to the discharge of their functions, water authorities and the appropriate minister are required *inter alia* to have regard to the desirability of preserving natural beauty and of conserving flora, fauna and geological or physiographical features of special interest (see also 7.4.30-32 and 39).

7.4.43 Prevention of Oil Pollution Act 1971

In addition to covering oil pollution from ships (see 7.1.25) the Act makes it an offence to discharge oil from a place on land into the territorial sea or inland waters navigable by sea-going ships.

Netherlands Legislation

7.4.44 Ministerial Decree on Limit Values for Mercury, 25 April 1986

This Decree implements EEC Directives 82/176 (7.4.18) and 84/156 (7.4.14).

7.4.45 Regulations on Limit Values for Cadmium, 2 August 1985

This Regulation implements EEC Directive 83/513/EEC (7.4.15).

7.4.46 Royal Decree Concerning Water Quality Standards, 3 November 1983

This Decree gives effect to a number of EEC Directives notably 75/440 (7.4.28), 79/869 (7.4.20) and 79/923 (7.4.19). The Decree provides for rules relating to the establishment of quality standards and measurement and only applies within the territorial sea.

Italian Legislation

7.4.47 Urgent Measures against Eutrophication of Waters, Law No. 283, 1989

One on several legislative measures introduced in 1988 and 1989, this law contains urgent measures against the phenomenon of eutrophication of the coastal waters of the

Adriatic Sea. To this end, it regulates the collection and elimination of algae and organic materials, the adaptation of coastal purification plants, the reduction of effluents into the sea and the monitoring of eutrophication in the Adriatic Sea.

7.4.48 Environmental Impact Assessment, Decree No. 377, 1988

This Decree implements EEC Directive 85/337 (see 7.4.13).

7.4.49 Bathing Waters, Law-Decree No. 155, 1988

This law relates to the implementation in Italy of EEC Directive 76/160 on the quality of bathing water (see 7.4.26).

7.4.50 Urgent Measures against Phenomena of Eutrophication of the Waters, Law No. 7, 1986

This law contains provisions against eutrophication of the lakes and seas. The measures provide for the reduction and control of agricultural and industrial discharges containing phosphorus. The law also prohibits the production, importation and trade of synthetic detergents containing phosphorus above specified fixed limits.

7.5 POLLUTION BY RADIOACTIVE SUBSTANCES
(excluding dumping — see 7.8)

International

7.5.1 International Atomic Energy Agency Regulations for the Safe Transport of Radioactive Materials 1973 (as amended 1979)

These Regulations provide the basis for class 7 of the IMDG Code which deals with the way in which radioactive substances should be carried at sea.

Regional

7.5.2 1974 Helsinki Convention (see 7.1.14)

The list of substances which contracting parties agree should be subject to measures of control that will minimise pollution from land includes radioactive materials.

7.5.3 1974 Paris Convention (see 7.4.10)

Under the Convention contracting parties undertake to adopt measures to forestall or eliminate pollution of the sea from sources on land by radioactive substances, including wastes.

7.5.4 European Atomic Energy Community Treaty 1957 / EURATOM Treaty

The Treaty sets out basic standards for the discharge of radioactive waste. Contracting parties must carry out monitoring to ensure that these are complied with and make periodic returns to the European Commission, which must also be informed about a state's plans for disposal so that it can determine whether these will lead to contamination of the land, water or air space of another Member State. The Commission's opinion or recommendations are, however, not binding on the state concerned. The Treaty also requires the Community to establish "basic standards for the protection of the health of workers and of the general public." These include maximum permissible doses, maximum permissible levels of exposure and basic principles for medical supervision of workers based on ICRP recommendations.

UK Legislation

7.5.5 The Ionising Radiations Regulations 1985 (SI 1333/1985)

These Regulations, which came into operation in January 1986, provide that, when a radioactive substance is used as a source of ionising radiation, the substance must, when reasonable, be in the form of a sealed source. Tests must be carried out to detect any leakage of radioactive substances. A number of safety measures apply in cases of transporting radioactive substances.

7.5.6 Control of Pollution (Radioactive Waste) Regulations 1985

These Regulations were drawn up under the Control of Pollution Act (COPA 74, see 7.4.41) and came into operation on 6 June 1985. They bring discharges of radioactive waste within the scope of the provisions of COPA II, relating to the control of pollution of rivers and coastal waters and to the control of discharges of trade and sewage effluents, etc. into such waters. While the discharges are subject to COPA II, no account is to be taken of the radioactive properties of such waste, which remains subject to control under the Radioactive Substances Act 1960.

7.5.7 Radiological Protection Act 1970

The Act establishes the National Radiological Protection Board (NRPB) to take over the functions of the Radioactive Substances Advisory Committee appointed under the Radioactive Substances Act 1948 and the Medical Research Council's Radiological Protection Service, together with some activities previously conducted by the Atomic Energy Authority (AEA). The Board's functions are to carry out research into radiological protection and to provide information and advice on the subject to those having responsibilities in the field, including government departments.

7.5.8 Nuclear Installations Act 1965

The Act gives the Nuclear Installations Inspectorate the right to exercise controls over the discharge of waste from a site licensed under the Act. In licensing the site the NII can attach conditions regulating discharges and requiring the effective monitoring of radiation levels. The Act also deals with the carriage of nuclear matter, the reporting of incidents and liability in the event of damage to people or property.

7.5.9 Radioactive Substances Act 1960

Under the Act disposals of radioactive waste are permissible only when authorised under Section 6 and are based on the following three criteria:

- to ensure, irrespective of cost, that no member of the public shall receive more than the relevant ICRP dose limits. (The International Commission on Radiological Protection ICRP, is a small group of scientists, chosen by the International Congress of Radiology on the basis of their individual scientific reputations, which recommends basic radiation standards);

- to ensure, irrespective of cost, that the whole population of the country shall not receive an average of more than one-fifth of the ICRP limit relating to genetic risk; and

- to do what is reasonably practicable, having regard to cost, convenience and the national importance of the subject, to reduce doses far below these levels.

Although not specifying the basic radiation doses, the Act empowers the Minister to make regulations on this point.

7.5.10 Nuclear Installations (Licensing and Insurance) Act 1959

The Act creates the Nuclear Installations Inspectorate (NII) to oversee the design and planning of plants and processes and to approve their operation on licensed nuclear sites. While AEA sites do not require statutory licensing, their safety arrangements are made in consultation with the NII, which was transferred from the Department of Energy to the Health and Safety Executive (HSE) in 1975.

7.5.11 Atomic Energy Authority Act 1954

This Act established the UK Atomic Energy Authority (AEA) to take over both the defence and civil projects of the Ministry of Supply with a mandate to "produce, use and dispose of atomic energy and carry out research into any matters connected therewith."

7.6 POLLUTION FROM OFFSHORE INSTALLATIONS AND PIPELINES

International

7.6.1 United Nations Convention on the Law of the Sea (UNCLOS) 1982

States must take measures to minimise pollution from installations operating in the marine environment, to ensure the safety of their operation and the design, construction, operation and manning of such installations (Art. 194(3)(c)(d)). See also 7.8.1.

7.6.2 MARPOL 73/78 (see 7.1.2)

With the entry into force of Annex V (see 7.3.2), the disposal of all garbage (but not sewage) into the sea from offshore installations is prohibited, except that food wastes may be disposed of provided the installation is more than 12 nautical miles from land and the waste is comminuted.

Regional

7.6.3 1985 Nairobi Convention (see 7.1.6)

7.6.4 1983 Cartagena Convention (see 7.1.8)

7.6.5 Bonn Agreement 1983 (see 7.1.7)

The provisions of this Agreement for active cooperation in the event of serious pollution or threats of pollution extend to cover spills from offshore installations.

7.6.6 1982 Jeddah Convention (see 7.1.9)

7.6.7 1981 Lima Convention (see 7.1.10)

7.6.8 1981 Abidjan Convention (see 7.1.11)

7.6.9 1978 Kuwait Convention (see 7.1.12)

Each of these conventions contains provisions similar to those of 7.6.5.

7.6.10 1976 Barcelona Convention (see 7.1.13)

The Convention outlines measures to prevent and abate pollution from exploration and exploitation of the continental shelf and seabed.

UK Legislation

7.6.11 Petroleum Production (Seaward Areas) Regulations 1988 (SI 1213/1988)

Like earlier Petroleum Production Regulations, these Regulations impose an obligation on those holding a licence to explore the UK's territorial sea or continental shelf for petroleum, or to produce petroleum therefrom, to prevent the escape of any petroleum into the sea.

7.6.12 The Petroleum Act 1987

The Act updates arrangements regulating oil and gas installations and their operations on the UK Continental Shelf. Part I establishes a regime for the removal of oil and gas installations at the end of their production life. The Secretary of State may also make provision for the prevention of pollution, inspection and safety requirements in respect of installations and pipelines not wholly removed.

7.6.13 Oil and Pipelines Act 1985

This Act repealed most of Part I of the Petroleum and Submarine Pipelines Act 1975 (7.6.17) establishing the British National Oil Corporation (BNOC). Under the new Act the BNOC is replaced by the Oil and Pipelines Agency, the members of which are appointed by the Secretary of State. The Agency's principles are:

* to assume responsibility for the management of over 1,000 miles of pipeline and a number of storage installations;

* to be responsible for the disposal of royalty in kind oil, of which 260,000 barrels are lifted daily;

* to maintain the rights and obligations to lift petroleum should this ever be necessary; consent for this rests with the Government and will be given only if there is an existing or imminent threat to the security of the nation's oil supplies.

7.6.14 The Prevention of Oil Pollution Act 1971 (Application of Section 1) Regulations 1984 (SI 1684/1984)

In implementation of UK obligations under the Convention for the Prevention of Marine Pollution from Land-Based Sources 1974, those Regulations which entered into force on 22 November 1984, made it an offence to discharge oil or an oil mixture into the sea from a pipeline or as a result of operations for the exploration or exploitation of the natural resources of the continental shelf.

7.6.15 Merchant Shipping (Prevention of Oil Pollution) Regulations 1983

The Regulations apply the provisions of MARPOL 73/78 to installations, including platforms, used in connection with the processing of offshore seabed mineral resources. Installations must be fitted, as far as is practicable, with oily water separating equipment and with storage tanks for oily residues that cannot be dealt with otherwise. A record of all oily or oily mixture discharges must be kept in an approved form.

7.6.16 Offshore Installations (Emergency Procedures) Regulations 1976

These Regulations apply to manned, fixed and mobile offshore installations concerned with underwater exploration and exploitation of mineral resources in UK waters. Responsibility for dealing with any spills that may occur rests in the first instance with the operator concerned, who is required to keep an emergency procedures manual on each manned installation detailing action to be taken in specified emergencies, including blow-outs and spills. All spills must be reported to both the DoE and the Coastguard.

7.6.17 Petroleum and Submarine Pipelines Act 1975

The Act includes provision for the creation of the British National Oil Corporation, which is empowered to search for and extract petroleum anywhere in the world and to deal in petroleum and its derivatives. Authorisation is required from the Secretary of State for the construction and use of submarine pipelines in territorial waters and for the construction and extension of refineries. In choosing the route of a pipeline operators are required to take into account the interests of fishermen and any factors which may affect the integrity of the line, such as possible movement of the seabed (see also 7.6.13).

The Act also permits the Secretary of State to exempt from the provisions of POPA 1971 (7.1.25) any discharge of crude oil, or mixture containing crude oil, produced as a result of operation for the exploration or exploitation of the resources of the seabed. Such exemptions are issued subject to conditions imposed following consultations with the various government departments concerned, and taking into consideration the requirements of the Paris Convention (see 7.4.10).

7.6.18 The Prevention of Oil Pollution Act 1971

This Act consolidated a number of earlier statutory provisions and, in particular, replaced that part of the Continental Shelf Act 1964 which dealt with oil pollution from pipelines and offshore installations. The Act makes it an offence to discharge oil or oily water as a result of exploration or exploitation of seabed resources but in practice only a 'best practicable' standard is demanded. Discharges from pipelines are also prohibited. Though strict liability applies to deliberate discharges, it is a defence, where the discharge consists of an escape of oil, to prove that reasonable care has been exercised.

7.6.19 Continental Shelf Act 1964

Under the Act any rights exercisable by the UK outside territorial waters with respect to the seabed, subsoil and their natural resources (excluding coal) are vested in the crown and regulated by the appropriate sections of the Petroleum (Production) Act 1934 (7.6.20). The Act makes it an offence for any act or emission taking place on, under or above an offshore installation or the surrounding designated area in the same way as if the act occurred in the UK. The Act also makes it an offence to break or damage a pipeline.

7.6.20 Petroleum (Production) Act 1934

Under the Act any petroleum in natural condition under the strata of Great Britain or beneath UK territorial waters becomes the property of the Crown, which has the exclusive right of searching, boring for and getting such petroleum.

Netherlands Legalisation

7.6.21 Ministerial Regulations Concerning the Discharge of Substances from Mining Installations, 1987

This regulation, based on the Continental Shelf Mining Regulations 1967, contains standards for the discharge of substances from mining installations and sets standards for oil-based muds and cuttings as well as drainage waters.

7.6.22 Decree on Environmental Impact Assessment, 1987

In this decree, activities are appointed which are only allowed to be exercised after an environmental impact assessment has been made. Among these activities are the extraction of oil and gas on the Continental Shelf and the laying of pipelines for the transport of gas.

Italian Legislation

7.6.23 Rules for the Exploration and Exploitation of Deep-Seabed Mineral Resources, Decree No. 200, 1988

This Decree contains regulations implementing Law No. 41 of 1985 (see 7.6.25). It regulates the terms and conditions for granting, extending, suspending and revoking permits for the exploration of deep-seabed mineral resources. To ensure consistency of exploitation with adequate protection of the marine environment, an environmental impact statement is to be included in applications for permits. The holder of a licence is requested, *inter alia*, to monitor the consequences of his activities on the marine environment and to draft an annual report. The Decree provides also for controls and inspections as well as for penalties for infractions.

7.6.24 Rules on Exploration and Exploitation of Goethermal Resources, Law No. 896, 1986

This law relates to goethermal resources found in the Italian territory, the territorial sea and Continental Shelf. Any licence application should be accompanied by a study identifying the impacts of the planned activities on the environment.

7.6.25 Rules for the Exploration and Exploitation of Deep-Seabed Mineral Resources, Law No. 41, 1985

The purpose of these Rules is to set up basic principles in order to regulate the exclusive competence of the State to grant licences for the exploration and exploitation of deep-seabed mineral resources. Its provisions are to be considered as an interim legal regime until the 1982 UN-CLOS enters into force. Under this law, any licence holder who commits serious infringements of the provisions relating to the protection and preservation of the marine environment shall have their licence withdrawn.

7.6.26 Offshore Exploration, Research and Development of Hydrocarbons Decree, 20 May 1982

This Decree regulates the activities of licence holders for research and development of hydrocarbons in the territorial sea and the Continental Shelf and provides for the equipment of platforms and assisting vessels.

7.7 POLLUTION FROM THE ATMOSPHERE

International

7.7.1 United Nations Convention on the Law of the Sea (UNCLOS) 1982

Articles 212 and 222 of the UN Convention provide that states shall prescribe and enforce legalisation to prevent marine pollution from the atmosphere, applicable to the air space under their sovereignty and to their vessels and aircraft. Article 212(3) calls on states to establish global and regional rules to prevent atmospheric pollution.

7.7.2 Convention on Long-Range Transboundary Air Pollution 1979, Helsinki Protocol on Sulphur Dioxide Emissions 1985 and Protocol on Nitrogen Oxides Emissions 1988

The most significant attempt so far to develop international rules dealing with atmospheric pollution is the UN Economic Commission for Europe's Convention on Long-Range Transboundary Air Pollution, 1979. Although the obligation imposed by the Convention on parties gradually to reduce and prevent air pollution is not accompanied by any precise details as to how and when this is to be done, precise provisions for reducing emissions from two pollutants have been laid down by subsequent protocols to the Convention. The first of these, the Helsinki Protocol of 1985, provides that parties to it are to reduce emissions of sulphur dioxide by 30 per cent of 1980 levels by 1993. The second Protocol, the Protocol on Nitrogen Oxides Emissions of 1988, commits its parties to restrict emissions of nitrogen oxides to their 1987 levels by 1994, to apply national emission standards for major new sources, and by 1996 to have adopted control policies based on the critical loads which the environment can tolerate.

7.7.3 Convention for the Prevention of Marine Pollution by Dumping of Wastes and Other Matter 1972 (London Dumping Convention, LDC)

In 1978 the Contracting Parties resolved to amend the Convention to permit and regulate incineration at sea. In issuing special permits, states have to apply Regulations for the Control of Incineration of Wastes and Other Matter at Sea. The amendments entered into force on 11 March 1979.

7.7.4 Nuclear Test Ban Treaty 1963

One of the consequences of this Treaty was the ending of pollution of the sea by radioactive fallout from nuclear weapons testing in the atmosphere.

Regional

7.7.5 Convention for the Prevention of Marine Pollution from Land-Based Sources 1974 (Paris Convention) and Protocol of 1986

The Paris Convention, as a result of a Protocol of 1986, which entered into force in September 1989, is now concerned with atmospheric pollution. The Paris Commission (PARCOM) has established a monitoring programme to try to discover the significance of various atmospheric pollutants for the marine environment, although it has not yet adopted any regulatory measures.

7.7.6 Convention for the Prevention of Marine Pollution by Dumping from Ships and Aircraft 1972 (Oslo Convention)

The Oslo Convention (see 7.8.12) regulates the incineration of waste at sea. A Protocol containing mandatory rules on incineration came into force on 1 September 1989.

EEC Directives

7.7.7 89/429/EEC on the Reduction of Air Pollution from Existing Municipal Waste Incineration Plants

This Directive provides that existing municipal waste incineration plants must meet the same emission limit values as new plants by 1 December 1996 in the case of plants with nominal capacity equal to or more than 6 tonnes of waste per hour and by 1 December 2000 in the case of other plants.

7.7.8 89/369/EEC on the Prevention of Air Pollution from New Municipal Waste Incineration Plants

This Directive sets emission limit values for dust, heavy metals, hydrochloric acid, hydrofluoric acid and sulphur dioxide from new plants incinerating domestic, commercial, trade and similar refuse. National authorities are to lay down emission values for other pollutants when they consider this to be appropriate because of the composition of the waste to be incinerated and the characteristics of the incineration plant. The Directive has to be implemented by 1 December 1990.

7.7.9 88/609/EEC on the Limitation of Emissions of Certain Pollutants into the Air from Large Combustion Plants

This Directive requires a reduction in overall annual emissions from existing large combustion plants (50MW or more) of sulphur dioxide from 1980 levels in three stages (by 1993, 1998 and 2003 respectively), the percentage reductions specified varying from Member State to Member State. Similarly, emissions of nitrogen oxides must be reduced in two stages (by 1993 and 1998), again by varying percentages. New plants are subject to emission limit values in respect of sulphur dioxide, nitrogen oxides and dust.

UK Legalisation

7.7.10 Health and Safety (Emissions into the Atmosphere) Regulations 1989 (SI 319/1989) and Control of Industrial Air Pollution (Regulation of Works) Regulations 1989 (SI 318/1989)

These Regulations implement EEC Directive 88/609 on the Limitation of Emissions of Certain Pollutants into the Air from Large Combustion Plants (7.7.9).

7.7.11 The Air Quality Standards Regulations 1989 (SI 317/1989)

These Regulations implement various EEC Directives (80/779, 82/884 and 85/203) setting air quality standards for sulphur dioxide, smoke, lead and nitrogen dioxide. The Regulations place a duty on the Secretary of State to take any necessary measures to ensure that concentrations of the substances mentioned do not exceed the limits prescribed in the directives.

7.8 DUMPING

International

7.8.1 United Nations Convention on the Law of the Sea (UNCLOS) 1982

Dumping is defined in this convention as "any deliberate disposal of wastes or other matter from vessels, aircraft, platforms or other man-made structures at sea" and also "any deliberate disposal of vessels, aircraft, platforms or other man-made structures at sea" (Art.1(5)(a)(i)(ii)).

States must adopt laws and regulations to prevent, reduce and control pollution of the marine environment by dumping. These must ensure that dumping is not carried out without the permission of the competent authorities of states and no dumping is permitted in a state's territorial waters, EEZ or onto the Continental Shelf without its express prior authorisation. National laws and regulations must not be less effective than the global rules and standards (Art.210).

7.8.2 Convention for the Prevention of Marine Pollution by Dumping of Wastes and Other Matter 1972 (London Dumping Convention, LDC)

The Convention, which entered into force in September 1975, governs dumping on a global basis. Contracting states are required to apply a licensing system to any waste dumped at sea from vessels or aircraft: before authorisation is given authorities should give careful consideration to environmental effects. Normal discharges from vessels are excluded, but all other conventional wastes are included, though special exemption is available for abnormal circumstances, eg. due to stress of weather or when safety of human life or of ship or aircraft is threatened.

The convention prohibits the dumping, except in trace quantities, of certain substances deemed to be toxic, persistent and bioaccumulative, eg. crude fuel and lubricating oils, radioactive waste and industrial wastes. Licences to dump waste at sea must contain information of the quality and quantity of the wastes to be dumped, and may also specify the site and method of dumping. Fishery departments carry out spot checks on dumping operations to ensure that licensing conditions are being met; they also survey the major dumping grounds at intervals to determine the effects of waste disposal.

Regional

7.8.3 Protocol to the Noumea Convention concerning the Prevention of Pollution of the South Pacific Region by Dumping 1986

This Protocol to the 1986 Noumea Convention (see 7.1.5) reflects the provisions of both the London Dumping Convention and UNCLOS.

7.8.4 1985 Nairobi Convention (see 7.1.6)

7.8.5 1983 Cartagena Convention (see 7.1.8)

7.8.6 1982 Jeddah Convention (see 7.1.9)

7.8.7 1981 Lima Convention (see 7.1.10)

7.8.8 1978 Kuwait Convention (see 7.1.12)

7.8.9 1981 Abidjan Convention (see 7.1.11)

The above Conventions (7.8.4-7.8.9) all contain similar provisions to those of the Barcelona Convention (7.8.10).

7.8.10 1976 Protocol to Barcelona Convention (see 7.1.13) concerning Dumping

Dumping of 'black list' substances is prohibited completely, while 'grey list' substances may be disposed of at sea only under special licence from the relevant national authority. All other dumping must have a general permit.

7.8.11 1974 Helsinki Convention (see 7.1.14)

The Convention prohibits dumping in the Baltic Sea area except when human life is at risk, or a ship or aircraft at sea is threatened by complete destruction or total loss. Dumping of dredged spoils that do not contain significant quantities of any listed substances is permissible by authorisation of the appropriate national authority.

7.8.12 Convention for the Prevention of Marine Pollution by Dumping from Ships and Aircraft 1972 (Oslo Convention)

The Convention, which entered into force in April 1974, is similar to the 1972 LDC (7.8.2) in its regulation of dumping, but is restricted to the North Sea and the North-East Atlantic.

The Convention, which is administered by the Oslo Commission (OSCOM) with representatives of the 13 signatory states of north-western Europe, operates through a licensing system. Substances with high potential for damaging the environment on account of their toxicity, persistence and bioaccumulation are placed on a 'black list' and may not be dumped (except when 'trace' quantities only are present); substances with only two of these characteristics may be dumped only when special precautions are taken. A **Decision for the Reduction and Cessation of the Dumping of Industrial Waste at Sea** entered into force for the North Sea on 31 December 1988 and will enter into force for other parts of Convention waters from 31 December 1995.

EEC Directives

7.8.13 89/428/EEC on Procedures for Harmonising the Programmes for the Reduction and Eventual Elimination of Pollution caused by Waste from the Titanium Dioxide Industry

This Directive provides that after the end of 1989 waste from the titanium dioxide industry produced by the sulphate or chlorine processes shall not be dumped in inland or marine waters, and solid, strong acid and treatment wastes from the sulphate process and solid waste and strong acid waste from the chloride process shall not be discharged into inland or marine waters. In addition, the Directive sets emission limits for other kinds of waste which must be reached by the end of 1989 or 1992, depending on the type of waste: alternatively, instead of complying with emission standards, a Member State may use quality objectives provided that these achieve the same effect. The Directive also sets atmospheric emission limits for the titanium dioxide industry.

7.8.14 78/176/EEC on Waste from the Titanium Dioxide Industry (see 7.4.23)

Under the Directive all dumping is made subject to prior authorisation by the competent authority and such authorisation is for a limited period only.

7.8.15 76/403/EEC on PCBs and PCTs (see 7.4.25)

The Directive aims at prohibiting the discharge, discarding and unsupervised dumping of PCBs and PCTs.

UK Legislation

7.8.16 Environmental Protection Bill 1989 (see 7.4.30)

Clause 111 amends certain of the provisions of the Food and Environment Protection Act 1985 dealing with the dumping of waste at sea. The effect of the amendments is to extend the scope of the 1985 Act to include the dumping or incineration of waste within UK continental shelf limits by foreign vessels which have loaded the waste to be dumped or incinerated in foreign ports. Enforcement officers' powers are also to be increased to enable them to bring into port vessels suspected of illegal dumping, and the maximum fine on summary conviction is increased from £2,000 to £50,000.

7.8.17 Food and Environment Protection Act 1985

This Act repealed and replaced the Dumping at Sea Act 1974 (see below). It prohibits all unlicensed deposits in the sea, except where these are specifically exempted by the Minister concerned through statutory instruments. Provisions referring to dumping materials are extended to include "those under the seabed". The provisions are also extended to include foreign vessels within British fishery limits (200 miles).

7.8.18 Dumping at Sea Act 1974 (DASA)

This Act, under which provisions of the LDC (7.8.2) and Oslo Convention (7.8.12) were implemented in the UK, has now been repealed and replaced by the Food and Environment Protection Act 1985 (see above).

7.8.19 Radioactive Substances Act 1960

A person wishing to dump radioactive waste at sea must have authorisation from the Secretary of State for the Environment f(or Scotland or Wales) in addition to licences under later legislation. For wastes from nuclear establishments in England this authority is exercised jointly with the Minister of Agriculture, Food and Fisheries.

Netherlands Legislation

7.8.20 Royal Decree Concerning the Dumping of Substances in the Sea, 12 December 1983

The Decree modifies a 1975 Decree which prohibited dumping of a number of designated substances. This new Decree designates a number of substances which are ex-

cluded from this prohibition.

7.8.21 Marine Pollution Act 1975/81

This Act implements the LDC (7.8.2) and the Oslo Convention (7.8.12).

Italian Legislation

7.8.22 Decree of the Minister for Merchant Shipping, 6 July 1983

This Decree contains additions to the list of harmful substances whose dumping is prohibited according to the Provisions of the Protection of the Sea 1982 (7.8.24).

7.8.23 Law No. 305, 2 May 1983

This law implements the LDC (7.8.2), which came into force for Italy on 30 May 1984.

7.8.24 Provisions for the Protection of the Sea 1982

Among its numerous provisions, the discharge at sea from merchant vessels of certain listed substances is prohibited.

7.9 SHIP SAFETY — VESSEL CONSTRUCTION, CREW TRAINING AND SALVAGE

International

7.9.1 Salvage Convention 1989

The Convention recognises the contribution made by salvage services to the protection of the marine environment and requires the owner and master of an endangered vessel to take timely and reasonable action to arrange for salvage and use their best endeavours to prevent or minimise environmental damage, thus limiting their freedom to contract for salvage. The salvor is under a duty to prevent or minimise damage to the environment and the coastal state may give directions to the salvor.

"Special compensation" is payable when the salvor fails to save the maritime property but succeeds in minimising environmental damage.

7.9.2 United Nations Convention on the Law of the Sea (UNCLOS) 1982

* coastal states may adopt laws and regulations, in conformity with rules of international law in respect of the safety of navigation and the regulation of maritime traffic and require states to exercise their right of innocent passage through its designated sea lanes and traffic separation schemes (Arts. 21(1)(a). 22(1));

* vessels during transit passage must comply with generally accepted international regulations, procedures and practices for safety at sea, including the International

Regulations for Preventing Collisions at Sea, and must navigate through designated sea lanes and traffic separation schemes (Arts. 41, 39(2)(a));

- installations or structures which are abandoned or disused shall be removed to ensure safety of navigation, taking into account any generally accepted international standards established by the competent international organisation. Appropriate publicity must be given to the depth, position and dimensions of any installations or structures not entirely removed (Art. 60(3));

- coastal states may establish safety zones around artificial islands and installations to ensure the safety both of navigation and of the installations (Art. 60(4));

- flag states must take measures for their ships to ensure safety at sea regarding the construction, equipment and seaworthiness of ships; the manning conditions and the training of crews; and the use of signals, the maintenance of communications and the prevention of collisions (Art. 94(3)(a)(b)(c));

- coastal states must promote the establishment, operation and maintenance of an adequate and effective search and rescue service regarding safety on and over the sea (Art. 98(2));

- during the exercise of their powers of enforcement, states must not endanger the safety to navigation or create any hazard to a vessel (Art. 225);

- laws and regulations of the coastal state in relation to the exercise of innocent passage must not apply to the design, construction, manning or equipment of foreign ships unless they are giving effect to generally accepted international rules and standards (Art. 21(2));

- a coastal state may adopt laws and regulations on the design, construction, manning and equipment of foreign vessels in special areas under its jurisdiction adopted in conformity with this Convention (Art. 211(6)(c));

- flag states must prevent their vessels from sailing if they do not comply with the international rules and standards relating to their design, construction, manning and equipment (Art. 217(3)).

7.9.3 International Convention on Maritime Search and Rescue 1979 (SAR)

The main purpose of the Convention, which entered into force in June 1985, is to facilitate cooperation between search and rescue organisations and between those participating in search and rescue operations at sea by establishing the legal and technical basis for an international SAR plan. Contracting parties are required to ensure that arrangements are made for the provision of adequate maritime SAR services off their coasts and are encouraged to enter into SAR agreements with neighbouring states involving the establishment of maritime SAR regions, the pooling of facilities, the establishment of common procedures, cooperation, training and liaison. Further provisions concern the establishment of preparatory measures and ship reporting systems.

7.9.4 International Convention on Standards of Training, Certification and Watchkeeping for Seafarers 1978 (STCW)

The Convention, which entered into force in April 1984, is the first attempt to establish global minimum professional standards for seafarers. It prescribes minimum standards which contracting parties are obliged to meet or exceed. It contains provisions for deck, engine and radio departments, spells out principles on keeping navigational and engineering watches, and lays down minimum requirements for the certification of masters, chief engineers, mates, radio officers, radio operators and other members of the crew. There are also provisions on maintaining and updating the necessary knowledge. Special rules concern the training of masters, officers and ratings of chemical and liquefied natural gas carriers.

7.9.5 ILO Merchant Shipping (Minimum Standards) Convention (No. 147) 1976

The Convention, which entered into force in November 1981, follows a 1936 Convention concerned with minimum professional requirements for masters and officers of merchant ships. The 1976 Convention is concerned with minimum standards for merchant ships, including standards of competency and manning, and contains provisions for port state enforcement.

7.9.6 Convention for the Safety of Life at Sea 1974 (SOLAS) (as amended in 1981) and the 1978 Protocol

The Convention, which entered into force in May 1980, establishes minimum standards for the construction, equipment and operation of ships, eg. provisions regulating fire-protection, lifesaving apparatus, safety of navigation etc. States are responsible for ensuring that ships under their flags comply with the requirements of the Convention and various certificates are prescribed as proof that this has been done. Contracting governments can inspect ships of other contracting states if there is reason to believe that a ship does not comply with the requirements of the Convention. The 1978 Protocol, which entered into force in May 1981, strengthens the provisions for ship inspection by including unscheduled inspections and mandatory annual surveys. The 1981 Amendments, which entered into force in September 1984, involve major changes in technical aspects covered by the original Convention which are designed to strengthen its effectiveness. A second set of Amendments, adopted in 1983, came into force on 1 July 1986. The Amendments affect five chapters of the Convention:

- Chapter II — 1: Subdivision and Stability, Machinery and Electrical Installations. These changes are largely editorial in content, made necessary by the changes made to Chapter III;

- Chapter II — 2: Construction — Fire Protection, Fire Detection and Fire Extinction. The new Amendments are a further improvement on the regulations that had been changed in the 1981 Amendments. Regulation 56 has been completely re-written, particularly in respect of combination carriers;

- Chapter III — Life-Saving Appliances and Arrangements: This chapter has been completely revised: there is great attention to detail and the regulations are much more precise. It also requires cargo ships to carry at least one rescue boat;

- Chapter IV - Radiotelegraphy and Radiotelephony. Amendments are made to Regulations 14 bis and 14 bis I, which deal with survival craft EPIRBs and two-way radiotelephone apparatus for survival craft;

- Chapter VII — Carriage of Dangerous Goods. The revised chapter now refers also to bulk dangerous goods by means of two codes, developed by IMO.

7.9.7 International Regulations for Preventing Collisions at Sea 1972 (as amended in 1981)

These Regulations entered into force in 1977. They provide for the mandatory observance of traffic separation schemes adopted by IMO, observance of which had previously not been binding.

Rule 10 of the Regulations specifies rules to be observed by ships proceeding through waters where such schemes are in operation. A 1981 amendment to these rules came into force in June 1983.

7.9.8 International Convention Load Lines 1966 as amended in 1971, 1975, 1979 and 1983

Under the Convention, which entered into force in July 1968, it is a requirement that a ship's master shall be supplied with sufficient information in approved form to enable him to arrange loading or ballasting of his ship without creating unacceptable structural stresses. Administrations are given discretion in applying this requirement. The Convention is now one of those taken into consideration in applying the Paris Memorandum of Understanding on Port State Control (see 7.9.9).

Regional

7.9.9 Paris Memorandum of Understanding (MOU) on Port State Control (PSC) 1982

The Memorandum aims to establish a harmonised and co-ordinated system for the inspection of foreign ships visiting ports of the maritime members of the European Community (including Spain and Portugal) together with Finland, Norway and Sweden. Inspection by surveyors is based on a quota system and ships failing to reach required standards may be detained until faults are put right.

7.9.10 1974 Helsinki Convention (see 7.1.14)

Recommendations adopted in 1980 and 1981 concern ac-

ceptance by the Baltic Sea States of International Instruments on Maritime Safety, while a further 1980 Recommendation concerns a position reporting system for ships in the Baltic Sea area.

7.9.11 The Hague Memorandum of Understanding between Certain Maritime Authorities on the Maintenance of Standards on Merchant Ships 1978

The Memorandum, adopted in 1978 by the eight north-western European countries bordering the North Sea, makes provision for more stringent surveillance of sea-going ships with a view to maintaining adequate standards. The Memorandum is not binding and the parties to the arrangement are not states but rather their authorities. In practice the Memorandum has now been largely superseded by the 1982 Paris Memorandum of Understanding on Part State Control (7.9.9).

UK Legislation

7.9.12 The Merchant Shipping (closing of openings in enclosed superstructures and bulkheads above the bulkhead deck) Regulations 1988 (SI 317/1988 and 642/1988)

SI 317/1988, made in the wake of the *Herald of Free Enterprise* disaster, applies new regulations regarding the closing of openings (doors etc.) to UK ships.

SI 642/1988 has the same content as SI 317/1988, but applies to non-UK ships in UK waters.

7.9.13 The Merchant Shipping (Passenger Boarding Cards) (Amendments) Regulations 1988 (SI 191/1987 and SI 641/1988)

These Regulations are made with the intentions of increasing safety on ro-ro passenger ferries. These Regulations, which came into force on 1 March 1988, are intended to tighten up the issue of boarding cards on UK ships. Breaches are punishable as a criminal offence.

SI 641/1987 is as SI 191/1987, but applies to non-UK vessels in UK waters.

7.9.14 Pilotage Act 1987

The Act transfers, as from 1 October 1988, responsibility for pilotage from the existing pilotage authorities to certain harbour authorities referred to in the Act as competent harbour authorities (CHAs) which already have responsibilities for the regulation of shipping movements and the safety of navigation. CHAs will have a duty to provide any pilotage services necessary to secure the safety of ships using their ports and approaches.

7.9.15 The Merchant Shipping (Passenger Ship Construction) (Amendments) Regulations 1987 (SI 1886/1987 and SI 2238/1987)

These Regulations came into force in February 1988. They concern the installation of vehicle-deck light indicators to show whether the bow doors are closed on the bridge of any UK passenger ship.

SI 2238/1987 These Regulations are as SI 1886/1987, but refer to non-UK ships in UK waters.

7.9.16 The Merchant Shipping (Fees) Regulations 1987 (SI 63/1987)

These Regulations, which came into force in February 1987, revoke and replace the Merchant Shipping (Fees) Regulations 1985 as amended. The Regulations prescribe an increase in fees of around 10 per cent for surveys and inspections and tonnage measurements except for fees for radio and navigational equipment surveys to which an increase of 18 per cent is applied.

7.9.17 The Offshore Installations (Lifesaving Appliances and Fire-Fighting Equipment) (Amendment) Regulations 1987 (SI 129/1987)

These Regulations, which came into force in March 1987, substitute the 1977 and the 1978 Regulations on Lifesaving Appliances in respect of examinations carried out by the Ministry.

7.9.18 The Merchant Shipping (Indemnification of Shipowners) Order 1987 (SI 220/1987)

This Order, which came into force in May 1987, revokes and replaces the Merchant Shipping (Indemnification of Shipowners) Order 1985, as amended. This amendment includes the amendments made in 1983 to the Safety at Sea Convention 1974, among the Conventions with which a shipowner must comply if he is to get indemnification in the circumstances set out in Article 5 of the Convention on the Establishment of an International Fund for the Compensation of Oil Pollution Damage 1971 (7.10.3).

7.9.19 The Merchant Shipping (Submersible Craft Operations) Regulations 1987 (SI 311/1987)

These Regulations, which came into force in October 1987, relate to the operation of manned submersible crafts in the UK waters, or elsewhere when the craft is operated from, or comprises, a UK ship. These Regulations impose safety requirements and provide for the reporting of casualties and other accidents which may occur in the course of such operations. See also the Merchant Shipping (Submersible Craft) (Amendment) Regulations 1987 (SI 306/1987).

7.9.20 Safety at Sea Act 1986

This is an Act intending to promote the safety of fishing and other vessels and prescribes the kind of lifesaving equipment that must be on board. The Act also extends the jurisdiction of the Secretary of State to include the safety and health on foreign ships and vessels while they are within a port in the United Kingdom. The Merchant Shipping Act 1971 is amended accordingly and these measures include any provisions of international agreements ratified by the United Kingdom.

7.9.21 The Merchant Shipping (Indemnification of Shipping Owners) (Amendment) Order 1986 (SI 296/1986)

See 7.10.9

7.9.22 The Merchant Shipping (Fire Protection) (Non-United Kingdom) (Non-SOLAS Ships) Rules 1986 (SI 1248/1986)

These Rules, which came into operation in September 1986, apply to certain non-United Kingdom ships when in a UK port. The Rules apply to ships which are not required to comply with the International Conventions for the Safety of Life at Sea (SOLAS) of 1960 and 1974 or the 1978 Protocol to the 1974 Convention because they are expressly excluded from these conventions by reason of size of type of ship. These Rules require that such ships follow the requirements of:

* the Merchant Shipping (Fire Protection) (Ships Built before 25 May 1980) Regulations 1985, as amended;

* the Merchant Shipping (Fire Appliances) Regulations 1980, as amended; or

* the Merchant Shipping (Fire Protection) Regulations 1984, as amended as the requirements apply to United Kingdom ships of similar size and tonnage.

7.9.23 Dangerous Vessels Act 1985

A harbourmaster is empowered to require the removal from his harbour of any vessel he considers poses a grave and imminent danger to the safety of any person or property, or which by sinking or foundering in the harbour may prevent its use by other vessels. Only the Secretary of State has the power to issue directions overruling such a decision.

7.9.24 Regulations (1985) Implementing 1981 Amendments to SOLAS

For the 1985 Regulations designed to implement the 1981 Amendments to SOLAS 74/78, see *ACOPS 1985-6 Yearbook* (pp. 36-37).

7.9.25 Regulations for improving crew standards and implementing STCW 1978 (7.9.4) include:

* **Merchant Shipping (Certification of Marine Officers) Regulations 1980;**

* **Merchant Shipping (Certification of Deck Officers)**

Regulations 1980;

- **Merchant Shipping (Certification and Watchkeeping) Regulations 1982;**

- **Merchant Shipping (Tankers — Officers and Ratings) Regulations 1984.**

7.9.26 Merchant Shipping (Tankers) (EEC Requirements) Regulations 1981

These Regulations, which were laid before Parliament in August 1981, give effect to certain provisions of the EEC Council Directive 79/11/EEC of December 1979 concerning minimum requirements for certain tankers entering or leaving Community ports. The master of a tanker to which the Regulations apply is required, prior to entering any British port, to report *inter alia* any defects which may arise and which may affect the safe manoeuvrability of the tanker and the safety of other vessels in the vicinity, or which may constitute a hazard to the marine environment or to persons or property on land in the vicinity.

7.9.27 Merchant Shipping Act 1979

The Act deals with a series of issues including carriage of passengers and luggage at sea, and safety and health on ships.

It enables the UK to ratify the 1978 Protocol to the 1974 SOLAS Convention, and establishes a Pilotage Commission to keep under consideration the organisation of pilotage services at ports and in waters off the coasts of the UK, to monitor the adequacy of pilotage services and to investigate whether pilotage should be made compulsory in more areas. Subsequent regulations made under the Act include the following, which give effect to relevant provisions of SOLAS 74/78 (7.9.6):

- **Passenger Ship Construction Regulations 1980,** which include provisions for survey and the issue of a Passenger Ship Safety Certificate;

- **Cargo Ship Safety Equipment Survey Regulations 1981,** which make similar provisions for cargo vessels;

- **Cargo Ship Construction and Survey Regulations 1981,** which apply to hulls, machinery and electrical equipment: and the

- **Merchant Shipping (Distress Signals and Prevention of Collisions) Regulations 1983** which give effect to the International Regulations for Preventing Collisions 1972 (7.9.7).

7.9.28 Merchant Shipping (Load Lines) Act 1967

The Act implements the International Convention on Load Lines 1966 (7.9.8) and makes an offence any infringement of the following:

- **Merchant Shipping (Load Lines) Rules 1968** Which lay down provisions for surveying and marking of ships;

- **Merchant Shipping (Load Lines) (Deck Cargo) Regulations 1968,** which lay down the requirements for

securing packaged goods carried as deck cargo.

Italian Legislation

7.9.29 Merchant Shipping Safety Decrees 1987

The Minister for Merchant Shipping issued several Decrees in 1987 to implement the 1978 STCW (7.9.4). They provide for the organisation of training courses for the safety of oil and chemical tankers, firefighting courses and survival and rescue training courses.

7.10 LIABILITY AND COMPENSATION

International

7.10.1 United Nations Convention on the Law of the Sea (UNCLOS) 1982

- flag states bear international responsibility for any loss or damage to a coastal state resulting from the non-compliance by a warship or government ships operated for non-commercial purposes in respect of the laws and regulations concerning innocent passage (Art. 31);

- in the case of the seizure of a ship on suspicion of piracy without adequate grounds the seizing state is liable to the state of nationality of the ship for any loss or damage caused (Art. 106);

- states are liable for damage or loss attributed to them arising from enforcement measures when such measures are unlawful or exceed those reasonably required in the light of available information (Art. 232);

- states are liable for their activities in the Area; state parties or international organisations acting together must bear joint and several liability (Art. 139);

- the provision of this Convention regarding responsibility and liability for damage are without prejudice to the application of existing rules regarding responsibility and liability under international law (Art. 304);

- states are responsible for the fulfilment of their international obligations in relation to pollution and are liable according to international law. Therefore, they must ensure that there is available resource in respect of damage caused by pollution and adequate compensation (Art. 235).

7.10.2 Convention of Limitation of Liability for Maritime Claims (LLMC) 1976

This Convention, adopted in 1976, entered into force on 1 December 1986. The Convention replaces a previous convention that related to the limitation of liability of owners of seagoing ships, adopted in 1957.

The most important effect of the LLMC Convention is to raise the amount of compensation available for loss of life or personal injury and for property damages. With regard to personal claims, liability for ships not exceeding 500 tons is limited to US $400,000. Additional amounts may be added to this figure for larger ships. For other claims, the limit of liability is US $200,000 for ships under 500 tons, with additional amounts for larger ships.

At December 1986, the Convention had been ratified by 13 countries.

7.10.3 International Convention on the Establishment of an International Fund for Oil Pollution Damage 1971 (Fund Convention, FC) and the 1984 Protocol

The Convention, which entered into force in 1978 with the establishment of the IOPC Fund, provides compensation for pollution damage additional to the sums available under CLC (7.10.6) so that, where CLC liability ends, Fund provision begins. Compensation is limited to 60 million SDR (US $78 million) per incident, including the sum paid by the shipowner under the CLC Convention. (7.10.6).

Liability may also arise under the Fund in certain cases where CLC does not apply, eg. pollution damage resulting from natural phenomena of an exceptional kind.

The scope of the Fund is governed by CLC provisions in that the Convention relates only to persistent oil carried in bulk, to damage occurring in the territory, including the territorial seas, of a contracting state, and to preventive measures. The Convention indemnifies shipowners for part of the clean-up costs incurred under CLC. Indemnification may be denied if it can be proved that the polluting ship, with the owner's knowledge, did not comply with relevant international conventions on safety and oil pollution.

Changes introduced by the 1984 Protocol include:

- application of the revised definitions and of the geographical area of the CLC (7.9.6) to the FC;

- removal of the indemnification to shipowners for increased liability imposed in excess of the 1957 Limitation Convention so that it is kept in rough proportion to that of the cargo-owner;

- the Fund will still compensate damage where exoneration applies under CLC, where the damage exceeds the CLC liability limit, or where the shipowner is insolvent;

- the Fund acquires the rights of those it compensates and is itself a legal entity able to enter into litigation;

- the Fund's organisation is streamlined through dissolution of the Executive Committee;

- introduction of a 'two-tier' coverage scheme for raising the present limit of US $47.1 million.

7.10.4 Tanker Owners Voluntary Agreement concerning Liability for Oil Pollution (TOVALOP)

This is an agreement amongst tanker owners, including bareboat charterers. The basis of the Agreement is that when a participating tanker spills, or threatens to spill, persistent oil being carried as cargo or fuel, the owner or bareboat charterer will either undertake to combat the pol-

lution or will voluntarily accept the responsibility for reimbursing any public authority or individual who incurs reasonable clean-up costs or who suffers pollution damage.

Since coming into effect in October 1969, TOVALOP has been amended on a number of occasions. The most recent amendments, which came into force on 20 February 1987, resulted, *inter alia*, in an increase in the maximum possible compensation available under TOVALOP to US $70 million through the addition of a Supplement. The terms of the TOVALOP Supplement apply worldwide but only to cases where the participating tanker is carrying a cargo owned by a member of CRISTAL. If this condition is not met, only the terms and limits of the TOVALOP Standing Agreement (the version existing prior to 20 February 1987) are available to claimants.

7.10.5 Contract Regarding an Interim Supplement to Tanker Liability for Oil Pollution, 1971 (CRISTAL)

Like the TOVALOP scheme, CRISTAL was originally devised as a voluntary, interim scheme pending the entry into force of the IOPC Fund. The oil companies party to CRISTAL comprise in excess of 90 per cent of total world and fuel oil cargoes.

When the IOPC Fund entered into force in 1978, CRISTAL was substantially revised in order to complement the other scheme more fully. Its scope is to supplement other compensation schemes to ensure adequate compensation for individuals and governments suffering oil pollution damage. It also supplies indemnification for tanker owners for partial liability under other compensation schemes. The agreement is applicable in the territory or territorial sea of any state regardless of where the spill originally occurred. The polluting vessel must be owned or bareboat (demise) chartered to a participant in the TOVALOP scheme. Oil must be owned by a party to CRISTAL.

The agreement covers all pollution damage, including threat removal, not recoverable from the tanker owner or other source. Liability may also arise under CRISTAL in certain cases where the Fund does not apply, eg. intentional act or omission by third party and negligence of governments. The amount of liability was raised in 1986 to a maximum amount of US $135 million. In countries where the IOPC scheme operates, the Fund will take precedence and the voluntary schemes would apply only if there were uncompensated schemes above its limits.

7.10.6 International Convention of Civil Liability for Oil Pollution Damage 1969 (CLC Convention) and the 1984 Protocol

The Convention, which entered into force in June 1975, applies only to persistant oils, as defined in 7.1.4. Under it a shipowner is strictly liable, subject to certain limited exceptions, for oil pollution damage caused by oil which has escaped from any tanker carrying oil in bulk. In return for strict liability, and the the absence of fault or privity on his part, the owner may limit his liability to a sum set by reference to the size of the vessel concerned. Ships carrying more than 2,000 t of oil in bulk must be insured against their maximum liability and a claimant is able to sue the insurer direct rather than the owner. Jurisdiction is restricted to the courts of the contracting party suffering the damage, and judgments are enforceable in other contracting states.

The Convention applies only to pollution damage caused in the territory, including the territorial sea, of a contracting state, and to reasonable measures taken to prevent and minimise such damage after pollution has occurred. The Convention does not apply to unidentifiable sources of pollution or to measures taken to remove the threat of pollution.

Changes introduced by the 1984 Protocol include:

- an amended definition of 'ship' to include combination carriers and unladen vessels not clean of oil residues;

- the definition of a 'pollution incident' is broadened to include the 'threat of pollution';

- the concept of 'pollution damage' includes compensation for impairment of the environment as long as this is limited to the costs of reasonable measures of reinstatement actually undertaken or to be undertaken;

- the area covered by the Convention is extended to the EEZ of a state or a comparable area;

- the limits of coverage are increased.

Regional

7.10.7 The Offshore Pollution Liability Agreement (OPOL)

A voluntary agreement exists between oil companies that operate or intend to operate offshore installations and pipelines to guarantee funds to meet certain claims for pollution damage arising out of the oil spillage or escape of oil and to pay for the costs of remedial measures. It also provides a mechanism for settling such claims on the basis of strict and limited liability. It covers direct loss or damage from contamination arising out of the escape or discharge of crude oil or natural gas liquids from one or more offshore facilities which is/are located within the jurisdiction of a country designated under the Agreement — the UK, Denmark, West Germany, France, Eire, the Netherlands and Norway. The liability limit is an overall maximum of US $60 million.

7.10.8 The Convention on Civil Liability for Oil Pollution Damage from Offshore Installations

This Convention, signed in London, in 1977 by the UK, France, West Germany, Denmark, Sweden, Norway, the Netherlands and Belgium, deals with compensation for pollution damage, and remedial measures taken in respect of pollution, from offshore installations in terms similar to those of OPOL. The Convention has not as yet been ratified by any contracting party and thus is not yet in force.

UK Legislation

7.10.9 Merchant Shipping Act 1988

Section 34 and schedule 4 implement the 1984 Protocols to the CLC Convention (7.10.6) and the Fund Convention (7.10.3).

7.10.10 The Merchant Shipping (Indemnification of Shipowners) (Amendment) Order 1986 (SI 296/1986)

This Order, which came into operation in August 1986, amends the Merchant Shipping (Indemnification of Shipowners) Order 1985 (see 7.10.11) to take account of the Decision of the Assembly of the International Fund for Compensation for Oil Pollution Damage made at its Eighth Session in October 1985. The decision was to replace references to Safety of Life at Sea Convention 1974 (7.9.6), as amended, and the International Convention for the Prevention of Pollution at Sea 1973 (7.1.2), as amended. A court may now exonerate the Fund wholly or partly from indemnifying shipowners in respect of their liability under the Merchant Shipping (Oil Pollution) Act 1971 (7.10.14) if it is proven that, as a result of the actual fault or privity of the shipowner, the ship did not comply with the requirements of the 1985 Order.

7.10.11 The Merchant Shipping (Indemnification of Shipowners) Order 1985 (SI 1665/1985)

This Order, made under the Merchant Shipping Act 1974, came into operation December 1985. The IOPC Fund indemnifies shipowners in respect of their liability under the Merchant Shipping (Oil Pollution) Act 1971 (7.10.14).

Under the Merchant Shipping Act 1974 the Fund may be wholly or partly exonerated if the shipowner does not comply with requirements prescribed by the Secretary of State. The present Order prescribes these requirements as now comprising specified sections of MARPOL 73/78; SOLAS 74, the 1978 Protocol and the 1981 amendments; the 1966 Load Lines Convention; and the 1972 International Regulations for Preventing Collisions at Sea and the 1981 amendments.

7.10.12 Merchant Shipping (Indemnification of Shipowners) Order 1978

The Order specifies the standards to be met by a shipowner as a condition of his qualifying for an indemnity from the Fund for part of his liability under the FC. If the ship did not comply with the statutory requirements set out in the Order, the Fund may be wholly or partly relieved of its obligation to indemnify him.

7.10.13 Merchant Shipping Act 1974

Part I of the Act incorporates the provisions of the 1971 Fund Convention (7.10.3) by rendering the Fund liable for pollution damage in the UK if the person suffering the damage has been unable to obtain full compensation under the Merchant Shipping (Oil Pollution) Act 1971 (7.10.12). No compensation is recoverable if the source of pollution cannot be identified. Statutory Instruments came into operation in October 1978 to implement the provisions of the FC.

7.10.14 Merchant Shipping (Oil Pollution) Act 1971

The Act gives effect to the 1969 CLC Convention (7.10.6) by making the owner of a vessel carrying oil in bulk as cargo liable without fault for oil pollution damage, including clean-up expenses, up to £55 per limitation ton or £5.8 million per incident. The Act applies to the area of the UK and to other Convention countries. A provision requires vessels carrying more than 2,000 tonnes of oil in bulk as cargo to carry on board certificates of financial responsibility in respect of the liability created by the 1969 Convention.

Sections of the Act dealing with liability came into force in September 1971 and the remainder of the Act on the coming into force of the 1969 Convention in 1975. The Act, which provides for compensation for oil pollution damage in civil proceedings, does not preclude the possibility of compensation through the application of fines imposed following prosecution under the Prevention of Oil Pollution Act 1971 (7.1.25) which creates criminal liability for oil pollution offences.

7.11 REGISTRATION

International

7.11.1 United Nations Convention on Conditions for Registration of Ships 1986

In February 1986, after 10 years of negotiations, the United Nations Conference on Conditions for Registration of Ships adopted a Convention that for the first time defines in an international instrument the elements of the 'genuine link' that should exist between a ship and the state whose flag it flies.

The central provisions of the 1986 Convention (Arts. 8, 9 and 10) provide for participation by nations of the flag state in the ownership, manning and management of ships on its register, thus establishing what are widely regarded as the key and necessary economic criteria that give meaning to the concept. A distinctive feature of the Convention is that it gives states a choice between complying with the provisions on ownership or with those on manning (Art. 7).

The convention will enter into force 12 months after it has been ratified by 40 states, representing 25 per cent of gross registered tonnage of ships used in international seaborne trade of 500 gross registered tonnage and above.

Considering that an important early objective of the exercise for the majority of countries was the phasing out of open registries (flags of convenience), it was emphasised that the Convention would, in time, bring about the aboli-

tion of certain abuses arising from inadequate links between a ship and its flag state. Nevertheless, the Convention has been criticised by INTERTANKO for its imprecise language and by the International Transport Workers Federation as perpetuating the institution of open registries.

UK Legislation

7.11.2 Merchant Shipping Act 1988

Part I of the Act allows orders to be made to tighten up the conditions under which ships can be registered in UK dependent territories, to ensure that their registers cannot be used for substandard ships. Such orders have so far been made for Bermuda and the Isle of Man (SI 2251/1988).

7.11.3 The Merchant Shipping (Fees) (Amendment) Regulations 1987 (SI 548/1987)

These Regulations, which came into force in April 1987, increase the fees payable in respect of the registration of ships by 3.6 per cent.

7.11.4 The Merchant Shipping (Light Dues) (Amendment) Regulations 1986 (SI 334/1986)

These Regulations, which came into operation in April 1986, provide for an increase of about 5 per cent in the scale of light dues payable under the Merchant Shipping (Mercantile Marine Fund) Act 1898 as set out in the Merchant Shipping (Light Dues) Regulations 1981 (as amended) and revokes previous amended regulations. The last change took effect in June 1983.

7.12 SPECIALLY PROTECTED AREAS

International

7.12.1 The United Nations Convention on the Law of the Sea (UNCLOS) 1982

Part XII of this Convention on the Protection and Preservation of the Marine Environment provides, in Article 211 regarding pollution from vessels, that a state may define as special areas parts of their EEZs and take special mandatory measures for the prevention of pollution. This must have the approval of the competent international organisation and is applicable only if the existing international rules are inadequate to meet special circumstances. Additional regulations may relate to discharges or navigational practices.

Coastal states also have the right to adopt and enforce non-discriminatory laws and regulations for the prevention, reduction and control of pollution from vessels in ice-covered areas within the limits of the EEZ. This applies only when severe climatic conditions create obstructions or exceptional hazards to navigation, and pollution of the marine environment could cause major harm or irreversible disturbance of the ecological balance (Art. 234).

7.12.2 Regulation 10 of Annex I of the International Convention of Pollution from Ships 1973 as amended in 1978 (MARPOL 73/78)

Annex I of this Convention, that came into force in October 1983, includes regulations for the prevention of pollution by oil. According to Regulation 10 stricter rules must be followed in special areas with minimal discharges of oil and no discharge of chemicals or other substances hazardous to the marine environment. According to the Regulation, the special areas are the Mediterranean Sea area, the Baltic Sea area, the Black Sea area, the Red Sea area, and the 'Gulf area'.

7.12.3 Amendments to the International Convention for the Prevention of Pollution of the Sea by Oil, 1954, concerning the Protection of the Great Barrier Reef 1971

These amendments were adopted in October 1971 but never came into force. Their purpose was to protect the Great Barrier Reef of Australia from pollution from ships.

Curiosity caught the seal on Pribilof Island, Alaska. Scientists have noted that the fur seal population has been declining rapidly. It is estimated that 30,000 northern fur seals die every year from entanglement. Photo: CEE

Regional

7.12.4 Protocol to the Convention for the Protection, Management and Development of the Marine and Coastal Environment of the Eastern African Region, 1985 concerning Protected Areas and Wild Fauna and Flora

The contracting parties to this Protocol can establish protected areas in areas under their jurisdiction to safeguard the natural resources and protect and preserve the areas. This includes a wide range of activities and species and it might also include the establishment of a 'buffer' zone where stricter measures could apply. Annex I refers to the protected species of wild flora in the area and Annex II to species of wild fauna requiring special protection. Annex III is a list of harvestable species of wild fauna requiring protection and Annex IV includes the protected migratory species.

7.12.5 Protocol to the Convention for the Protection of the Mediterranean Sea against Pollution, 1976, concerning Mediterranean Specially Protected Areas 1982

For the purpose of this Protocol, specially protected areas are limited to the territorial waters of the parties. These areas can be established to safeguard: sites of biological and ecological value; breeding grounds and population levels of marine species and habitats; and sites of particular importance because of their scientific, aesthetic, historical, archaeological, cultural or educational character.

UK Legislation

7.12.6 The Lundy (Bristol Channel) Marine Nature Reserve Order 1986 (SI/1986)

This is the only order so far made under Section 36 of the Wildlife and Countryside Act 1981 (7.12.7).

7.12.7 The Wildlife and Countryside Act, 1981

Section 36 of this Act permits the establishment of marine nature reserves to protect and preserve endangered species or the marine ecosystem. Marine nature reserves can be established within three miles. The Territorial Sea Act 1987, provides specifically in Section 3(2) that the 12-mile extent of the territorial sea does not extend the jurisdiction of the Wildlife Act.

Italian Legislation

7.12.8 Safeguarding Measures in the Gulf of Orosei, Decree of the Minister for the Environment No. 344, 28 July 1987

This Decree was issued according to Art. 7.2 of Law No. 59 (below) as an interim measure before the transformation of the area into a marine reserve. Fishing and tourist shipping is prohibited within 2 km of the coast in the Gulf of Orosei. The provisions were issued in order to protect the last few monk seals which inhabit the area.

7.12.9 Urgent and Temporary Measures for the Operation of the Ministry for the Environment, Law No. 59, 1987

Law No. 349 of July 1986 established a Ministry for the Environment to undertake the task of ensuring the enhancement, conservation and restoration of environmental conditions corresponding to the interests of the community and the quality of life. Law No. 59 empowers the Minister for the Environment, in conjunction with the Minister for Merchant Shipping, to adopt measures for the protection of areas proposed as marine reserves. Use and transformation of these areas, as well as fishing, can be prevented.

7.12.10 Urgent Measures Concerning the Protection of Areas of Special Interest, Law No. 431 1985

According to the provisions of this Law, coastal areas up to 300 metres from the waterline, certain listed waterways and wetlands are included in a list of specially protected areas. The regions are requested to draft rules and plans for the protection of areas within their boundaries.

7.13 DECOMMISSIONING OF OFFSHORE PLATFORMS

International

7.13.1 IMO Guidelines and Standards for the Removal of Offshore Installations and Structures on the Continental Shelf and in the Exclusive Economic Zone

These guidelines and standards, adopted by the IMO Assembly in November 1989, constitute the "generally accepted international standards" for removal of abandoned installations or structures called for by Art. 60(3) of the UN Convention on the Law of the Sea 1982 (7.13.2). Removal is required except where non-removal or partial removal is consistent with these detailed guidelines and standards.

7.13.2 The United Nations Convention on the Law of the Sea (UNCLOS) 1982

Under Art. 60(3), abandoned or disused installations or structures on the continental shelf or in the EEZ have to be removed to ensure safety of navigation, taking into account any generally accepted international standards established by IMO (see 7.13.1). Removal must also have due regard to fishing, the protection of the marine environment and the rights and duties of other states. By acknowledging the permissibility of partial removal, the Convention departs from the absolute removal rule laid down in the Geneva Convention on the Continental Shelf 1958 (7.13.4).

7.13.3 Convention for the Prevention of Marine Pollution by Dumping of Wastes and Other Matter 1972 (London Dumping Convention, LDC) (see 7.8.2)

This convention includes the deliberate disposal of platforms in its definition of dumping and a special permit would be required where the dumping would present a serious obstacle to fishing or navigation. See Arts. III and IV.

7.13.4 The Geneva Convention on the Continental Shelf 1958

Under Art. 5(5), any continental shelf installations which are abandoned or disused must be entirely removed.

Regional

7.13.5 Convention for the Prevention of Marine Pollution by Dumping from Ships and Aircraft 1972 (Oslo Convention) (see 7.8.12)

It is likely that a permit from national authorities would be required for the dumping of abandoned or disused installations, see Art. 6 and Annex II.

UK Legislation

7.13.6 Petroleum Act 1987

This Act of 9 April 1987 updates the arrangements regulating oil and gas installations and their operations on the UK Continental Shelf. It amends the Petroleum Act 1934 and the law relating to submarine pipelines; Sections 34 to 39 of the Petroleum and Submarine Pipelines Act 1975 are repealed.

Part I establishes a regime for the removal of oil and gas installations and submarine pipelines at the end of their production life. Part II deals with licensing and enables the licensing regime to be extended to the territorial waters of Northern Ireland; Part III provides for the automatic establishment of safety zones round offshore installations.

Clause II gives provision for the making of regulations relating to the abandonment of offshore installations and pipelines. Regulations may be made prescribing inspection and safety requirements in the removal and disposal of installations and pipelines, and making provision for the prevention of pollution. See also *ACOPS' 1986-87 Yearbook*, pp 40-41.

Mudflats at Cardiff Bay, which may become a 500-acre, barrage-held lake. The RSPB have proposed an alternative scheme, including a nature reserve and 2,500 new homes. See 8.2 for a report on estuarine barrages. Photo: Andrew Wiard

8. Scientific and Technical

8.1 ACOPS' SURVEYS OF MARINE OIL POLLUTION AROUND THE COAST OF THE UNITED KINGDOM

8.1.1   1988

There has been an increase in the number of reported pollution incidents, for the third year in succession. The slight decrease in spills from offshore oil installations in the North Sea has been more than offset by the increase in port spills and reported slicks on the open sea; increased, perhaps by the 45 incidents reported from the Department of Transport surveillance flights, from which, unfortunately no UK prosecutions have resulted.

In general terms beach pollution has decreased and in-port pollution has increased. The largest spill, excluding the *Piper Alpha* disaster in July, was a 100 tonne spill in Milford Haven which resulted in the largest clean up costs and a fine of £4,250 on the Master. The greatest-increase in port incidents was in Sullom Voe.

An overall view of oil pollution around the UK can be obtained from Map 1. It is interesting to compare this map with one of the first such maps published by ACOPS in 1979, when in a comparable survey only 236 incidents were reported.

ACOPS was notified of 559 cases, 12% more than the total for 1987, with a mean annual increase of 15% since 1985. In 1988 a total of 154 spillages in ports was recorded by members of the British Ports Federation and other reporting organisations, 17% more than the mean annual total for the five year period ending in 1987. This was due, in part, to 15% higher response rates, but a real increase was also evident in ports throughout the United Kingdom, including the major oil terminals.

Excluding spills from North Sea oil installations, the largest number of oil slicks was detected at sea since 1983,

and 28% more than the mean annual total between 1983 and 1987. The primary cause was the introduction on 1 April 1988 of Department of Transport surveillance flights, using remote sensing techniques, over the main shipping lanes. These are in addition to similar flights instigated in 1986 by the Department of Energy over the North Sea oilfields. Consequently, an extra 45 spills, including possible sightings, which would not have been otherwise identified, were entered into the survey data base.

Although the Department of Transport was unable to obtain any successful prosecutions in the survey area, there was evidence that considerable efforts had been made to identify and punish offenders responsible for illegal discharges in international waters within the survey area. Photographic and other evidence was submitted to foreign flag states on several occasions and suspected vessels boarded and inspected by the appropriate authorities under the terms of current Port State Control arrangements.

Expenditure on clean-up reached record levels during 1988, particularly following the accidents in the North Sea, and including the £350,000 bill for the clean-up operation and compensation claims for the *El Omar* spill in Milford Haven. The costs of surveillance flights by the Departments of Energy and Transport have not been included in the survey data.

The locations of all cases of oil pollution which occurred during 1988, and the boundaries of the survey area are shown in map form before the individual enumeration area reports. Several reported incidents have not been included in the data because they were located outside the survey area, primarily to the south of the median line in the English Channel and to the west of the British Isles.

Data

The considerable volume of data collected is best shown in diagram form.

Figure 1 compares the 1988 annual total with those of the previous 8 years. The most noticeable features are: first,

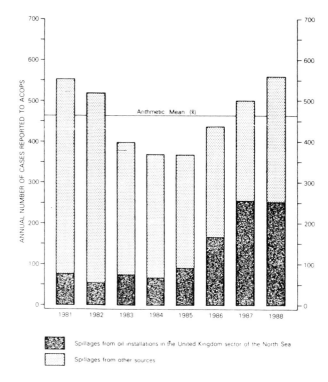

Fig. 1. Bar chart showing the number of cases of oil pollution in the waters around the UK 1981-1988.

the 1988 total was the largest recorded during the time-series; and second, it represented an increase in the annual total of cases reported to ACOPS for the third successive year. For the 3-year period ending in 1988, the mean annual increase was 15% although there is evidence to suggest that the rate of increase on the previous year may be levelling off, from 19% in 1986 to 12% in 1988. However, it is interesting to note the first increase in the total number of oil spills from sources other than offshore installations since 1984.

The total of 340 spills for the open sea was the largest number since the survey began. It was 71% higher than the 5-year mean annual total to 1987. This can probably be explained by the more accurate reporting of spillages from oil installations on the United Kingdom Continental Shelf over the last 2 years, and by the introduction of airborne surveillance patrols for oil spills from ships as well as offshore rigs and platforms.

From 1 April 1988 the Department of Transport's Marine Pollution Control Unit (MPCU) introduced regular surveillance patrols over the shipping lanes around the UK coastline using a Cessna 402 aircraft equipped with side-ways-looking-radar (SLAR) and infrared (IR) and ultraviolet line scanners. This was in addition to the flights undertaken by the Department of Energy. The aircraft flew a total of 115 hours in 1988; a total of 45 possible slicks were observed.

Comparisons of the port spill data between 1987 and 1988 revealed 22% more cases in England, 45% and 41% more in Wales and Scotland respectively. In Scotland much of the increase was due to the location of incidents in the vicinity of the Cromarty Firth. During 1987 most occurred outside the port area and were therefore classified as pre-dominantly estuarine or nearshore events, whereas the reverse was true during 1988.

Analysis of the information concerning amounts and extent of oil pollution incidents gave inconsistent results. The estimated median spill volume for the survey period was 328 litres (n = 362), 52 litres less than the corresponding 1987 statistic. However, there were 5 more spills over 1 tonne (1,180 litres) during the same time interval. At sea, 18% of the 188 oil slicks observed in which the dimensions were known were greater than 1.6 km in length, representing a steady increase in the proportion within this category over recent years. A similar trend was apparent with beach pollution data. During 1988, 42% (n = 64) of incidents were greater than 1.6 km in liner distance, compared with a mean of 33% for 1986 and 1987. Likewise, the total linear distance of shoreline polluted amounted to 333 km, 23% more than the 1987 total. Extensive pollution of south coast beaches by heavy fuel oil in December, including the entire Dorset coastline, was important in this context.

The source of pollution was identified on 428 occasions, representing 77% of all incidents reported. Excluding spills from offshore oil installations, 57% were identified, slightly less than the mean of 60% for 1983 to 1987. The percentage was generally highest in those areas where port spills were prominent.

Comparing 1988 with 1987, there were a few differences in the relative frequencies of the different polluting oils identified. No reports were received from the Department of Transport relating to any prosecutions initiated or successfully completed against offenders within the survey area. However, photographic and other evidence of possible illegal discharges was submitted to foreign flag states on several occasions during 1988, including Denmark, Liberia, Eire and the Federal Republic of Germany.

Fourteen convictions were obtained in United Kingdom ports with fines totalling £20,650 and another 13 cases had not been called or concluded at the time of reporting. The median fine was £1,000 and the Milford Haven Port Authority appeared to be the most vigilant with 7 prosecutions.

There were 120 clean-up operations undertaken during 1988, 15 more than the 1987 total. Although information was not available concerning cases arising from offshore oil installations in the North Sea, spraying vessels dealt with localised patches of oil following the *Piper Alpha* disaster on 6 July. Almost 67% of clean-ups used 25,000 litres of chemical dispersants. This was the largest volume reported used since 1982, and the mean volume per treatment was highest in area 7 (Irish Sea), 1,363 litres, followed by area 2 (Eastern England), 644 litres. Only 16% of the reports suggested the oil was removed from the sea.

The 1988 total expenditure bill for both reporting and non-reporting organisations was approximately 250% more than 1987. This was due in part to major incidents, including a final bill of £350,000 for clean-up costs and compensation claims following the *El Omar* spill in Milford Haven on 3 December. Furthermore, the MPCU submitted claims totalling over £119,000 for their involvement in anti-pollution measures, including surveillance flights, following the *Piper Alpha* disaster and *Ocean Odyssey* blow-out.

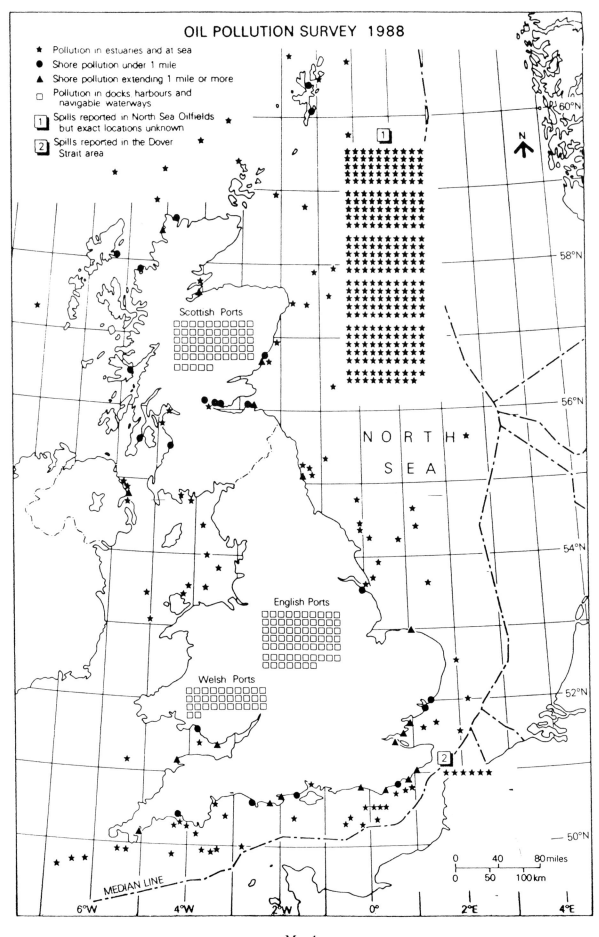

OIL POLLUTION SURVEY 1988

★   Pollution in estuaries and at sea

●   Shore pollution under 1 mile

▲   Shore pollution extending 1 mile or more

□   Pollution in docks, harbours and navigable waterways

1   Spills reported in North Sea Oilfields but exact locations unknown

2   Spills reported in the Dover Strait area

Scottish Ports

English Ports

Welsh Ports

N O R T H
S E A

MEDIAN LINE

0 40 80 miles
0 50 100 km

Map 1

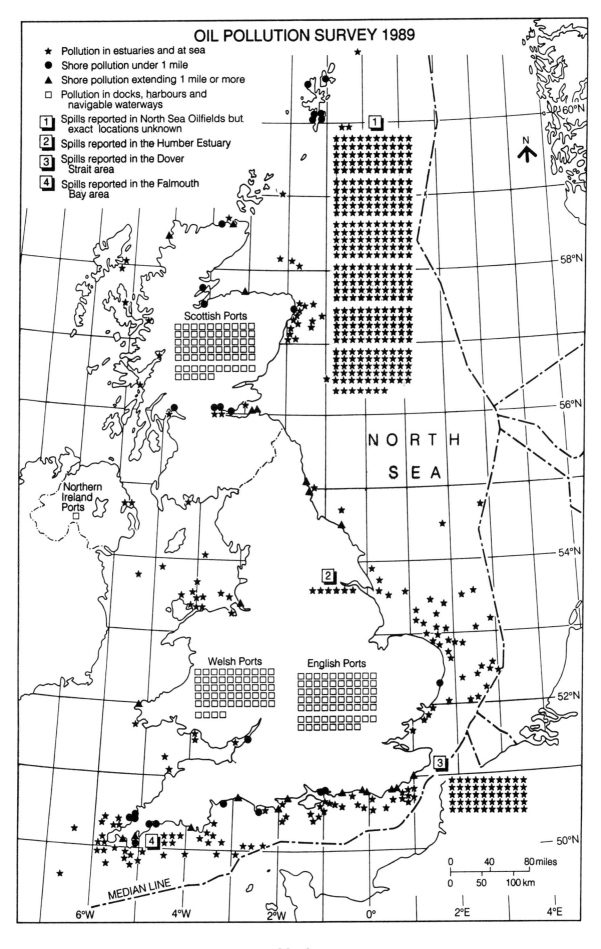

OIL POLLUTION SURVEY 1989

★ Pollution in estuaries and at sea
● Shore pollution under 1 mile
▲ Shore pollution extending 1 mile or more
□ Pollution in docks, harbours and navigable waterways
1 Spills reported in North Sea Oilfields but exact locations unknown
2 Spills reported in the Humber Estuary
3 Spills reported in the Dover Strait area
4 Spills reported in the Falmouth Bay area

Scottish Ports

Northern Ireland Ports

NORTH SEA

Welsh Ports

English Ports

MEDIAN LINE

0   40   80 miles
0   50   100 km

60°N
58°N
56°N
54°N
52°N
50°N

6°W   4°W   2°W   0°   2°E   4°E

Map 2

Large Spills Reported in and around the United Kingdom 1988

A summary of the largest spills, of 2 tonnes or more, is given below.

24 December
> 1,500 t crude oil after the Floating Storage Unit broke loose in bad weather in the Fulmar Field of the North Sea.

6 July
> 750 t crude oil following a series of explosions and a fire which destroyed the *Piper Alpha* platform in the North Sea.

3 December
> 100 t light Iranian crude from the tanker *El Omar* in Milford Haven.

7 November
> Approximately 20 t diesel oil accidentally discharged into the Exe Estuary.

21 May
> 12 t diesel oil from a Norwegian carrier which sank off the Orkneys.

15 May
> 11 t heavy fuel oil overflowed from a coastal tanker into Milford Haven.

10 February
> 10 t crude oil lost after a loading arm failure at ICI's North Tees No.4 jetty.

5 July
> 10 t oil-based product following a terminal spill at Milford Haven.

17 October
> 10 t crude oil from a tanker at Sullom Voe when loading cargo.

8 December
> 10 t oil based product accidentally discharged into Ardrossan Harbour after an equipment failure.

8 January
> 9 t fuel oil from a coaster which sank in the North Sea off the Humber Estuary.

28 June
> 8 t crude oil from a tanker in Milford Haven.

8 May
> 5 t fuel oil into Ramsgate Harbour from a ferry discharging ballast.

28 March
> 4 t of oil from a commercial premises into Whitley Bay.

10 March
> 3 t oil residues washed ashore in the Cromarty Firth from an unknown source.

March
> 2.5 t of fuel oil into Heysham Harbour when a non-tanker vessel was refuelling.

5 April
> 2.5 t fuel oil escaped into Portsmouth Harbour from a non-tanker vessel.

8 December
> 2.5 t crude oil lost from a tanker at Teesport's Phillips No.2 Berth.

April
> 2 t oil based product from a leaking storage tank at

an industrial premises escaped into the sewage system, and by a local river into Stonehaven Bay.

30 June
> Approximately 2 t fuel oil from a hose failure into Aberdeen Harbour.

21 September
> 2 t heavy fuel oil into the River Medway at Kenthole Reach from a tanker.

22 December
> 2 t fuel oil spilled when a tanker was discharging its cargo at Kenthole Reach in the River Medway.

In addition there were another 33 spills of 2 t or more reported by the Department of Energy from fixed platforms and mobile rigs on the United Kingdom Continental Shelf during 1988.

Source of Data

At total of 817 reports were processed to provide the 1988 statistics, 105 more than the previous year. The overall response rate for those organisations despatching and receiving questionnaires from their members was 77%, compared with 73% for 1987. The 15% improved response rate for the British Ports Federation over the previous year is a significant figure.[1]

8.1.2   1989

The ACOPS survey of coastal oil pollution during 1989 has been completed, although the final report was still being prepared at the time of going to press. The report shows a marked increase in the number of oil-related incidents around UK coasts compared with 1988 (Map 2). The total number of reported incidents has increased by 37% from 559 to 764. This may be due in part to increased public awareness on pollution in general, and oil pollution in particular, and the fact that many relatively small incidents have been reported. There has also been a large increase in reports of oil on the open sea: 489 compared to 340 in 1988; an increase of 44%. Many of these reports are from Government aircraft surveillance flights. Most of these slicks disperse naturally and do not pollute the coastline where, despite the public interest, the increase in reports was smaller: 48 compared to 40 in 1988 (20%).

Incidents reported in the North Sea oil field went up from 269 in 1988 to 307 in 1989, a 12% increase; incidents in ports increased by 22% (from 154 to 188). The pollution from vessels broke down to 67 tankers and 101 other vessels. One in three incidents were reported to be the direct result of human error.

This is the fourth successive year that the number of reported incidents has increased.[1]

8.2   Estuarine Barrages

The general consensus being reached by scientists internationally that global warming will result in significant rises in mean sea level over the next century has provided additional stimulus to the investigation of estuarine barrages in the UK. Many of the 'recreational and amenity' barrages being proposed are now being sold as flood protection structures against storm surges. Meanwhile tidal power barrages are being promoted on the basis of producing energy free of carbon dioxide emissions, as well as their potential for storm surge flood protection.

A further factor stimulating interest in tidal barrages in the UK has been the inclusion of the non-fossil fuel obligation (NFFO) for the purchase of electricity by the privatised area electricity supply companies in the UK Government's electricity industry privatisation programme. A wide variety of generating proposals have applied for inclusion in the NFFO, including tidal barrages, wind farms, land-fill gas power stations and combined heat and power units. Of the schemes applying, however, individual tidal barrage schemes could satisfy a large proportion of the NFFO.

Internationally, major tidal flood control barrier schemes are being proposed or constructed in Thailand and South Korea. Detailed engineering and scientific studies, related to the Venice Lagoon Barrier are also continuing prior to scheme implementation and the Netherlands, ambitious flood defence barriers are now in place. Studies related to tidal power generation in tropical Australia and Pakistan have also been proposed.

With regard to tidal power, a study was completed during 1989 for the Energy Technology Support Unit (ETSU) of the Department of Energy on the UK potential for tidal energy from small estuaries. A total of 118 west coast estuaries were examined, but only a small proportion were found to provide a prospect for economic energy production. Nevertheless, detailed feasibility studies are being carried out with regard to two small schemes on the Conwy and Wyre Estuaries. It is likely that other small schemes will be the subject of feasibility studies over the next couple of years.

The major tidal energy proposals for the Severn and Mersey Estuaries were the subject of the Third Conference on Tidal Power on 28–29 November 1989. One of the Sessions of the Conference dealt with environmental effects. The timing of the conference was particularly appropriate,

since it concluded with the completion of the £4¼m Severn Barrage Development Project, while the Mersey Barrage Project was about three quarters of the way through the Feasibility Study. In the final discussion session of the conference, it was generally agreed that a Severn Barrage would be unlikely to be built within the next quarter of a century, but that a Mersey scheme might commence construction within the next five years.

Further recreational and amenity barrages continue to be promoted, and a contract has been let for the construction of the Tawe Barrage. Nevertheless, a number of schemes have been, or are on the point of being, abandoned. The Bill promoting the controversial Cardiff Bay Barrage continues its passage through the UK Parliament, and agreements are still lacking between the promoters and the National Rivers Authority and other petitioners against the Bill, on a number of environmental issues. Whilst a majority of barrage proposals have come from South Wales, a barrage is being promoted for the Tees Estuary.

It is clear from experience both in the UK and internationally that there will be increasing pressure on the coastal and estuarine environment; whether it be from tidal and wave power generation or flood protection schemes. Thus, it will be necessary to consider and understand the cumulative effect of these developments on fisheries and other natural resources and sensitive ecosystems, such as those supporting migrating birds. To achieve this, it is necessary that planning policies that encourage the management of the overall coastal zone - including marine, estuarial and coastal, terrestrial environments are adopted, so that the conflicting pressures on coasts and estuaries can be resolved and accommodated with minimal environmental impact.[2]

Notes and References

1. The ACOPS surveys were compiled by A J Dixon and T R Dixon (Buckinghamshire College) and edited by J Wardley Smith (Vice Chairman). Maps and graphs by R Fry.
2. Report by John Towner.

9. Reference Section

9.1 INDEX OF FURTHER INFORMATION

Listed below in alphabetical order, are references to organisations to be found in previous issues of the Yearbook. Up to and including 1982 ACOPS produced an Annual Report, which was superseded in 1983 by the Yearbook. Years up to 1982 are therefore references to the Annual Report of that year; dates after this time refer to the appropriate ACOPS Yearbook. See 9.4 for acronyms.

Tanker Advisory Center: 1979 p 42, 1981 p 44, 1982 p 39, 1983 p 46, 1984 p 50, 1985-86 p 78, 1986-87 p 70,97, 1987-88 p 91.

Turkish Legislation: 1987-88 p 59.

UK Government Reports: 1985-86 p 41, 1986-87 p 42, 1987-88 p 60.

UK Legislation: 1978 p 14, 1979 p 19, 1980 p 15, 1982 p 26, 1983 p 8, 1984 p 28, 1985-86 p 34, 1986-87 p 38, 1987-88 p 49.

UK Offshore Operators Association: 1980 p 47, 1981 p 46, 1982 p 40, 1983 p 51, 1985-86 p 77, 1986-87 p 69, 1987-88 p 91.

UK Petroleum Industry Association: 1980 p 47, 1981 p 46, 1982 p 40, 1983 p 50, 1984 p 49, 1985-86 p 78, 1986-87 p 70, 1987-88 p 90.

United Nations Environment Programme: 1977 p 9, 1978 p 63, 1979 p 33, 1980 p 24, 1981 p 23, 1982 p 18, 1983 p 20, 1984 p 13, 1985-86 p 19, 1986-87 p 20, 1987-88 p 21.

Venezuelan Legislation: 1987-88 p 58.

Warren Spring Laboratory: 1986-87 p 70, 1987-88 p 91.

Water Authorities Association: 1985-86 p 70, 1986-87 p 60, 1987-88 p 83.

World Commission of Environment and Development: 1 986-87 p 57.

World Health Organization: 1979 p 34, 1980 p 26, 1981 p 27, 1982 p 21, 1983 p 24, 1984 p 16, 1985-86 p 22, 1986-87 p 24, 1987-88 p 25.

World Maritime University: 1986-87 p 70.

World Meteorological Organisation: 1987-88 p 26.

World Wide Fund for Nature: 1987-88 p 79.

9.2 GLOSSARY AND UNITS

9.2.1 Glossary of Terms

Advanced Gas-cooled Reactor (AGR): a development of the Magnox reactor, using enriched uranium oxide fuel in stainless steel containers.

Biological oxygen demand (BOD): the amount of oxygen required to degrade the organic material and oxidise reduced substances in a water sample; used as a measure of the oxygen requirement of bacterial populations in serving as an index of water pollution.

Bulk cargoes: large consignments of loose cargo usually travelling in full ship loads between two named ports and normally arranged under the terms of a charter party.

Clean ballast tanks (CBT): the reserving of certain cargo tanks solely for ballast water.

Collective effective dose equivalent (collective dose): the quantity obtained by multiplying the average effective dose equivalent by the numbers of persons exposed to a given source of radiation. Expressed in man-sievert (man-Sv).

Combined carrier: a general-purpose merchant vessel designed to carry bulk homogenous and non-homogenous types of cargo (eg. iron ore and oil); the ship may be volume or deadweight limited for tonnage, according to the type of cargo it carries.

Continental margin: the submerged prolongation of the land mass of a coastal state, consisting of the seabed and subsoil of the continental shelf (qv.), the slope and the rise, but not the deep ocean floor or its subsoil.

Continental shelf: the seabed and subsoil of the submarine areas that extend beyond a state's territorial sea (qv.) throughout the natural prolongation of its land territory to the outer edge of the continental margin (qv.) or to a distance of 200 mm from the baselines from which the territorial seas is measured if this is greater.

Crude oil washing (COW): the cleaning or washing of crude oil tanks by dislodging residues of waxy and asphaltic substances using high pressure jets of crude oil.

Dose equivalent: the quantity obtained by multiplying the absorbed dose by a factor to allow for the different effectiveness of the various ionising radiations in causing harm to tissue. Unit sievert, symbol Sv. The factor for gamma rays, X-rays, and beta particles is 1, for neutrons 10, and for alpha particles 20.

Emission standard: limit on the emission or discharge of a contaminant from a particular source.

Environmental quality objective (EQO): a statement of quality to be aimed for in a particular aspect of the environment; usually expressed in qualitative terms.

Environmental quality standard (EQS): more specific than an EQO and expressed in quantitative terms; it is often part of a regulatory framework and has some legal basis; it specifies the maximum permissible level of a contaminant, or the minimum permissible level of a positive environmental attribute.

Estuary: arm of the sea at the mouth of tidal river where tidal effect is influenced by the river current.

Eutrophication: the process of nutrient enrichment of water which leads to enhanced organic growth but which, if carried too far (hypertrophication), causes undesirable effects.

Exclusive economic zone (EEZ): the area beyond and adjacent to the territorial sea (qv.) over which the coastal state has sovereign rights for the purpose of exploring and exploiting, conserving and managing the natural resources of the waters, seabed and subsoil.

Halflife: the time taken for the activity of a radionuclide to reduce by half through radioactive decay.

High seas: all parts of the sea not included in the exclusive economic zone, the territorial sea (qv.), or in the archipelagic waters of an archipelago.

Hydrocarbon: compound of hydrogen and carbon with nothing else, eg. petroleum, coal, asphalt and natural waxes.

Inert gas systems: replacement of flammable gas content in cargo tanks by inert gas from the ship's own boiler flue gas.

Innocent passage: foreign shipping has a right of innocent passage through the territorial sea; such passage must not prejudice the peace, good order or security of the coastal state.

Internal waters: all waters lying landward of the baseline ie, bays, rivers, harbours and inland waterways; there is no right of innocent passage for foreign ships.

Littoral zone: narrow area along a coast that extends from the low-tide mark to the high-tide mark and is characterised by special flora and fauna.

Load lines: lines on ship's side to mark limits to which ships may be loaded to provide safe freeboard in different areas of the sea during various seasons.

Load line zones: the oceans of the world are divided into seasonal zones and seasonal periods, eg. the tropical zone and the North Atlantic winter seasonal zone: ships entering these zones must have the appropriate minimum freeboard.

Load on top (LOT): method of washing out oil tankers; oil washings from all cargo tanks are pumped into a single slop tank where oil and water separate; the water is drained into the sea, oil remains on board and new oil is loaded 'on top' of it.

Magnox reactor: a thermal reactor named after the magnesium alloy in which the uranium metal fuel is contained. The moderator is graphite and the coolant carbon dioxide gas.

Ore-bulk oil (OBO) carriers: a variation of a bulk carrier which can carry ore, other bulk cargoes and oil.

Pressurised Water Reactor (PWR): a thermal reactor using water both as a moderator and coolant. Uses enriched uranium oxide fuel.

Salvage: the rescue of ships, lives or cargo from danger is a salvage service; the reward paid to the salvor in respect of the successful performance of a salvage service is also called salvage.

Segregated ballast tanks (SBT): special tanks on an oil tanker reserved for the carriage of sea water ballast; required by the 1973 MARPOL Convention, which came into force in 1984.

Sievert: see dose equivalent.

Territorial seas: traditionally, the width was 3 nm. Under the 1982 Law of the Sea Convention every state has the right to establish the width of its territorial sea up to a limit not exceeding 12 nm measured from baselines determined in accordance with the Convention.

Uniform emission standard (UES): a system whereby the same limit is placed on all emissions of a particular contaminant.

Very large crude oil carrier (VLCC): a very large crude oil carrier which is subdivided into a number of tanks; oil may be discharged by use of under deck cargo pipelines, or by a 'free flow' system where the vessel is trimmed by the stern and oil flows through valves in the tank bulk-heads into one big tank before being pumped ashore.

9.2.2   Units of Measurement

Conversion factors

Distance
1 nautical mile = 1.852 kilometres = 1.15 statute miles
1 kilometre = 0.62 statute miles = 0.54 nautical miles
1 statute mile = 0.87 nautical miles = 1.6093 kilometres
Depth
1 fathom = 1.829 metres = 6 feet
1 metre = 3.281 feet = 0.547 fathoms
1 foot = 0.166 fathoms = 0.305 metres
1 shackle = 15 fathoms
1 cable = 608 feet (approx. 100 fathoms)

10 cables = 1 nautical mile
Weight
1 tonne = 0.984 long tons = 1.102 short tons
1 long ton = 1.120 short tons = 1.016 tonnes
1 short ton = 0.907 tonnes = 0.893 long tons
Crude oil (based on average gravity)
1 barrel = 0.136 tonnes = 35.6 gallons (Imp)
1 tonne = 256 gallons (Imp) = 7.33 barrels
1,000 gallons (Imp) = 28.6 barrels

Tonnage definitions

Tonnage is a means of measuring a ship, and expressing its 'size' and earning capacity. It is used in assessing harbour and port dues and charges for the services rendered to a ship. Frequently there is confusion over ship tonnages, arising from the use of measurement tons (at 100 cubic ft to the ton) and weight tons (2,240 lb) to express ship characteristics. The following are some of the more frequently used forms of tonnage:

Registered tonnage: the internal capacity of a ship, donoting the cubic capacity under the tonnage deck; a registered ton is 100 cubic feet of internal measurement.

Gross registered tonnage (GRT): the registered tonnage plus the measured tonnage 'over deck', such as the spaces for the bridge, accommodation, etc. GRT is a normal unit for passenger ships and cargo liners and is used as a basis for safety requirements and manning.

Net registered tonnage (NRT): a measure of the earning capacity of a vessel, expressed in units of 100 cubic feet. It is the GRT less the volumes of certain spaces not used for the carriage of cargo (engine room, certain water tanks, etc). NRT is frequently the basis on which harbour dues and pilotage fees are levied. It is a normal unit for passenger ships and cargo liners.

Displacement tonnage: the actual weight of a ship; it equals the weight of water displaced by a floating vessel.

Light displacement: the weight of the ship when in an unloaded state, but including the weight of water in boilers and any permanent ballast.

Load displacement: the light displacement weight plus the weight of cargo, fuel, stores, fresh water and water ballast.

Deadweight tonnage (DWT): the difference between light and load displacements; it is a measure of the carrying capacity of a vessel and is the weight of cargo, fuel, fresh water and stores that it is able to carry (normal unit for bulk carriers and tankers).

Freight tonnage: freight rates are assessed on weight or volume of cargo; if an item takes up more than 40 cubic feet per ton weight then the shipper pays on volume.

New tonnage rules: under a new convention, which came into force in July 1982, the gross tonnage is based on the volume of all enclosed spaces, and net tonnage is the volume of cargo spaces plus the volume of passenger spaces, multiplied by a coefficient. The gross and net tonnage under the new convention are no longer expressed in tons of 100 cubic feet but in cubic metres.

Radiological Units

Radioactivity

New SI Unit and symbol	Old Unit and symbol
becquerel (Bq)	curie (Ci)
Terabecquerel (TBq)	

Conversion data

1 Ci = 3.7 x 1010 Bq
1 Bq = 2.7 x 10-11 Ci = 27 pCi
1TBq = 1012 Bq = 27 Ci

Dose equivalent

New SI Unit and symbol	Old Unit and symbol
sievert (Sv)	rem (rem)

Conversion data

1 rem = 10-2 Sv = 10 mSv
1 Sv = 102 rem

Miscellaneous

cm	centimetre
m	metre
m^3	cubic metre
MW	mega watts
nm	nautical mile
ppm	parts per million

Sources

House of Commons Environment Committee, First Report (HMSO, 1986).
MAFF Directorate of Fisheries Research: *Aquatic Environment Monitoring Report, No. 12, 1985.*
Tenth Report of the Royal Commission on Environmental Pollution (HMSO, Cmnd 9149, 1984).
The Times Atlas of the Oceans (Times Books, London, 1983).

9.3 DIRECTORY

Advisory Committee on Pollution of the Sea (ACOPS)
57 Duke Street, London WIM SDH, UK. tel: 071 493 3092; fax: 071 493 3092

African Ministerial Conference on the Environment (AMCEN) Secretariat
See UNEP.

African NGO's Environment Network (ANEN)
Red Cross House, 2nd Floor, St John's Gate/Parliament Road, Nairobi, Kenya. tel: 28138

Aktionskonferenz Nordsee e.V.
Kreuzstrasse 61, D-2800 Bremen 1, FRG. tel: 0421 77675

Asociacion Uruguaya de Derecho Ambiental (AUDA)
Echevarriarza 336, Montevideo, Uruguay. tel: 72 10 24

Association of County Councils (ACC)
66a Eaton Square, London SW1W 9BH, UK. tel: 071 235 1200

Association of District Councils (ADC)
9 Buckingham Gate, London SW1E 6LE, UK. tel: 071 828 7931

Association of Metropolitan Authorities (AMA)
35 Great Smith Street, London SW1P 3BJ, UK. tel: 071 222 8100

Association of Sea Fisheries Committees of England and Wales (ASFCEW)
11 Clive Avenue, Lytham St Annes, Lancashire FY8 2RU, UK. tel: 0253 721848

Basle Convention, Interim Secretariat
Villa Belle Rive, 266 Rue de Lausanne, Chambesy, Geneva, Switzerland. tel: 022 758 2510

British Hotels, Restaurants and Caterers Association (BHRCA)
40 Duke Street, London W1M 6HR, UK. tel: 071 499 6641

British Ports Federation (BPF) Ltd
Commonwealth House, 1-19 New Oxford Street, London WC1A 1DZ, UK. tel: 071 242 1200

British Resorts Association (BRA)
PO Box 9, Margate, Thanet, Kent CT9 1XZ, UK. tel: 0843 225511

British Trust for Ornithology
Beech Grove, Tring, Hertfordshire, UK. tel: 044 282 3461; fax: 044 282 8455

Caribbean Community (CARICOM)
c/o UNEP 14-20 Port Royal St, Kingston, Jamaica tel: 809 92 29267; fax: 809 92 29292

Center for Environmental Education
1725 DeSales Street NW, Washington DC 20036, USA. tel: 202 429 5609

Coastwatch
c/o Farnborough College of Technology, Boundary Road, Farnborough, Hampshire GU14 6SB, UK. tel: 0252 515511; fax: 0252 549682

Commission of the European Communities (CEC)
200 Rue de la Loi, B-1049 Brussels, Belgium. tel: 02 235 1111

Commonwealth Secretariat
Marlborough House, Pall Mall, London SW1Y 5HX, UK. tel: 071 839 3411

Convention of Scottish Local Authorities (COSLA)
Rosebery House, 9 Haymarket Terrace, Edinburgh EH12 5XZ, UK. tel:031 346 1222

Council of Europe
BP 431 R6, 67006 Strasbourg Cedex, France. tel: 88 61 49 61

Council of European Municipalities and Regions (CEMR)
41 Quai d'Orsay, 75007 Paris, France. tel: 1 45 51 40 01; fax: 1 47 05 97 43

Council for the Protection of Rural England (CPRE)
4 Hobart Place, London SW1W 0HY, UK. tel: 071 235 9481

Council for the Protection of Rural Wales (CPRW)
Ty Gwyn, 31 High Street, Welshpool, Powys SY21 7JP, UK. tel: 0938 2525

CRISTAL (Services Ltd)
Staple Hall, Stonehouse Court, 87-90 Houndsditch, London EC3A 7AB, UK. tel: 071 621 1322

Department of Agriculture and Fisheries for Scotland (DAFS)
Dover House, Whitehall, London W1, UK. tel: 071 212 3434

Department of the Environment (DoE)
Romney House, 43 Marsham Street, London SW1 3PY, UK. tel: 071 212 3434

Department of the Arts, Sport, the Environment, Tourism and Territories
GPO Box 787, Canberra, ACT 2601, Australia. tel: 616 274 1111; fax: 616 274 1123

Directorate of Water, Pollution and Prevention (DEPPR)
Ministry of Environment, 14 Bvd du General Leclerc, 92524 Neuilly-sur-Seine, France. tel: 147 58 12 12; fax: 47 45 04 79

English Tourist Board
Thames Tower, Black's Road, London W6 9EL, UK. tel: 081 846 9000

Environmental Problems Foundation of Turkey (EPFT)
33/3, 06660 Kavaklidere, Ankara, Turkey. tel: 041 255 508; fax: 041 185 118

Estuarine and Brackish Water Sciences Association
c/o Dr N V Jones, Institute of Estuarine and Coastal Studies, University of Hull, Hull HU6 7RX, UK.

European Council of Chemical Manufacturers' Federations (CEFIC)
250 Avenue Louise, BTE 71, B-1050 Brussels, Belgium. tel: 02 640 2095

European Environmental Bureau (EEB)
Rue du Luxembourg, 20, B-1040, Brussels, Belgium. tel: 02 514 1250; fax: 02 514 0937

Federal Ministry for the Environment, Nature Conservation and Nuclear Safety
Postfach 12 06 29, 5300 Bonn 1, FRG. tel: 0228 205 2524; fax: 0228 305 3524

Field Studies Council Research Centre (FSC)
Fort Popton, Angle Pembroke, Dyfed SA71 5AD, UK. tel: 0646 641404; fax: 0646 641425

Finnish Association for Nature Conservation (SLL)
Suomen Luonnonsuojelun Tuki Oy, Nervanderinkatu 11, SF-00100 Helsinki, Finland. tel: 0 406 262

Food and Agriculture Organization (of the UN) (FAO)
Via delle Terme di Caracalla, 00100 Rome, Italy. tel: 06 57971

Foundation for Sea Use Management Studies (SEA)
167 Oosteinde, NL-2611 VD Delft, The Netherlands. tel: 015 12 4552

Fridtjof Nansen Institute
Fridtjof Nansens vei 17, PO Box 326, N-1324 Lysaker, Norway. tel: 02 538 912; fax: 02 125 047

Friends of the Earth UK (FOE)
26-28 Underwood Street, London, UK. tel: 071 490 1555

Friends of the Earth International (FOEI)
as FOE UK

Green Alliance
60 Chandos Place, London WC2N 4HG, UK. tel: 071 836 0341

Greenpeace, UK
30-31 Islington Green, London N1 8XE, UK. tel: 071 354 5100

Greenpeace International
124 Cannon Workshops, West India Docks, London, E14 9SA, UK. tel: 071 515 0275

Hellenic Marine Environment Protection Association (HELMEPA)
5 Pergamou Street, 171 21 Nea Smyrni, Athens, Greece. tel: 934 3088

Helsinki Commission (HELCOM)
Mannerheimintie 12A, SF 00100, Helsinki 10, Finland. tel: 0 602366

Indian Ocean Marine Affairs Co-operation Commission (IOMAC)
BMICH Building, Bauddhaloka Manatha, Colombo 7, Sri Lanka. tel: 01 599 691

Institute for European Environmental Policy (IEEP)
3 Endsleigh Street, London WC1H ODD, UK. tel: 071 388 2117; fax: 071 388 2826

Institute of International Law and International Organizations
University of Parma, 12 43100 Parma, Italy.

Institute of Petroleum (IP)
61 New Cavendish Street, London, E1M 8AR, UK. tel: 071 636 1004; fax: 071 255 1472

Intergovernmental Oceanographic Commission (IOC) of UNESCO
7 Place de Fontenoy, 75700 Paris, France. tel: 45 68 10 00; fax: 40 56 93 16

International Association of Independent Tank Owners (INTERTANKO)
Gange-Tolvs gatte 5, N-0273 Oslo 2, Norway. tel: 47 2 44 03 40; fax: 47 2 44 03 40

International Association of Ports and Harbors (IAPH)
Kotohira-Kaikan Bldg, 1-2-8, Toranomon, Minato-ku, Tokyo 105, Japan. tel: 03 591 4261; fax: 03 580 0364

International Atomic Energy Agency (IAEA)
Wagramerstrasse 5, PO Box 100, A-1400 Vienna, Austria. tel: 43 1 2360; fax: 43 1 234564

International Centre for Ocean Development (ICOD)
5670 Spring Garden Rd, 9th floor, Halifax, Nova Scotia, Canada B3J 1H6. tel: 902 426 1512; fax: 902 426 4464

International Chamber of Shipping (ICS)
30-32 St. Mary Axe, London, EC3A 8ET, UK. tel: 071 283 2922; fax: 010 44 1 626 8135

International Conference on the Protection of the North Sea, Secretariat
PO Box 5807, 2280 HV Riswijk, The Netherlands. tel: 70 394 9500; fax: 70 390 0691

International Co-operative Alliance (ICA)
Ronte Morions 15, 1218 Caconnex, Geneva, Switzerland.

International Council for the Exploration of the Sea (ICES)
Palaegade 2-4, DK-1261 Copenhagen K, Denmark. tel: 33 15 42 25; fax: 33 93 42 15

International Institute for Environment and Development (IIED)
3 Endleigh Street, London WC1H 0DD, UK. tel: 071 388 2117; fax: 071 388 2826

International Institute for Transportation and Ocean Policies Studies
1236 Henry Street, Halifax, Nova Scotia, Canada B3H 3J5. tel: 902 424 3879

International Juridical Organisation (IJO)
Via Barbarini 3, 00187 Rome, Italy.

International Labour Office (of the UN) (ILO)
4 Routes des Morillons, CH-1211 Geneva 22, Switzerland. tel: 22 996111

International Maritime Bureau (IMB)
Maritime House, 1 Linton Road, Barking, Essex IG11 8HG, UK. tel: 081 591 3000; fax: 081 594 2833

International Maritime Law Institute (IMLI)
PO Box 20, St Julian's, Malta. tel: 319 342

International Maritime Organization (of the UN) (IMO)
4 Albert Embankment, London SE1 7SR, UK. tel: 071 735 7611

International Ocean Institute (IOI)
1321 Edward Street, Halifax, Nova Scotia, Canada B3H 3H5. tel: 902 424 2034; fax: 902 424 2319

International Oil Pollution Compensation Fund (IOPC Fund)
4 Albert Embankment, London SE1 7SR, UK. tel: 071 582 2606; fax: 071 587 3210

International Organization of Consumers Unions (IOCU)
Emmastraat 9, 2595 EG The Hague, The Netherlands. tel: 70 47 63 31; fax: 83 49 76

International Petroleum Industry Environmental Conservation Association (IPIECA)
1 College Hill (1st Floor), London EC4R 2RA, UK. tel: 071 248 3447/8; fax: 071 489 9067

International Salvage Union (ISU)
Central House, 32-66 High Street, London E15 2PS, UK. tel: 081 519 4872; fax: 081 519 5483

International Tanker Owners Pollution Federation Ltd (ITOPF)
Staple Hall, Stonehouse Court, 87-90 Houndsditch, London EC3A 7AX, UK. tel: 071 621 1255; fax: 071 621 1783

International Transport Workers' Federation (ITF)
133-135 Great Suffolk Street, London SE1 1PD, UK. tel: 071 403 2733; fax: 071 357 7871

International Union for the Conservation of Nature and Natural Resources (IUCN)
now The World Conservation Union
Avenue de Mont Blanc, CH1196 Gland, Switzerland. tel: 22 64 91 14; fax: 22 642 926

International Union of Local Authorities (IULA)
PO Box 90646, 2509 The Hague, The Netherlands. tel: 70 21 44 032; fax: 70 32 46 916

International Whaling Commission (IWC)
The Red House, Station Road, Histon, Cambridge CB4 4NP, UK. tel: 0223 233971; fax: 0223 232876

Lega Navale Italiana
Sezione di Agrigento, Bia Diodoro Siculo, 1-Casella Postale 92100 Agrigento, Italy. tel: 41 01929 27

Local Government International Bureay

35 Great Smith St, London SW1P 3BJ. tel: 071 222 1636; fax: 071 233 2179

Marine Conservation Society
4 Gloucester Road, Ross-on-Wye, Herefordshire HR9 5BU, UK. tel: 0989 66017

Marine Forum for Environmental Issues
80 York Way, London N1 9AG, UK. tel: 071 837 5359

Marine Pollution Control Unit (of the DTp) (MPCU)
Sunley House, 90 High Holborn, London WC1V 6LP, UK. tel: 071 405 6911

Memorandum of Understanding on Port State Control (Secretariat)
PO Box 5817, 2280 HV Rijswijk, The Netherlands. tel: 070 94 94 20; fax: 070 99 62 74

Meteorology & Environmental Protection Administration (MEPA)
PO Box 1358, Jeddah, Saudi Arabia. tel: 671 0448

Ministry of Agriculture, Fisheries and Food (MAFF)
Whitehall Place, London SW1, UK. tel: 071 233 3000

National Federation of Fishermen's Organizations (NFFO)
Mersden Road, Fish Docks, Grimsby DN31 3SG, UK. tel: 0472 352141; fax: 0472 242486

National Research Coucil (US)
Georgetown Facility Room HA 250, 2001 Wisconsin Avenue NW, Washington DC 20418, USA. tel: 202 334 3119; fax: 202 334 2620

National Rivers Authority (NRA)
30-34 Albert Embankment, London SE1 7TL, UK. tel: 071 820 0101; fax: 071 820 1603

National Union of Marine Aviation and Shipping Transport Officers (NUMAST)
Oceanair House, 750-760 High Road, Leytonstone, London E11 3BB, UK. tel: 081 989 6677

National Union of Seamen (NUS)
Maritime House, Old Town, London SW4 OJP, UK. tel: 071 622 5581/8

Nature Conservancy Council (NCC)
Northminster House, Peterborough PE1 1UA, UK. tel: 0733 40345

Netherlands Institute for the Law of the Sea (NILOS)
Janskerhof 3 3512 BK, Utrecht, The Netherlands. tel: 39 30 60; fax: 39 30 73

North East River Purification Board
Greyhope House, Greyhope Road, Torry, Aberdeen AB13RD, UK. tel: 0224 248338

North Sea Working Group (NSWG)
Vossiusstraat 20-111, 1071 AD Amsterdam, The Netherlands. tel: 20 76 14 77

Oceanic Society
1536 16th Street NW, Washington DC 20036, USA. tel: 202 328 0098

Oceans and Coastal Areas Programme Activity Centre (OCA/PAC) of UNEP
PO Box 30552, Nairobi, Kenya. tel: 2 333930

Oil Companies European Organization for Environmental and Health Protection (CONCAWE)
Madouplein 1, B-1030 Brussels, Belgium. tel: 2 2203111; fax: 2 21944646

Oil Companies International Marine Forum (OCIMF)
6th Floor, Portland House, Stag Place, London SW1E 5BH, UK. tel: 071 828 7696/6283

Oil Industry International Exploration and Production Forum (E & P Forum)
25/28 Old Burlington Street, London W1X 1LB, UK. tel: 071 437 6291

Organization for Economic Co-operation and Development (OECD)
2 Rue Andre Pascal, 75775 Paris, France. tel: 1 45 02 7700; fax: 1 45 24 7876

Oslo Commission (OSCOM)
New Court, 48 Carey Street, London WC2A 2JE, UK. tel: 071 242 9927

Overseas Development Administration (ODA)
Eland House, Stag Place, London SW1E 5DH, UK. tel: 071 273 3000

Paris Commission (PARCOM)
New Court, 48 Carey Street, London WC2A 2JE, UK. tel: 071 242 9927

Permanent Commission of the South-East Pacific (Comision Permanente del Pacifico Sur, CPPS)
Casilla 16638, Agencia 6400-9, Santiago 9, Chile. tel: 726652-4

Petroleum and Industry Association (P & I Association)
1 Pepys Street, London EC3N 4AL, UK. tel: 071 480 7272

Regional Marine Pollution Emergency Response Centre for the Mediterranean Sea (REMPEC)
Manoel Island, Malta. tel: 33 72 96-8; fax: 33 99 51

Royal Society for the Prevention of Cruelty to Animals (RSPCA)
Causeway, Horsham, West Sussex RH12 1HG, UK. tel: 0403 64181

Royal Society for the Protection of Birds (RSPB)
The Lodge, Sandy, Bedfordshire SG19 2DL, UK. tel: 0767 80551; fax: 0767 292365

Scottish Fishermen's Federation (SFF)
Bonn Accord Crescent, Aberdeen AB1 2DE, tel: 0224 582583

Scottish River Purification Boards' Association (SRPBA)
City Chambers, Glasgow G2 1DU. tel: 041 227 4190

Seas at Risk
Vossiusstraat 20-111, NL 1071 AD Amsterdam, The Netherlands. tel: 20 754336; fax: 20 753806

Seatrade Annual Awards for Achievement
Fairfax House, Causton Road, Colchester CO1 1RJ, UK. tel: 0206 45121; fax: 0206 45190

Society for the Environmental Protection of the German North Sea Coast (SDN)
Weserstrasse 78, D 2940 Wilhelmshaven, FRG. tel: 04421 44316; fax: 04421 43146

Society of International Gas Tanker and Terminal Operators Ltd (SIGTTO)
London Liaison Officer, Staple Hall, Stonehouse Court, 87-90 Houndsditch, London EC3A 7AX, UK. tel: 071 621 1422

South Pacific Commission (SPC)
PO Box D5, Noumea Cedex, New Caledonia. tel: 26 20 00

South Pacific Forum Fisheries Agency (FFA)
PO Box 629, Honiara, Solomon Islands. tel: 21124; fax: 23995

Swedish International Development Authority (SIDA)
S 105 25 Stockholm, Sweden. tel: 010 46 8 728 5100; fax: 08 32 21 41; fax: 08 32 21 41

Tanker Advisory Center
217 East 85th Street, Suite 259, New York, NY 10028, USA. tel: 212 628 7686

Tidy Britain Group
The Pier, Wigan, Lancashire WN3 4EX, UK. tel: 0942 824620

United Kingdom Centre for Economic and Environmental Development (UKCEED)
12 Upper Belgrave Street, London SW1X 8BA, UK. tel: 071 245 6440/1; fax: 071 235 5478

United Kingdom Offshore Operators Association Ltd (UKOOA)
3 Hans Crescent, London SW1X 0LN, UK. tel: 071 589 5255

United Kingdom Petroleum Industry Association Limited (UKPIA)
9 Kingsway, London WC2B 6XH, UK. tel: 071 240 0289

United Kingdom Pilots' Association (UKPA)
Transport House, Smith Square, London SW1P 3JB, UK. tel: 071 828 7788

United Nations Educational, Scientific and Cultural Organization (UNESCO)
7 Place de Fontenoy, 75007 Paris, France. tel: 1 56 81000

United Nations Development Programme (UNDP)
1 United Nations Plaza, New York, NY 10017, USA. tel: 212 906 5000

United Nations Environment Programme (UNEP)
PO Box 30552, Nairobi, Kenya. tel: 2542 333930; fax: 2542 520561

Warren Spring Laboratory
Gunnels Wood Road, Stevenage, Hertfordshire SG1 2BX, UK. tel: 0438 741122; fax: 0438 360858

Water Research Centre (WRC)
PO Box 16, Marlow, Buckinghamshire SL7 2HD, UK. tel: 0491 571531; fax: 0491 579094

Water Services Association
1 Queen Anne's Gate, London SW1H 9BT, UK. tel: 071 222 8111; fax: 071 222 1811

World Commission on Environment and Development (WCED)
Palais Wilson, 52 Rue de Paquis, 1201 Geneva, Switzerland. tel: 022 32 71 17

World Health Organisation (of the UN) (WHO)
CH-1211 Geneva 27, Switzerland. tel: 791 21 11; fax: 791 07 46

World Maritime University (WMU)
PO Box 500, S-201 24 Malmo, Sweden. tel: 4640 123223

World Resources Institute
1709 New York Avenue, NW, Washington DC 20006, USA. tel: 202 638 6300; fax: 202 638 0036

World Wide Fund for Nature (WWF) UK
Panda House, Weyside Park, Catteshall Lane, Godalming, Surrey GU7 1XR, UK. tel: 0483 426444; fax: 0483 426409

World Wide Fund for Nature (WWF) International
Avenue de Mont Blanc, 1196 Gland, Switzerland. tel: 022 647181

9.4 ACRONYMS AND ABBREVIATIONS

AKN	Aktionskonferenz Nordsee
ALARA/P	As low as reasonably achievable/practicable
AMCEN	African Ministerial Conference on Environment
ANEN	African NGOs Environment Network
ASEAN	Association of South-East Asian Nations
ASFC	Association of Sea Fisheries Committees
ATS	Antarctic Treaty System
BCH	Code Code for the construction and equipment of ships carrying dangerous chemicals in bulk
BOD	Biological Oxygen Demand
BPEO	Best Practicable Environmental Option
CARICOM	Caribbean Communities
CCAMLR	Convention for the Conservation of Antarctic Marine Living Resources
CDT	Carriage of Dangerous Goods
CEC	Commission of the European Community
CEDRE	Centre du Documentation de Recherche et d'Experimentation sur les Pollutions Accidentelles des Eaux
CEE	Center for Environmental Education
CEED	Centre for Economic and Environmental Education
CEFIC	Council of European Chemical Manufacturers' Federation
CEHI	Caribbean Environmental Health Institute
CEMR	Council of European Municipalities and Regions
CEN	European Committee for Standardisation
CEPNET	Caribbean Environmental Programmes Information Network
CIDA	Canadian International Development Agency
CITES	Convention on the International Trade in Endangered Species
CLC	International Convention on Civil Liability for Oil Pollution Damage 1969
CLRAE	Standing Conference of Local and Regional Authorities of Europe
CMEA	Council for Mutual Economic Assistance
CMSWA	Convention on Migratory Species and Wild Animals
COBSEA	Coordinating Body of the Seas of East Asia
CONCAWE	Oil Companies' European Organisation for Environmental and Health Protection
CONPACSE	Coordinated Programme to Research and Monitor Marine Pollution in the South-East Pacific Region
CPPS	Permanent Commission for the South-East Pacific
CRAMRA	Convention on the Regulation of Antarctic Mineral Activities 1988
CWI	Clean World International
DASA	Dumping at Sea Act 1974
DoE	Department of Environment (UK)
DTp	Department of Transport (UK)
DWT/dwt	deadweight (tonnage)
EC	European Communities
ECOSOC	Economic and Social Council (of the UN)
ECU	European Currency Unit
EEB	European Environmental Bureau
EEC	European Economic Community
EEZ	European Economic Zone
EFTA	European Free Trade Association
EIA	Environment Impact Assessment
EP	European Parliament
EPA	(US) Environmental Protection Agency
EPFT	Environmental Problems Foundation of Turkey
EQO	Environmental Quality Objective
EQS	Environmental Quality Standard
FAO	Food and Agriculture Organisation (of the UN)

FC	International Convention on the Establishment of an International Fund for Compensation of Oil Pollution Damage 1971 (Fund Convention)
FFA	Forum Fisheries Agency (of the South Pacific)
FOEI	Friends of the Earth International
FSCRC	Field Studies Council Research Centre (UK)
GEEP	IOC/IMO/UNEP Working Group on the Environmental Effects of Pollution
GESAMP	Joint Group of Experts on the Scientific Aspect of Marine Pollution (of the UN)
GIPME	Global Investigation of Pollution in the Marine Environment
GRT	Gross registered tonnage
HCH	Hexachlorocyclohexane (Lindane)
HELCOM	Helsinki Commission
HELMEPA	Hellenic Marine Environment Protection Association
HNS	Hazardous and Noxious Substance (other than oil)
IAEA	International Atomic Energy Agency (of the UN)
IAPH	Internationl Association of Ports and Harbors
IBC Code	International Code for the construction and equipment of ships carrying dangerous chemicals in bulk
ICES	Internationl Council for the Exploration of the Sea
ICLOS	Indonesian Center for the Law of the Sea
ICOD	International Centre for Ocean Development
ICA	International Cooperative Alliance
ICS	International Chamber of Shipping
IEEP	Institute for European Environmental Policy
IIED	International Institute for Environment and Development
IITOPS	International Intsitute for Transportation and Ocean Policy Studies
IJO	International Juridical Organisation
ILO	International Labour Office (of the UN)
IMB	International Maritime Bureau
IMDG Code	International Maritime Dangerous Goods Code
IMO	International Maritime Organisation
INTERTANKO	International Association of Independent Tanker Owners
IOC	Intergovernmental Oceanographic Commission (of UNESCO)
IOCU	International Organisation of Consumer Unions
IOI	International Ocean Institute
IOMAC	Indian Ocean Marine Affairs Cooperation
IOPC	Fund International Oil Pollution Compensation Fund
IOPP (Certificate)	International Oil Pollution Prevention (Certicifate)
IP	Institute of Petroleum
IPCC	Intergovernmental Panel on Climate Change
IPIECA	International Petroleum Industry Environmental Conservation Association
ISO	International Standards Organisation
ITF	International Transport Workers' Federation
ITOPF	International Tanker Owners Pollution Federation Ltd
IUCN	International Union for the Conservation of Nature and Natural Resources (now the World Conservation Union)
IULA	International Union of Local Authorities
IWC	International Whaling Commission
IWIC	International Waste Identification Code
KAP	Kuwait Action Plan
LDC	Convention for the Prevention of Marine Pollution by Dumping of Wastes and Other Matter 1972 (London Dumping Convention)
m	metre
MAFF	Ministry of Agriculture, Fisheries and Food (UK)
MARPOL 73/78	International Convention for the Prevention of Pollution from Ships 1973 (as amended by the 1978 Protocol)
MCS	Marine Conservation Society

MEP	Member of the European Parliament
MEPA	Meteorology and Environmental Protection Administration (Saudi Arabia)
MEPC	Marine Environment Protection Committee (of IMO)
MF	Marine Forum for Environmental Issues (UK)
MOU	(Paris) Memorandum of Understanding on Port State Control
MPCU	Marine Pollution Control Unit (of the UK DTp)
MSC	Maritime Safety Committee (of IMO)
MV	motor vessel
NCC	Nature Conservancy Council (UK)
NFFO	National Federation of Fishermen's Organisations (UK)
NGO	Non-governmental organisation
NILOS	Netherlands Institute for the Law of the Sea
NORAID	Norwegian Agency for International Development
NRA	National Rivers Authority (UK)
NRC	(US) National Research Council
nrt/NRT	Net registered tonnage
NSTF	North Sea Task Force
NSWG	North Sea Working Group
NUS	National Union of Seamen (UK)
OAU	Organisation for African Unity
OCA	Oceans and Coastal Areas (of UNEP Programme)
OCIMF	Oil Companies International Marine Forum
ODA	Overseas Development Administration (UK)
OECD	Organisation for Economic Cooperation and Development
OILPOL	International Convention for the Prevention of Pollution of the Sea by Oil 1954
OPRU	Oil Pollution Research Unit (Orielton) (UK)
OSCOM	Oslo Commission
OSIR	Oil Spill Intelligence Report
PAC	Programme Activity Centre (of UNEP)
PARCOM	Paris Commission
PCB	Polychlorinated biphenols
PCP	Polychorinated phenols
PCT	Polychlorinated terphenols
PERSGA	Programme for the Environment of the Red Sea and Gulf of Aden
POLREP	Common reporting system (of pollution incidents)
POPA 71	Prevention of Oil Pollution Act 1971 (UK)
PSC	Port State Control
RCU	Regional Coordinating Unit
REMPEC	Regional Marine Pollution Emergency Response Centre for the Mediterranean Sea
ROPME	Regional Organisation for the Protection of the Marine Environment (of the Gulf)
RSP	Regional Seas Programme (of UNEP)
RSPB	Royal Society for the Protection of Birds (UK)
SCAR	Scientific Committee on Antarctic Research
SEA	Foundation for Sea Use Management Studies
SI	Statutory Instrument
SIDA	Swedish International Development Authority
SIGTTO	Society of International Gas Tanker and Terminal Operators Ltd
SOLAS	International Convention for the Safety of Life at Sea 1974
SPA	Special Protection Area
SPAEI	South Pacific Association of Environmental Institutions
SPC	South Pacific Commission
SRPBA	Scottish River Purification Boards' Assoication
SSSI	Site of Special Scientific Interest
STCW	International Convention on Standards of Training, Certification and Watchkeeping for Seafarers 1978

t	ton(ne)s
TAC	Tanker Advisory Center
TBG	Tidy Brtiain Group
TBT	Tri-butyl tin
TOVALOP	Tanker Owners Voluntary Agreement concerning Liability for Oil Pollution
TSPP	Tanker Safety and Pollution Prevention
UKOOA	United Kingdom Offshore Operators' Association
UKPIA	United Kingdom Petroleum Industry Association
UN	United Nations
UNCLOS	United Nations Convention on the Law of the Sea 1982
UNDP	United Nations Development Programme
UNEP	United Nations Environment Programme
UNESCO	United Nations Educational, Scientific and Cultural Organisation
USCG	United States Coast Guard
UWIST	University of Wales Institute of Science and Technology
WCED	World Commission on Environment and Development
WHO	World Health Organisation (of the UN)
WMO	World Meteorological Organisation
WMU	World Maritime University
WRI	World Resources Institute
WSL	Warren Spring Laboratory (UK)
WWF	World Wide Fund for Nature
WSA	Water Services Association (of England and Wales)

ANNEX I

Conclusions of a Special Meeting of Vice Presidents of ACOPS, held on 4 October 1989, concerning:

Protecting the Environment from Hazardous Substances

1 States should ratify the Basle Convention forthwith;

2 Prior to ratification, the provisions of the Basle Convention should be applied as widely as possible by governments and industry on a voluntary basis with immediate effect;

3 It is most desirable that the consultations which the Secretary-General of the Organisation for African Unity was mandated to undertake in accordance with OAU Resolution 1225 result in a recommendation for the rapid acceptance of the Basle Convention. They also endorse the decision of the OAU to proceed with the adoption of a more stringent regional agreement;

4 Protocols to the relevant international agreements on the control and regulation of the transport of toxic wastes should be adopted;

5 They welcome the EC's proposals: to amend the present Directive on Waste (75/442/EEC); to replace Directive 78/319/EEC with a new Directive on hazardous wastes; to ensure greater freedom of public assess to environmental information; to impose strict liability on those responsible for the pollution of the environment; and they also welcome the waste strategy policy of the EC. Other states are urged to consider these proposals.

6 Pursuant to Resolutions of the Basle Convention, the Assembly and other bodies of IMO and the Meeting of Contracting Parties to the LDC should be urged to augment and to ensure proper monitoring of any dumping of waste at sea;

7 Governments and industries must, as a matter of urgency, incorporate into their planning policies the use of clean technologies to ensure that less waste is generated worldwide and further that the Polluter Pays Principle must be applied effectively;

8 Industrialised countries which do not as yet have adequate facilities for the safe disposal of toxic waste generated by them should take immediate action to remedy the situation;

9 Donor agencies and industrialised countries should provide, as may be required, necessary assistance to developing countries to enable them to set up their own waste processing installation, training centres for personnel and training centres to promote public awareness, the development of sound management of hazardous wastes and the adoption of clean technologies pursuant to Article 10 of the Basle Convention; and

10 Measures should be taken on a global basis to protect developing countries by preventing the exploitation of their economic position by unscrupulous industrial concerns.

Signed by:
    The Rt Ho Lord Callaghan of Cardiff PC KG (President)
    Lord Clinton-Davis (Chairman)
    Mr J Wardley-Smith OBE (Vice Chairman)

Vice Presidents:
    Dr Arnoldo Jose Gabaldón (Venezuela)
    Mr Bengt Hamdahl (Sweden)
    Mr Phillip Muller (Western Samoa)
    Rear Admiral Michael L Stacey CB (UK)
    Mr Gaetano Zorzetto (Italy)
    Mrs Elisabeth Mann Borgese (Canada)
    Mrs Martine Remond-Gouilloud for Mr Louis le Pensec (France)
    Rear Admiral Sidney A Wallace for Mr Russell Train (USA)
on the 4 October 1989, London.
    The following Vice Presidents have also signified their approval of the Conclusions:
    Professor Alexander Yankov (Bulgaria)
    Dr Abdulbar Algain (Saudi Arabia)
    Mr Cheikh Cissokho (Senegal)
    Professor Gennady Polikarpov (USSR)

Note: Vice Presidents of ACOPS serve in their personal capacity and their views do not necessarily reflect those of their respective governments.

Appendix

OFFICERS

President
The Rt Hon Lord Callaghan of Cardiff KG

Vice Presidents
Dr Abdulbar Algain (Saudi Arabia)
Professor Elisabeth Mann Borgese (Canada)
Mr Cheikh Cissokho (Senegal)
Dr Arnoldo José Gabaldón (Venezuela)
Mr Bengt Hamdahl (Sweden)
Mr Phillip Muller (Western Samoa)
Mr Louis le Pensec (France)
Professor Gennady Polikarpov (USSR)
Professor Emil Salim (Indonesia)
Mr Chandrika Srivastava (India)
Rear Admiral Michael L Stacey CB (UK)
Mr Russell Train (USA)
Professor Alexander Yankov (Bulgaria)
Assessore Gaetano Zorzetto (Italy)

Chairman
Lord Clinton-Davis

Vice Chairman
Mr Wardley-Smith OBE

Executive Secretary
Dr Viktor Sebek

Honorary Treasurer
Mr Douglas Jack

Legal Advisor
Maxwell Bruce QC

Chairman, Fundraising Committee
The Rt Hon Viscount Hardinge (until July 1989)

STAFF

Executive Secretary
Dr Viktor Sebek

Administrator
Mrs Mary Thorold

Legal Officer
Mrs Jennie Holloway

Publications
Patricia Dent

INSTITUTIONAL MEMBERS

Association of County Councils
Association of District Councils
Association of Metropolitan Authorities
Association of Sea Fisheries Committees of England and Wales
Bermuda Biological Station for Research
British Hotels, Restaurants and Caterers Association
British Ports Federation
British Resorts Association
Clean World International
Convention of Scottish Local Authorities
Council of European Municipalities and Regions
Council for the Protection of Rural Wales
English Tourist Board
European Environmental Bureau
Field Studies Council
Fridtjof Nansen Institute
Friends of the Earth
International Association of Independent Tanker Owners
International Hotel Association
International Institute for Environment and Development
International Ocean Institute
International Transport Workers Federation
International Union of Local Authorities
National Federation of Fishermen's Organisations
National Rivers Authority
National Union of Seamen
Netherlands Institute for the Law of the Sea
Oceanic Society
Royal Society for the Prevention of Cruelty to Animals
Royal Society for the Protection of Birds
Scottish Fishermen's Federation
Scottish River Purification Boards' Association
Temple, Barker & Sloane Inc.
Universities Federation for Animal Welfare
Water Services Association

Observers

United Kingdom Offshore Operators' Association

INDIVIDUAL MEMBERS

Dr Anil Agarwal (India)
The Rt Hon Lord Campbell of Croy PC MC DL (UK)
Dr D Cormack (UK)
Senora Carmen Diez de Rivera Icaza MEP (Spain)
Professor E Gold (Canada)
Mr J Hume (UK)
Mrs Caroline Jackson (UK)
Lady Kennet (UK)
Senhor Manuel Medina Ortega MEP (Portugal)
Sir PN Miller (UK)
Mr H Muntingh MEP (The Netherlands)
Senhor Carlos Pimenta MEP (Portugal)
Mr C Pinto (Sri Lanka)

Commander MBF Ranken (UK)
Mme Martine Remond Gouilloud (France)
M F Roelants du Vivier MEP (Belgium)
Frau U Schleicher MEP (FRG)
Dr Han-Joachim Secker (FRG)
Mr Madron Seligman MEP (UK)
Professor H Smets (Belgium)
Mme A Spaak (Belgium)
Signora V Squarcialupi (Italy)
Mr Maurice Strong (Canada)
Mme Simone Veil MEP (France)
Rear Admiral S Wallace (USA)
Sir Frederick Warner FRS (UK)
The Rt Hon Baroness White (UK)

ACOPS LEGAL AND POLICY COMMITTEE

Chairman
Rear Admiral ML Stacey

UK Members

Dr Patricia Birnie
Professor ED Browm
Mr Maxwell Bruce
Mr B Eder
Mr M Forster

Lady Kennet
Dr V Sebek
Mrs J Holloway

ACOPS SCIENTIFIC AND TECHNICAL COMMITTEE

Chairman
Mr J Wardley Smith OBE

Members
Dr J Baker
Mr C Chubb
Mr J Corlett
Dr D Cormack

Mr T Dixon
Dr C Pattinson
Dr Portmann
Dr Towner

LEGAL AND SCIENTIFIC EXPERTS FOR VICE PRESIDENTS

France
Dr Rouchdy Kbaier
Professor M Remond-Gouilloud

South/Central America
Raul Branes (Chile)
Isabel de los Rios (Venezuela)
Beatriz Armada (Venezuela)
Deud Dumith (Venezuela)
Vicente Sanchez (Chile)
Enrique Laff (Mexico)
Pat Robinson (Jamaica)

Other Areas
Professsor Tullio Scovazzi (Italy)
Capt Saeed Yafai (People's Demo-
cratic Republic of Yemen)
Professor H Smets (Belgium)

USA
Rear Admiral Sid Wallace
Professor Philip Sorenson
Ms Anita Yurchyshyn
Virgil Keith
Joseph Porricelli
William Myhre

Africa
Mr Bakery Kante (Senegal)
Dr R Ojikutu (Nigeria)